Advances in Experimental Medicine and Biology

Protein Reviews

Volume 1111

Series editor
M. Zouhair Atassi, Houston, TX, USA

More information about this series at http://www.springer.com/series/14330

M. Zouhair Atassi
Editor

Protein Reviews – Purinergic Receptors

Volume 20

Editor
M. Zouhair Atassi
Biochem and Mol Biol
Baylor College of Medicine
Houston, TX, USA

ISSN 0065-2598   ISSN 2214-8019   (electronic)
Advances in Experimental Medicine and Biology
ISSN 2520-1891   ISSN 2520-1905   (electronic)
Protein Reviews
ISBN 978-3-030-14341-1   ISBN 978-3-030-14339-8   (eBook)
https://doi.org/10.1007/978-3-030-14339-8

This Springer imprint is published by the registered company Springer Nature Switzerland AG.
The registered company address is: Gewerbestrasse 11, 6330 Cham, Switzerland

Preface

Protein Reviews, as a book series, has been published by Springer since 2005. It has published nineteen printed volumes to date. To see the first 16 volumes please go to (http://www.springer.com/series/6876). From volume 17 on, in order to speed up the publication process and enhance accessibility, all articles have appeared online before they are published in a printed book. The book series appears as a subseries of Advances in Experimental Medicine and Biology (http://www.springer.com/series/14330). The books are published in volumes each of which will focus on a given theme or volumes that contain reviews on an assortment of topics, in order to remain up-to-date and to publish timely reviews in an efficient manner.

The aim of the *Protein Reviews* is to serve as a publication vehicle for reviews that focus on crucial contemporary and vital aspects of protein structure, function, evolution and genetics. Publications will be selected based on their importance to the understanding of biological systems, their relevance to the unravelling of issues associated with health and disease or their impact on scientific or technological advances and developments. Proteins linked to diseases or to the appearance and progress of diseases provide essential topics. Moreover, proteins that are, or can be, used as potential biomarkers or candidates for treatment, and/or for the design of distinctive new therapeutics, will receive high attention in this book series.

The issues may include biochemistry, biophysics, immunology, structural and molecular biology, genetics, molecular and cellular mechanisms of action, clinical studies and new pioneering therapies. A given volume may be focused on a particular theme, or may contain a selected assortment of different current topics.

The authors of the articles are selected from leading basic or medical scientists in academic or industrial organizations. The invited authors are nominated by the Editorial Board or by experts in the scientific community. However, interested individuals may suggest a topic for review and/or may propose a person to review a current important topic. Colleagues interested in writing a review or in guest-editing a special thematic issue are encouraged to submit their proposals and list of authors of the suggested chapters/topics to the Editor for consideration.

The manuscripts are reviewed and evaluated in the usual manner by independent outside experts in the field. The articles will be published online no later than 6 weeks after editorial review and acceptance.

Protein Reviews will publish all accepted review articles online before they appear in print. *Protein Reviews* imposes no page or color charges and has no page or color image limitations.

Volume 20 has ten chapters. The first five chapters deal with various aspects of membrane binding. The first chapter focuses on the phox-homology (PX) domain, which is a phosphoinositide-binding domain conserved in all eukaryotes and present in forty-nine human proteins. The next chapter deals with the modeling of PH domains/phosphoinositides interactions. This is followed by a chapter on BAR domain proteins regulate Rho GTPase signaling. The BAR (Bin–Amphiphysin–Rvs) domain is a membrane lipid binding domain present in a wide variety of proteins, often proteins with a role in Rho-regulated signaling pathways. The fourth article presents AP180 N-terminal homology (ANTH) and Epsin N-terminal homology (ENTH) domains and discusses their physiological functions and involvement in disease. The fifth article reviews the polyphosphoinositide-binding domains and presents insights from peripheral membrane and lipid-transfer proteins. This is followed by a chapter on the physiological functions of phosphoinositide-modifying enzymes and their interacting proteins in Arabidopsis, then by a chapter on the molecular mechanisms of Vaspin action in various tissues such as adipose tissue, skin, bone, blood vessels, and the brain. The eighth chapter deals with exceptionally selective substrate targeting by the metalloprotease anthrax lethal factor followed by an article on Salmonella, E. coli, and Citrobacter type III secretion system effector proteins that alter host innate immunity. The last chapter presents New techniques to study intracellular receptors in living cells, with insights into RIG-I-like receptor *signaling*. Volume 20 is intended for research scientists, clinicians, physicians and graduate students in the fields of biochemistry, cell biology, molecular biology, immunology and genetics.

I hope that this volume of *Protein Reviews* will continue to serve the scientific community as a valuable vehicle for dissemination of vital and essential contemporary discoveries on protein molecules and their immensely versatile biological activities.

Houston, TX, USA   M. Zouhair Atassi

Contents

The Phox Homology (PX) Domain 1
Mintu Chandra and Brett M. Collins

Modeling of PH Domains and Phosphoinositides Interactions
and Beyond 19
Jiarong Feng, Lei He, Yuqian Li, Fei Xiao, and Guang Hu

BAR Domain Proteins Regulate Rho GTPase Signaling 33
Pontus Aspenström

AP180 N-Terminal Homology (ANTH) and Epsin N-Terminal
Homology (ENTH) Domains: Physiological Functions and
Involvement in Disease 55
Sho Takatori and Taisuke Tomita

Polyphosphoinositide-Binding Domains: Insights from
Peripheral Membrane and Lipid-Transfer Proteins 77
Joshua G. Pemberton and Tamas Balla

Physiological Functions of Phosphoinositide-Modifying
Enzymes and Their Interacting Proteins in Arabidopsis 139
Tomoko Hirano and Masa H. Sato

Molecular Mechanisms of Vaspin Action – From Adipose
Tissue to Skin and Bone, from Blood Vessels to the Brain 159
Juliane Weiner, Konstanze Zieger, Jan Pippel, and John T. Heiker

Exceptionally Selective Substrate Targeting by the
Metalloprotease Anthrax Lethal Factor 189
Benjamin E. Turk

Salmonella, *E. coli*, and *Citrobacter* Type III Secretion System
Effector Proteins that Alter Host Innate Immunity 205
Samir El Qaidi, Miaomiao Wu, Congrui Zhu,
and Philip R. Hardwidge

**New Techniques to Study Intracellular Receptors in Living Cells:
Insights Into RIG-I-Like Receptor Signaling** 219
M. J. Corby, Valerica Raicu, and David N. Frick

**Correction to: Polyphosphoinositide-Binding Domains: Insights
from Peripheral Membrane and Lipid-Transfer Proteins** 241
Joshua G. Pemberton and Tamas Balla

Index 243

Adv Exp Med Biol - Protein Reviews (2019) 20: 1–17
https://doi.org/10.1007/5584_2018_185
© Springer Nature Singapore Pte Ltd. 2018
Published online: 23 March 2018

The Phox Homology (PX) Domain

Mintu Chandra and Brett M. Collins

Abstract

The phox-homology (PX) domain is a phosphoinositide-binding domain conserved in all eukaryotes and present in 49 human proteins. Proteins containing PX domains, many of which are also known as sorting nexins (SNXs), have a large variety of functions in membrane trafficking, cell signaling, and lipid metabolism in association with membranes of the secretory and endocytic system. In this review we discuss the structural basis for both canonical lipid interactions with the endosome-enriched lipid phosphatidylinositol-3-phosphate (PtdIns3P) as well as non-canonical lipids that promote membrane association. We also describe recent advances in defining the diverse mechanisms by which PX domains interact with other proteins including the retromer trafficking complex and proteins secreted by bacterial pathogens. Like other membrane interacting domains, the attachment of PX domain proteins to specific membranes is often facilitated by additional interactions that contribute to binding avidity, and we discuss this coincidence detection for several known examples.

Keywords

Endosome · Phosphoinositide · PX domain · Retromer · Sorting nexin · SNX

1 Introduction

The coordination of various cellular events including membrane trafficking and receptor signalling, requires both spatial and temporal regulation that is initiated by the recruitment of cytosolic protein complexes to specific membrane compartments. The cellular membranes of eukaryotic organelles typically contain specialized lipids that provide recruitment signals and allosteric regulation of these diverse peripheral membrane proteins. Perhaps the most important group of anchoring lipids are the phosphorylated derivatives of phosphatidylinositol, commonly known as phosphoinositides, which are dynamically regulated by lipid phosphatases and kinases and play a critical role in vesicular trafficking and cell signalling (Balla 2013; Viaud et al. 2016). To date, a number of highly conserved phosphoinositide binding modules have been identified, including the C2 (PKC conserved region 2), PH (pleckstrin homology), FYVE (Fab1, YOTB, Vac1 and EEA1), ENTH (epsin N-terminal homology) and the PX (Phox homology) domains (Hammond and Balla 2015; Kutateladze 2010; Lemmon 2008; Viaud et al. 2016). These domains engage specific

M. Chandra and B. M. Collins (✉)
The University of Queensland, Institute for Molecular Bioscience, St. Lucia, QLD, Australia
e-mail: b.collins@imb.uq.edu.au

phosphoinositide lipids to direct the proteins to discrete sites inside the cell and often serve as protein-protein interaction modules as well.

The PX domain was first identified in two subunits of the NADPH oxidase, p40[phox] and p47 [phox], and was shown to be present in a number of other proteins such as phospholipase D1 (PLD1) and sorting nexin 1 (SNX1) in humans and Vam7p and Mdm1p in yeast (Ponting 1996). Since then the PX domain has been found in diverse proteins across all eukaryotic species, and in humans there are 49 proteins known to contain this domain including class II phosphoinositide 3-kinases (PI3Ks), sorting nexins containing membrane-bending BAR (bin/amphiphysin/rvs) domains and many others (Fig. 1). In 2001 a series of key papers showed that PX domains were specifically able to recognise the endosome-enriched lipid phosphatidylinositol-3-phosphate (PtdIn3P), and this played a crucial role in membrane recruitment of proteins including yeast Vam7p, and human SNX3 and p40[phox] (Ago et al. 2001; Cheever et al. 2001; Ellson et al. 2001, 2002; Kanai et al. 2001; Sato et al. 2001; Song et al. 2001; Virbasius et al. 2001; Xu et al. 2001; Yu and Lemmon 2001). The first structure of a PX domain was the NMR solution structure of the domain from p47[phox] (Hiroaki et al. 2001), revealing a core fold consisting of three anti-parallel β-strands followed by three α-helices. Subsequently the crystal structure of the p40[phox] PX domain bound to PtdIns3P provided the first insight into the phospholipid recognition mechanism (Bravo et al. 2001). Since then numerous structures have been reported of PX domains from different proteins and from different species (Fig. 2), but to date only two other structures have been reported in the presence of the PtdIns3P headgroup (Pylypenko et al. 2007; Zhou et al. 2003).

As shown in Fig. 1 the PX domain proteins can be classified into a number of subgroups based on their structural architecture. Thus the proteins in this family cannot be functionally classified into a single pathway, rather they play diverse roles in membrane trafficking, membrane remodeling and cell signaling, depending on the activities of the other domains that are present. The single common link between the proteins is that they are recruited to and active on various compartments of the secretory and endocytic system. Several reviews have been published previously on the PX domain and SNX proteins (Cullen 2008; Ellson et al. 2002; Sato et al. 2001; Seet and Hong 2006; Teasdale and Collins 2012; Worby and Dixon 2002; Xu et al. 2001). In this chapter we focus on key aspects of PX domain structure and function including structural embellishments of the PX domain, the mechanisms of canonical and non-canonical lipid interactions, and the emerging understanding of its role in not only lipid but also protein-protein interactions.

2 The Structure of the PX Domain and Its Canonical Interaction with PtdIns3P

At the time of writing more than 50 structures of PX domains had been determined using either NMR spectroscopy or X-ray crystallography (Fig. 2). Structurally, the PX domain adopts a globular fold approximately 120 residues in length and composed of three anti-parallel β-strands followed by three α-helices (Fig. 3a). Sequence alignment of the PX domain across all human homologues (Seet and Hong 2006) indicates that it contains several well conserved regions, including a number of basic residues, and an extended region between helices α1 and α2 that we refer to as the PPK loop as it contains a conserved sequence with the general consensus $\Psi PxxPxK$ (Ψ = large aliphatic amino acids V, I, L and M) (Fig. 4). In most PX domains, the structure results in the formation of a shallow, positively charged surface pocket between β1, α1 and the extended PPK loop, which is generally considered to be the site of binding to negatively-charged phosphoinositide headgroups.

As discussed above, the first studies of PX domains showed they were able to specifically interact with the endosomal lipid PtdIns3P. An analysis of all PX domains from *Saccharomyces cerevisiae* suggested this was a common feature of the domain (Yu and Lemmon 2001), and a

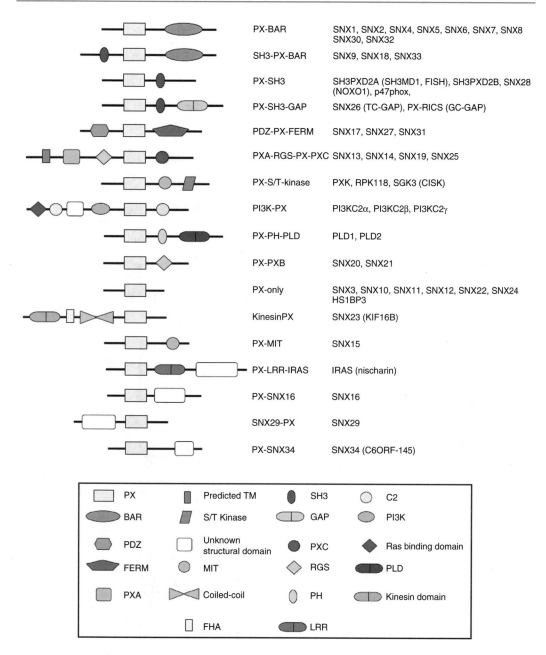

Fig. 1 The PX domain protein family. Schematic diagram of the domain structures of the known human PX domain-containing human proteins. The key/annotation for the various structures is shown in the right lower inset

survey of the literature found that PtdIns3*P* was also the most commonly reported lipid target of mammalian PX domain proteins (Teasdale and Collins 2012). The crystal structure of the p40[phox] PX domain bound to PtdIns3*P* provided the first molecular details on the mechanism of PtdIns3*P* recognition (Bravo et al. 2001), subsequently confirmed by structures of PtdIns3*P*-bound SNX9 (Pylypenko et al. 2007) and yeast Snx3p (Zhou et al. 2003) (Fig. 3b). It is evident from these structures that there are four key residues within these PX domains that determine

Fig. 2 Summary of the known PX domain structures. The figures lists each of the known human PX domain-containing proteins, classified into sub-families as in Fig. 1. The PDB codes for known NMR or crystal structures are shown on the right. The central panel highlights the four key side-chains required for coordination of PtdIns3*P* in the canonical binding pocket. Conservative substitutions in these side-chains are highlighted in grey and major alterations are indicated in black. The three yeast PX proteins with known structures are also shown for comparison

Sub-family	Protein	3-phosphate (R)	inositol (Y)	1-phosphate (K)	4,5-hydroxyls (R)	PDB Files
PX-BAR	SNX1		F			2I4K
	SNX2		F			
	SNX4					
	SNX5	Q	H		T	3HPB, 3HPC, 5TGI, 5TGJ, 5TGH
	SNX6	Q	H		T	
	SNX7					3IQ2
	SNX8					
	SNX30					
	SNX32	Q	H		T	
SH3-PX-BAR	SNX9					2RAI, 2RAJ, 2RAK, 3DYT, 3DYU
	SNX18					
	SNX33					4AKV
RGS-PX	SNX13					
	SNX14				K	4BGJ, 4PQO, 4PQP
	SNX19					4P2I, 4P2J
	SNX25		L		S	5XDZ, 5WOE
PX-FERM	SNX17					3FOG, 3LUI
	SNX27					4HAS
	SNX31					
PX-only	SNX3					2YPS, 5F0J, 5F0L
	SNX10					4ON3, 4PZG
	SNX11					4IKB, 4IKD
	SNX12					2CSK
	SNX22					2ETT
	SNX24					4AZ9
	HS1BP3	K			K	
PX-SH3	SH3PXD2A					
	SH3PXD2B					
	SNX28	S	W	L		2L73
	p40phox					1H6H, 2DYB
	p47phox		F	P		1O7K, 1KQ6, 1GD5
PX-S/T kinase	PXK					
	RPS6KC1					
	SGK3					1XTE, 1XTN, 4OXW
PX-SH3-GAP	SNX26	S		P	L	
	PX-RICS	S		R	V	
PX-PI3-kinase	PIK3C2A	T	F	R		2AR5, 2REA, 2RED, 2IWL
	PIK3C2B	T	F	R		
	PIK3C2G	S	F	W		2WWE
PX-PH-PLD	PLD1	K	F	R		
	PLD2	K		R		
PX-PXB	SNX20					
	SNX21					
Kinesin-PX	SNX23					2V14
PX-MIT	SNX15			R		
PX-LRR-IRAS	IRAS					3P0C
PX-SNX16	SNX16					5GW0, 5GW1
SNX29-PX	SNX29					
PX-SNX34	SNX34		S	R		
Yeast	Bem1p	Y		P		2V6V, 2CZO
	Snx3p					1OCS, 1OCU
	Vam7p					1KMD

☐ no change   ▣ similar   ■ different

specificity and are required for coordination of the headgroup of PtdIns3P. With p40^{phox} as reference the first key residue is Arg58 at the end of the β1 strand, which forms a salt bridge with the 3-phosphate. The second essential residue is Tyr59, immediately following Arg58 and lying at the start of the α1 helix, which forms a stacking interaction with the inositol ring itself. The third residue is Lys92, which is part of the ΨPxxPxK sequence in the extended PPK loop between α1 and α2 and forms a salt-bridge with the PtdIns3P 1-phosphoryl group. Finally, the hydroxyl groups at the un-phosphorylated 4 and 5 positions of the inositol ring are both coordinated by Arg105. This interaction is critical to the specificity, as phosphorylation of either the 4 or 5 positions would lead to steric exclusion of the phosphoinositide headgroup. Analogous side chains in both SNX9 and Snx3p coordinate PtdIns3P in an identical manner, thus defining a key set of residues for lipid specificity (Fig. 3b). There are now many examples where mutations in any of these key residues have been shown to abolish PtdIns3P interaction *in vitro*, and prevent association with endosomal membranes in cells. As just a small sample, mutation of the Arg side-chain in p40^{phox} that coordinates the D3 phosphate prevents membrane binding *in vitro* and *in situ* (Bravo et al. 2001), as do similar mutations in SNX17 (Ghai et al. 2011) and SNX16 (Xu et al. 2017). In yeast, the Mdm1p protein tethers the endoplasmic reticulum to the vacuole through PX domain binding to PtdIns3P, and mutation of the PX domain leads to disrupted contacts between these organelles (Henne et al. 2015).

Based on the conserved and essential role of the four key side-chains required for canonical PtdIns3P interactions, we can compare the sequences and structures of other PX domains (Seet and Hong 2006; Teasdale and Collins 2012) to predict the ability of the various family members to coordinate PtdIns3P through this canonical binding mode. Of the 49 human PX proteins, only 30 possess all four key binding residues (Figs. 2 and 4). While some proteins possess relatively conservative substitutions at key sites, notable exceptions include the PI3KC2 and PX-SH3-GAP sub-families as well

as the homologous SNX5, SNX6 and SNX32 proteins that all have major changes in the canonical binding pocket. These proteins would be predicted to not bind PtdIns3P at all based on these structural considerations.

3 Non-canonical Lipid Binding by PX Domains

Although PtdIns3P is the most commonly reported phosphoinositide to bind to PX domain containing proteins, many other phosphoinositide interactions have also been reported in the literature (Teasdale and Collins 2012). While this supports the general model that PX domain proteins play diverse roles in trafficking and signaling events at different subcellular compartments (Cullen and Carlton 2012; Kutateladze 2010; Lemmon 2008; Viaud et al. 2016), we urge caution in interpreting some of the previous data on alternative phosphoinositide binding by PX domains. There are many experimental methods that have been used to measure phospholipid binding by protein domains *in vitro*, including liposome pelleting assays, isothermal titration calorimetry (ITC), surface plasmon resonance (SPR) and simple dot-blot experiments. Each method comes with its own advantages and disadvantages, with dot-blot assays in particular being noted for producing false readings (Narayan and Lemmon 2006). Ideally, specific phosphoinositide interactions should be confirmed by more than one method *in vitro*, and also tested in cells before a lipid-binding specificity can be confidently assigned. Nonetheless it is clear that PtdIns3P is not the only lipid able to bind to PX domains. Rather than detail all of the reported phosphoinositide interactions we believe it is more illustrative to highlight several well studied examples.

Early work on the structure of the p47^{phox} PX domain and the lipid-binding studies reveal that the PX domain of p47^{phox} has two distinct lipid-binding sites with different specificities (Kanai et al. 2001; Karathanassis et al. 2002) (Fig. 3c). The first site, which is analogous to the canonical PtdIns3P-binding pocket of p40^{phox},

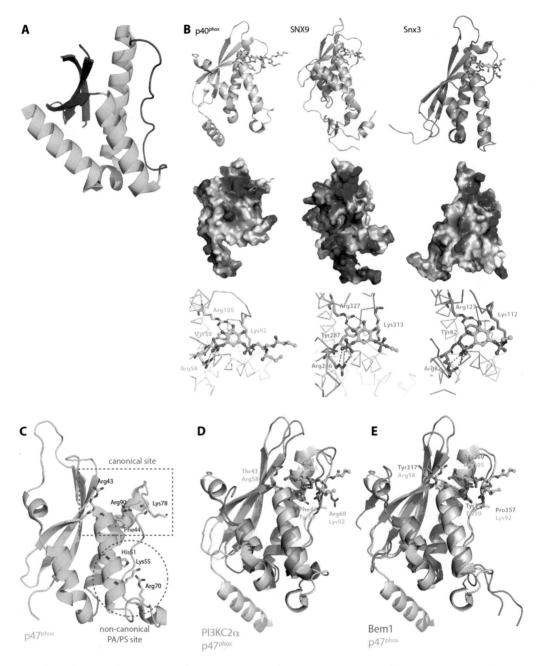

Fig. 3 Structure of the PX domain and canonical PtdIns3P interactions. (**a**) Cartoon representation of a representative PX domain from SNX17 (PDB code 3LUI) showing the core fold consisting of three anti-parallel β-strands (β1 – β3, shown in red) followed by three α-helices (α1 – α3, shown in cyan). The extended region between helices α1 and α2, also called polyproline loop (containing the ΨPxxPxK motif), is shown in magenta. (**b**) The known structures of PX domains bound to the canonical PtdIns3P lipid headgroup are shown. These are human p40phox (PDB code 1H6H), human SNX9 (PDB code 2RAK) and and yeast Snx3 (PDB code 1OCU). The PtdIns3P headgroup is shown in green ball-and-sticks with the top section showing the ribbon structure, the middle section showing the corresponding electrostatic surface, and the bottom section showing an enlarged view of the PtdIns3P binding site highlighting the four key side chains making direct contact with PtdIns3P. The hydrogen bonding interactions are shown as red dashed lines. (**c**) The crystal structure of

preferentially binds PtdIns(3,4)P_2; while the second site binds abundant anionic phospholipids including phosphatidic acid (PA) or phosphatidylserine (PS). The 'canonical' binding site in p47^{phox} possesses a conservative substitution of the key Tyr side-chain with a Phe. In the crystal structure a sulphate ion is coordinated in the same way as the 3-phosphate of PtdIns3P by p40^{phox}. It still remains to be determined as to the precise mechanism of PtdIns(3,4)P_2 engagement, although notably p47^{phox} also has weak affinities for other phosphoinositides suggesting that the binding pocket is perhaps just slightly altered from the typical coordination to be able to bind the doubly phosphorylated inositol ring. The second PA/PS binding site lies on the same surface of the domain as the phosphoinositide-binding pocket and is separated from it by a membrane insertion loop. This secondary binding site contains the side chains of Arg70, Lys55 and His51, giving it a basic character. In the crystal structure of p47^{phox} the guanidinium group of Arg70 forms hydrogen-bonding interactions with a bound sulfate ion and the backbone of Pro73 in the polyproline helix of the conserved ΨPxxPxK motif (residues 73–76; Fig. 3c). Again, the mechanism of PA/PS binding remains to be determined, but critically from a functional perspective, the simultaneous engagement of the two phospholipid-binding sites leads to a great synergistic increase in membrane affinity.

Two other clear examples of altered phospholipid specificity are PI3KC2α and the yeast protein Bem1p. The crystal structure of the PI3KC2α PX domain revealed a typical PX domain fold, but with a significantly altered canonical phosphoinositide-binding pocket (Stahelin et al. 2006) (Figs. 2, 3d and 4). Three of the four PtdIns3P-binding side chains are altered in PI3KC2α (and closely related homologues PI3KC2β and PI3KC2γ). SPR was used to test binding to liposomes containing different phosphoinostides, and showed clear binding to PtdIns(4,5)P_2 but not to PtdIns3P or other phospholipid membranes. Mutagenesis of side-chains around the canonical pocket disrupted PtdIns(4,5)P_2 interaction, but as of yet no co-crystal structures have been determined to show how this interaction occurs at the molecular level. PI3KC2α is known to associate with clathrin-coated vesicles where PtdIns(4,5)P_2 is enriched, and is involved in converting PtdIns4P into PtdIns(3,4)P_2 (Posor et al. 2013). It was also shown that the PI3KC2α C2 domain, which is adjacent to the C-terminal PX domain, is able to bind to many different phosphoinositide lipids (Stahelin et al. 2006), which suggests this enzyme is involved in a complicated network of phosphoinositide interactions and metabolism during the formation of these endocytic vesicles. In the case of Bem1p, there is also a significant alteration in the sequence of residues forming the canonical pocket (Figs. 2, 3e and 4). Similar to the PI3KC2α PX domain, the structure of Bem1p shows a binding pocket with major alterations that would preclude typical PtdIns3P interaction (Stahelin et al. 2007). Using similar SPR liposome binding experiments, Bem1p was found to bind specifically to PtdIns4P, although again the molecular details that underpin this specificity remain to be determined. Overall, despite the strong evidence for non-canonical phosphoinositide and phospholipid interactions mediated by various PX domains, there is still a notable lack of structural information on the mechanisms for these interactions.

Fig. 3 (continued) the PX domain of p47^{phox} (PDB code 1KQ6) shows two distinct lipid-binding sites. The first site, highlighted as blue dashed box, is analogous to the canonical PtdIns3P binding pocket of p40^{phox} and preferentially binds to PtdIns(3,4)P_2. The second site, highlighted as red dashed circle, binds to phosphatidic acid (PA) or phosphatidylserine (PS). The key residues thought to be involved in phosphoinositide binding to each of the respective sites are highlighted in stick representation. (**d**) Structural superposition of the PX domain of PI3KC2α (PDB code 2AR5) (shown as forest green colour) and PtdIns3P bound p40^{phox} (PDB code 1H6H) (shown as wheat colour) showing the presence of altered PtdIns3P-binding side chains in PI3KC2α. (**e**) Structural superposition of the PX domain of Bem1p (PDB code 2V6V) (shown as violet colour) and PtdIns3P bound p40^{phox} (PDB code 1H6H) (shown as wheat colour) showing the major alteration in the sequence of the residues at the canonical phospholipid binding pocket

```
                    β3      α1                              ΨPxxPxK                                                                      α2
                    RY                                                                                                                  R
SNX1_PX      QFAVKRRFSDFLGLYEKLSEKH--------------------SQNGFIVPPPPEKSLIGMTKVKVGKEDSS-------------------SAEFLEKRRAALERYLQRIVN
SNX2_PX      EFSVKRRFSDFLGLHSKLASKY--------------------LHVGYIVPPAPEKSIVGMTKVKVGKEDSS-------------------STEFVEKRRAALERYLQRTVK
SNX3_PX      ESTVRRRYSDFEWLRSELERES--------------------KVVVPPLPGKAFLRQLPFRGDDGIF-------------------DDNFIEERKQGLEQFINKVAG
SNX4_PX      TDSLWRRYSEFELLRSYLLVYY--------------------PHIVVPPLPEKRAEFVWHKLSADNM---------------------DPDFVERRRIGLENFLLRIAS
SNX5_PX      EFSVTRQHEDFVWLHDTLIETT--------------------DYAGLIIPPAPTKPDFDGPREKMQKLGEGEGSMTKEEFAKMKQELEAEYLAVFKKTVSSHEVFLQRLSS
SNX6_PX      EFSVVRQHEEFIWLHDSFVENE--------------------DYAGYIIPPAPPRPDFDASREKLQKLGEGEGSMTKEEFTKMKQELEAEYLAIFKKTVAMHEVFLCRVAA
SNX7_PX      EFEVRRRYQDFLWLKGKLEEAH--------------------PTLIIPPLPEKFIVKGMVERF-------------------NDDFIETRRKALHKFLNRIAD
SNX8_PX      KSSVYRRYNDFVVFQEMLLHKF--------------------PYRMVPALPPKRMLGA---------------------DREFIEARRRALKRFVNLVAR
SNX9_PX      NRSVNHRYKHFDWLYERLLVKF--------------------GSAIPIPSLPDKQVTGRF-------------------EEEFIKMRMERLQAWMTRMCR
SNX10_PX     TSCVRRRYREFVWLRQRLQSNA--------------------LLVQLPELPSKNLFFNMN---------------------NRQHVDQRRQGLEDFLRKVLQ
SNX11_PX     TSCVRRRYREFVWLRKQLQRNA--------------------GLVPVPELPGKSTFFGT---------------------SDEFIEKRRQGLQHFLEKVLQ
SNX12_PX     ESCVRRRYSDFEWLKNELERDS--------------------KIVVPPLPGKALKRQLPFRGDEGIF-------------------EESFIEERRQGLEQFINKIAG
SNX13_PX     MWKTYRRYSDFHDFHMRITEQF--------------------ESLSSSILKLPGKTFFNNM---------------------DRDFLEKRKKDLNAYLQLLLA
SNX14_PX     HWSVYRRYLEFYVLESKLTEFH--------------------GAFPDAQLPSKRIIGPK---------------------NYEFLKSKREEFQEYLQKLLQ
SNX15_PX     EVVVWKRYSDFRKLHGDLAYTH--------------------RNLFRRLEEFPAFPRAQVFGRF---------------------EASVIEERRKGAEDLLRFTVH
SNX16_PX     SWVVFRRYTDFSRLNDKLKEMF--------------------PGFRLALPPKRWFKDNY---------------------NADFLEDRQLGLQAFLQNLVA
SNX17_PX     VLHCRVRYSQLLGLHEQLRKEY--------------------GANVLPAFPPKKLFSL-----------------TPAEVEQRREQLEKYMQAVRQ
SNX18_PX     QVPVHRRYKHFDWLYARLAEKF--------------------PVISVPHLPEKQATGRF---------------------EEDFISKRRKGLIWWMNHMAS
SNX19_PX     YHTVNRRYREFLNLQTRLEEKP--------------------DLRKFIKNVKGPKKLFPDLPLGNM---------------------DSDRVEARKSLLESFLKQLCA
SNX20_PX     KAVLERRYSDFAKLQKALLKTF--------------------REEIEDVEFPRKHLTGNF---------------------AEEMICERRRALQEYLGLLYA
SNX21_PX     PAQISRRYSDFERLHRNLQRQF--------------------RGPMAAISFPRKRLRRNF---------------------TAETIARRSRAFEQFLGHLQA
SNX22_PX     RHTVPRRYSEFHALHKRIKKLY--------------------KVPDFPSKRLPNW-----------------RTRGLEQRRQGLEAYIQGILY
SNX23_PX     TWTVFRRYSRFREMHKTLKLKY--------------------AELAALEFPPKKLFGNK---------------------DERVIAERRSHLEKYLRDFFS
SNX24_PX     KHFVEKRYSEFHALHKKLKKCI--------------------KTPEIPSKHVRNW-----------------VPKVLEQRRQGLETYLQAVIL
SNX25_PX     NWTVPRRLSEFQNLHRKLSECV--------------------PSLKKVQLPSLSKLPFKSI---------------------DQKFMEKSKNQLNKFLQNLL-
SNX26_PX     SWPVLRSIDDFRSLDAHLHRCI--------------------FDRRFSCLPELPPPPE-----------------GARAAQMLVPLLLQYLETLSG
SNX27_PX     RQLCSKRYREFAILHQNLKREF--------------------ANFTFPRLPGKWPFSL-----------------SEQQLDARRRGLEEYLEKVCS
SNX28_PX     DTFVRRSWDEFRQLQKTLKKTF------------PVEAGLLRRSEQVLPKLPDAPLLTRRG---------------------HTGRGLVRLRLLDTYVQALLA
SNX29_PX     EWNIYRRYTEFRSLHHKLQNKY--------------------PQVRAYNFPPKAIGNK---------------------DAKFVEERRKQLQNYLRSVMN
SNX30_PX     EYSVRRRYQDFDWLRSKLEESQ--------------------PTHLIPPLPEKFVVKGVVDRF---------------------SEEFVETRRKALDKFLKRITD
SNX31_PX     FLFCRVRYSQLHGWNEQLRRVF--------------------GNCLPFFPPKYYLAM-----------------TTAMADERRDQLEQYLQNVTM
SNX32_PX     EFSVVRQHEEFIWLHDAYVENE--------------------EYAGLIIPPAPPRPDFEASREKLQKLGEGDSSVTREEFAKMKQELEAEYLAIFKKTVAMHEVFLQRLAA
SNX33_PX     ASPVYRRYKHFDWLYNRLLHKF--------------------TVISVPHLPEKQATGRF---------------------EEDFIEKKRRLILWMDHMTS
p40phox_PX   KYLIYRRYRQFYALQSKLEERF------------GPESKNSPFTCTLPTLPAKVYMA---------------------KQEIAETRIPALNAYMKNLLS
p47phox_PX   EKVVYRRFTEIYEFHKILKEMF------------PIEAGDINPENRIIPHLPAPRWFD---------------------GQRVAESRQGTLTEYCSALMS
PLD1_PX      KWQVKRKFKHFQEFHRELLKYKAFIRIPIPTRRHTFRRQNVREEPREMPSLPRSSENMI---------------------REEQFLGRRKQLEDYILTKILK
PLD2_PX      SWTTKKKYRHFQELHRDLLRHKVLMSLLPLARFAVAYSPARDAGNREMPSLPRAGPEG---------------------STRHAASKQKYLENYLNRLLT
HS1BP3_PX    QFLVSKKYSEIEEFYQKLSSRY--------------------AAASLPPYLPRKVLFV---------------------GESDIRERRAVFNEILRCVSK
SH3PXD2A_PX  SQTIYRRYSKFFDLQMQLLDKF------------PIEGGQKDPKQRIIPFLPGKILFRRSH---------------------IRDVAVRKLKPIDEYCRALVR
SH3PXD2B_PX  TEAIYRRYSKFFDLQMQMLDKF------------PMEGGQKDPKQRIIPFLPGKILFRRSH---------------------IRDVAVKLLIPIDEYCKALIQ
IRAS_PX      EWTVKHRYSDFHDLHEKLVAER--------------------KIDKNLLPPKKIIGKNSRSLVEKREKDN---------------------SRSLVEKREKDLEVYLQKLLA
Nischarin_PX EWTVKHRYSDFHDLHEKLVAER--------------------KIDKNLLPPKKIIGKN---------------------SRSLVEKREKDLEVYLQKLLA
PXK_PX       SWQIVRRYSDFDLLNNSLQIAG--------------------LSLLPPKKLIGNM---------------------DREFIEARQKGLQNYLNVITT
SGK3_PX      EWFVFRRYAEFDKLYNTLKKQF--------------------PAMAL-KIPAKRIFGDNF---------------------DPDFIKQRRAGLNEFIQNLVR
PI3KC2A_PX   PSFVFRTFDEFQELHNKLSIIF--------------------PLWKLPGFPNRMVLGRTH---------------------IKDVAAKRKIELNSYLQSLMN
PI3KC2B_PX   ATYIQRTFEEFQELHNKLRLLF--------------------PSSHLPSFPSRFVIGRSR---------------------GEAVAERREELNGYIWHLIH
PI3KC2G_PX   TSLTEKSFEQFSKLHSQLQKQF--------------------ASLTLPEFPHWWHLP-----------------FTNSDHRRFRDLNHYMEQILN
```

Fig. 4 **Alignment of amino acid sequences of human PX domains**. The amino acid sequences of the PX domains of human proteins are aligned within the phosphoinositide recognition region (strand β3 to helix α2). The secondary structure of the PX domain of p40[phox] is indicated on the top for reference. The four key residues involved in phosphoinositide recognition are shown in blue and highlighted at the top of the sequence. The polyproline loop (α1α2 loop), which includes the sequence motif ΨPxxPxK is is also highlighted with Ψ residues coloured in green and conserved Prolines in the motif coloured red. The Lysine of this motif is also part of the core tetrad of residues involved in PtdIns3P interaction. The alignment was performed using ClustalW and then manually adjusted to ensure accurate alignment of the key residues. Where available the precision of the sequence alignments was also confirmed by structural comparisons of the PX domains

4   Intermolecular Protein-Protein Interactions Mediated by PX Domains

Since the first reports of the ability of PX domains to interact with membrane phospholipids, the large majority of subsequent studies have focused on this key function of the domain. However, there are also a number of notable examples where PX domains act as protein-protein interacting modules (reviewed in (Teasdale and Collins 2012)). Indeed several of the first studies of PX domains noted the presence of the conserved ΨPxxPxK sequence in the extended α1-α2 loop, and postulated the ability of these sequences to mediate interactions with polyproline binding SH3 (Src homology 3) domains (Hiroaki et al. 2001; Ponting 1996). While this is an attractive idea, the evidence for such interactions still remains limited, with only one report showing a direct interaction of the PLCγ1 (phospholipase Cγ1) SH3 domain with the ΨPxxPxK sequence of PLD1 (phospholipase D1) and PLD2 PX domains (Jang et al. 2003), and high-throuput peptide binding experiments confirming the ability of SH3 domains to bind sequences from several PX domain proteins (Wu et al. 2007). As well as these intermolecular interactions a number of PX domain proteins possess one or more SH3 domains themselves, suggesting the possibility of intramolecular regulatory interactions. Again the evidence for this is limited, but the C-terminal SH3 domain of p47[phox] has been shown to bind its PX domain by pull-down assays and NMR experiments (Hiroaki et al. 2001). Despite the high degree of conservation of the ΨPxxPxK motif and the potential implications for co-regulation by SH3 domains and phosphoinositides binding to adjacent sites, SH3 domain interactions have remained remarkably poorly studied.

The first clear example of a direct PX domain-mediated protein interaction was revealed by the crystal structure of SNX3 with VPS26 and VPS35 subunits of the retromer endosomal trafficking complex (Lucas et al. 2016) (Fig. 5a). SNX3 is a member of the sub-family that consists of only a PX domain with some extended unstructured N and C-terminal sequences. It was first shown to interact with retromer in *S. cerevisiae* (Strochlic et al. 2007), and subsequently confirmed in human cells and with purified proteins (Harrison et al. 2014; Harterink et al. 2011). The crystal structure revealed that SNX3 interactions with the membrane and retromer occur on opposite sides of the SNX3 PX domain, consistent with SNX3 being a structural scaffold primed for recruiting retromer to endosomal membranes (Fig. 5a). SNX3 binding also involves contacts with retromer via an N-terminal extended structure, and the authors confirmed the importance of these interactions for retromer recruitment and Dmt1-II trafficking in cells via structure-based SNX3 mutagenesis. Thus, SNX3 integrates multiple binding sites within a single PX domain, enabling the recruitment of retromer to endosomal membranes.

Another clear example of protein-protein interactions mediated by PX domains is that of the SNX5 and SNX6 proteins with a protein from the bacterial pathogen *Chlamydia trachomatis*. *C. trachomatis* is an obligate intracellular pathogen, which hijacks host cell intracellular trafficking and lipid transport pathways to promote infection by generating a stable membrane enclosed vacuole termed the inclusion. The bacteria secrete a large array of short transmembrane inclusion proteins that decorate the membrane including the protein IncE, which was shown to bind directly to the SNX5 and SNX6 PX domains (Aeberhard et al. 2015; Mirrashidi et al. 2015). Subsequent crystal structures of the SNX5 PX domain in complex with IncE showed that IncE forms a long β-hairpin structure that binds within a complementary groove at the base of an extended α-helical insertion of the SNX5 PX domain and adjacent to the β-sheet sub-domain (Fig. 5b) (Elwell et al. 2017; Paul et al. 2017). An interesting finding was that the IncE protein binds to a very highly conserved surface of the SNX5 protein, suggesting that it is binding to site that is required for normal SNX5 function. While it is not yet clear what this function might be, it has been shown that the SNX5 and SNX6 PX domains can interact directly with other proteins including Homer1b (Niu et al. 2017) and PIPKIγ (Type Iγ phosphatidylinositol-5-kinase) (Sun et al. 2013). It now remains to be determined if these or other molecules may utilise the same binding pocket in this family of sorting nexin PX domains.

Other putative protein-protein interactions with PX domains include the binding of Vam7 to the HOPS tethering complex (Stroupe et al. 2006), PLD1 and PLD2 to dynamin (Lee et al. 2006), Munc18-1 (Lee et al. 2004) and PKCζ (Kim et al. 2005), p47[phox] with moesin (Zhan et al. 2004) and cPLA2 (Shmelzer et al. 2008), and SNX20 with PSGL1 (Schaff et al. 2008). Altogether, the above examples provide substantial evidence that PX domains act not only as lipid recognition modules, but play a key role in protein–protein interactions. The molecular details of some of these interactions have provided critical insights into their functional importance, but many more remain to be further charcaterised. A better understanding of the protein-binding partners of the PX domains will allow us to discern common features governing their association, lead to the identification of other interacting molecules, and will drive the design of

Fig. 5 Intermolecular protein-protein interactions mediated by PX domains. (**a**) Crystal structure of retromer subunits VPS26-VPS35(N) in complex with the PX domain of SNX3 and the cargo peptide sequence of the divalent cation transporter, Dmt1-II (PDB code 5F0J). The interactions with the membrane and retromer occur on opposite sides of the SNX3 PX domain. (**b**) Crystal structure of the SNX5 PX domain in complex with the IncE from *C. trachomatis* (PDB code 5TGI). The site of canonical PtdIns3*P* binding site is marked as dashed black box, however in SNX5 this site is completely altered. IncE forms a long β-hairpin structure that binds within a complementary groove at the base of the extended α-helical insertion of the SNX5 PX domain and adjacent to the β-sheet sub-domain

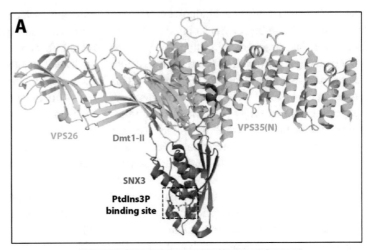

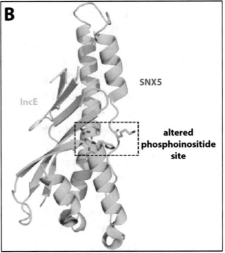

specific mutants aimed at dissecting the functional importance of PX domain protein recruitment.

5 Embellishments of the PX Domain

Although all of the PX domain structures determined to date share the core fold revealed in the very first structure of p47[phox] (Hiroaki et al. 2001), there are several members that possess unique features of both known and unknown function (Fig. 6). The best understood example of this is SNX5, which possesses a long α-helical

hairpin that protrudes from the core PX domain fold that is also found in close homologues SNX6 and SNX32 (Koharudin et al. 2009). This unique structural element is formed by an extended sequence between the common ΨPxxPxK motif-containing loop and the α2-helix, which is not present in any other PX domains. As discussed above this region is involved in binding to the bacterial effector IncE and is likely performing a conserved function in the normal activity of this protein in endosomal membrane trafficking (Elwell et al. 2017; Paul et al. 2017). Other embellished PX domain structures involve extensions at either the N- or C-termini. In the case of both SNX10 and SNX11, two additional

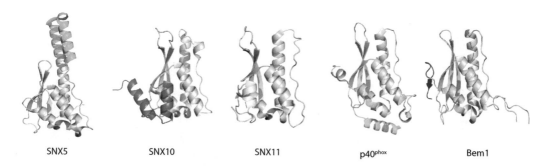

SNX5   SNX10   SNX11   p40phox   Bem1

Fig. 6 Structural embellishments of the PX domain and non-canonical lipid interactions. PX domains with structural extensions are shown in ribbon diagram with the core PX fold shown in wheat colouring and the extended structural elements highlights in alternative colours. The PDB codes are: SNX5 (3HPB), SNX10 (4ON3), SNX11 (4IKB), p40phox (1H6H), and yeast Bem1p (2V6V)

α-helices are present downstream of the conventional PX domain (Xu et al. 2013, 2014). This extended PX domain (referred to as the PXe domain) is found to be critical for the activity of these proteins in generating enlarged endocytic vacuoles upon over-expression (Qin et al. 2006; Xu et al. 2014). p40phox also possesses an additional α-helix, in this case at the N-terminus, the function of which is currently unknown (Bravo et al. 2001), and the yeast Bem1p protein has an extra β-strand at the C-terminus, again of unknown function (Stahelin et al. 2007). The final known example of an extended PX domain is that of SNX3. In this case the N-terminal region of SNX3, which is disordered on its own, forms a long and extended interface with the retromer subunits VPS26 and VPS35 when in complex as mentioned above (Fig. 5a) (Lucas et al. 2016).

6   Coordination of the PX Domain with Other Functional Modules

As is shown in Fig. 1, most PX domain proteins possess one or more other functional domains. While the PX domain tends to be thought of as a membrane-binding scaffold that recruits these proteins to where they need to be, there are now several examples of where the PX domain is clearly integrated into a larger structural and functional module. The best example of this is that of SNX9 (and its close homologues SNX18 and SNX33) (Fig. 7a). SNX9 has an N-terminal SH3 domain connected to the central PX domain, which is then followed closely by a C-terminal BAR domain. The BAR domain of SNX9 forms a coiled-coil consisting of three α helices which then dimerises forming a curved structure for membrane bending. The crystal structure of the SNX9 PX and BAR domains (Pylypenko et al. 2007; Wang et al. 2008) as well as the unpublished SNX33 PX-BAR structure (PDB ID 4AKV) reveals two intimately associated domains that are clasped together in a rigid architecture by a novel structure called the 'yoke' domain, which consists of short sequences both N- and C-terminal to the PX domain. The structure of SNX9 was also solved in the presence of PtdIns3P, showing that overall the PX-BAR unit is primed for both PtdIns3P membrane recruitment by the PX domain and for membrane tubulation by the BAR domain (Pylypenko et al. 2007).

Other examples where the PX domain is found to be intimately associated with additional structures are p40phox (Fig. 7b) and SNX16 (Fig. 7c). In p40phox, the N-terminal PX domain is followed by a central SH3 domain, with a C-terminal PB1 (phox and Bem1) domain. The crystal structure of full-length p40phox shows that the N-terminal PX and C-terminal PB1 domains in fact form an intimate interface with each other, while the central SH3 domain is in a flexible orientation (Honbou et al. 2007). This interface

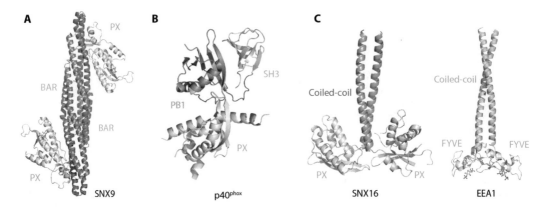

Fig. 7 Coordination of the PX domain with other functional domains. (a) The close association between the PX (wheat colour) and BAR domains (skyblue colour) as revealed in the crystal structure of SNX9 (PDB code 2RAI). (b) Crystal structure of full-length p40[phox] (PDB code 2DYB) shows that the N-terminal PX (wheat colour) and C-terminal PB1 domains (green) form an intimate interface with each other, while the central SH3 domain (light blue) is in a flexible orientation. (c) The structure of SNX16 PX domain with its C-terminal coiled-coil (CC) domain (PDB code 5GW0) shows that it forms a coiled-coil homodimer (salmon colour), where the two N-terminal PX domains (wheat colour) are positioned to bind to membrane phospholipids. The structure of PtdIns3P-bound EEA1 FYVE domain (wheat) (PDB code 1JOC) with its N-terminal coiled-coil dimerization domain (light green) is shown for comparison. Its dimeric structure with dual phosphoinositide binding domains is very similar in principle to the SNX16 protein

is found to be autoinhibitory for membrane recruitment, in the sense that the interaction between the PX and PB1 domains prevents interactions with PtdIns3P-containing membranes via steric occlusion. In contrast this interface does not interfere with the heterodimeric binding of the p40[phox] PB1 domain with the PB1 domain of p67[phox]. Therefore it is postulated that a pre-assembled phagocyte oxidase complex requires either an additional protein or post-translational modification of p40[phox] to release the autoinhibitory intramolecular interaction and allow membrane binding *in vivo*. A third recent example is SNX16, where the PX domain is adjacent to a C-terminal coiled-coil sequence. The structure of SNX16 shows that it forms a coiled-coil homodimer, where the two N-terminal PX domains are positioned to bind to membrane phospholipids (Xu et al. 2017) (Fig. 7c). Dimerisation of membrane associating domains is often thought to confer membrane-binding avidity, with an obvious analogue being that of the PtdIns3P-binding FYVE domain, which shares a similar coiled-coil driven dimerisation to that of SNX16 (Dumas et al. 2001)(Fig. 7c). This dimerization of SNX16 is also thought to

promote protein-protein interaction between the PX domain of SNX16 and the cytoplasmic domain of E-cadherin to promote endosomal recycling to the plasma membrane.

7 Coincidence Detection by PX Domain Proteins

Coincidence detection is a process whereby peripheral membrane proteins are recruited to specific cellular compartments required for their function by multiple coordinated molecular interactions (Carlton and Cullen 2005). This phenomenon can involve either the simultaneous recognition of phosphoinositides and other membrane lipids, the dual interaction with both membrane lipids and either transmembrane or other membrane-associated proteins, or the coordinated recognition of both membrane lipids and membrane shape (i.e. curvature).

The PX domain family provides instances of all of these different coincidence detection mechanisms (Fig. 8). As discussed above, the PX domain of p47[phox] recognizes both PtdIns$(3,4)P_2$ and PA and this simultaneous binding

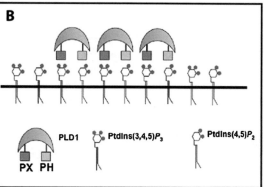

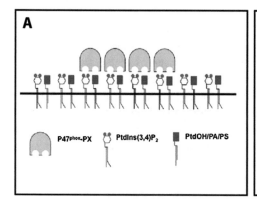

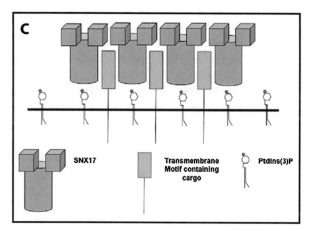

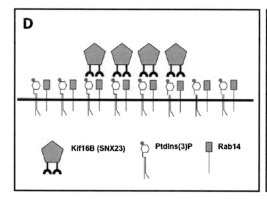

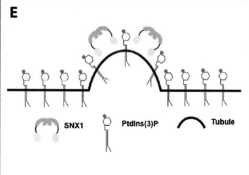

Fig. 8 Coincidence detection by PX domain containing proteins. Various mechanisms of membrane coincidence detection are exemplified by different members of the PX domain protein family. (**a**) Dual lipid recognition – Detection of phosphoinositides and other membrane lipids. Exemplified by the PtdIns(3,4)P_2- and PtdOH/PA/PS-dependent localization of p47^{phox}. (**b**) Two lipid binding domains – Detection of two different phosphoinositides. The PtdIns(3,4,5)P_3- and PtdIns(4,5)P_2 dependent localization of PLD1 through its PX and PH domains respectively. (**c**) Detection of phosphoinositides and transmembrane motif containing cargo. Exemplified by the transmembrane motif containing cargo and PtdIns3P-dependent localization of SNX17 to the endosomal membrane. (**d**) Detection of phosphoinositides and membrane tethered small monomeric GTPases. Exemplified by the Rab14 and PtdIns(3)P-dependent localization of Kif16B (SNX23). (**e**) Detection of phosphoinositides and geometric curvature. Exemplified by the PtdIns3P and curvature-dependent localization of SNX1 to endosomes

significantly enhances the avidity for membranes (Kanai et al. 2001; Karathanassis et al. 2002). PLD1 contains not only a PX domain but also a PH domain, able to bind respectively to PtdIns $(3,4,5)P_3$ (Lee et al. 2005; Stahelin et al. 2004) and PtdIns$(4,5)P_2$ (Hodgkin et al. 2000), while SNX27 possesses a second phospholipid binding site within its C-terminal FERM domain that regulates its recruitment to membranes of the T-cell immune synapse (Ghai et al. 2015). Both SNX27 and its homologue SNX17 are known to interact with PtdIns3P via the PX domain and with transmembane cargo proteins, via the FERM domain of either protein with cargos containing FxNPxY sequences, or via the PDZ domain of SNX27 with cargos containing PDZ binding motifs (Clairfeuille et al. 2016; Ghai et al. 2011, 2013). Membrane-attached small GTPases from the Rab family are common regulators of peripheral membrane proteins, and although to-date PX proteins have not been strongly implicated as direct Rab effectors, the kinesin Kif16B/SNX23 is partly endosome-associated via dual interaction with PtdIns3P and Rab14 (Ueno et al. 2011). Finally, PX domain proteins possessing C-terminal BAR domains have been found to preferentially associate with and function at phospholipid membranes with increased curvature. SNX1 was found to associate more strongly with membranes of higher curvature *in vitro* and is found on curved membrane domains in cells (Carlton et al. 2004). In addition SNX9 activity in promoting actin polymerization from membrane surfaces is potentiated by the presence of both PtdIns3P and by increased membrane curvature (Gallop et al. 2013). These are just some examples of membrane coincidence detection by PX domain proteins, and certainly many more family members will also be regulated by similar multivalent membrane interactions.

8 Conclusions and Future Perspectives

While the functions of many of the PX domain containing proteins are reasonably well understood, there are still many others whose function is either poorly characterized or essentially unknown. Similar to other phosphoinositide-binding modules such as the PH domain, the PX domain itself can serve different functions both in controlling subcellular localization or via regulation of protein-protein interactions. In this chapter we have summarized some of the key aspects of PX domain structure and function including the different mechanisms of membrane recognition and the emerging understanding of both the inter- and intra-molecular protein association by this small domain across diverse family members. Many questions remain however, in particular we still do not understand how alternative phosphoinositides are engaged by some PX domains, the potential role of SH3 domain interactions with the ΨPxxPxK sequence has received very little attention, and we have only just begun to scratch the surface regarding our understanding of different PX domain protein interactions. PX domain proteins are not only highly conserved proteins with fundamental roles in eukaryotic cell physiology, their dysfunction is also implicated in many human disorders including pathogen invasion, cancer and neurodegeneration (Teasdale and Collins 2012), and new insights into their diverse cellular activities may lead not just to new fundamental knowledge but also to the development of methods to target these pathways in disease. There remains a great deal yet for both cell biologists and biomedical scientists to discover about this important protein family.

Acknowledgements BMC is supported by an Australian National Health and Medical Research Council (NHMRC) Career Development Fellowship (APP1061574), and research in the lab is supported by an NHMRC Project Grant (APP1099114).

References

Aeberhard L, Banhart S, Fischer M, Jehmlich N, Rose L, Koch S, Laue M, Renard BY, Schmidt F, Heuer D (2015) The proteome of the isolated chlamydia trachomatis containing vacuole reveals a complex trafficking platform enriched for Retromer components. PLoS Pathog 11:e1004883

Ago T, Takeya R, Hiroaki H, Kuribayashi F, Ito T, Kohda D, Sumimoto H (2001) The PX domain as a novel phosphoinositide- binding module. Biochem Biophys Res Commun 287:733–738

Balla T (2013) Phosphoinositides: tiny lipids with giant impact on cell regulation. Physiol Rev 93:1019–1137

Bravo J, Karathanassis D, Pacold CM, Pacold ME, Ellson CD, Anderson KE, Butler PJ, Lavenir I, Perisic O, Hawkins PT et al (2001) The crystal structure of the PX domain from p40(phox) bound to phosphatidylinositol 3-phosphate. Mol Cell 8:829–839

Carlton JG, Cullen PJ (2005) Coincidence detection in phosphoinositide signaling. Trends Cell Biol 15:540–547

Carlton J, Bujny M, Peter BJ, Oorschot VM, Rutherford A, Mellor H, Klumperman J, McMahon HT, Cullen PJ (2004) Sorting nexin-1 mediates tubular endosome-to-TGN transport through coincidence sensing of high- curvature membranes and 3-phosphoinositides. Current Biol 14:1791–1800

Cheever ML, Sato TK, de Beer T, Kutateladze TG, Emr SD, Overduin M (2001) Phox domain interaction with PtdIns(3)P targets the Vam7 t-SNARE to vacuole membranes. Nat Cell Biol 3:613–618

Clairfeuille T, Mas C, Chan AS, Yang Z, Tello-Lafoz M, Chandra M, Widagdo J, Kerr MC, Paul B, Merida I et al (2016) A molecular code for endosomal recycling of phosphorylated cargos by the SNX27-retromer complex. Nat Struct Mol Biol 23:921–932

Cullen PJ (2008) Endosomal sorting and signalling: an emerging role for sorting nexins. Nat Rev Mol Cell Biol 9:574–582

Cullen PJ, Carlton JG (2012) Phosphoinositides in the mammalian endo-lysosomal network. Subcell Biochem 59:65–110

Dumas JJ, Merithew E, Sudharshan E, Rajamani D, Hayes S, Lawe D, Corvera S, Lambright DG (2001) Multivalent endosome targeting by homodimeric EEA1. Mol Cell 8:947–958

Ellson CD, Gobert-Gosse S, Anderson KE, Davidson K, Erdjument-Bromage H, Tempst P, Thuring JW, Cooper MA, Lim ZY, Holmes AB et al (2001) PtdIns(3)P regulates the neutrophil oxidase complex by binding to the PX domain of p40(phox). Nat Cell Biol 3:679–682

Ellson CD, Andrews S, Stephens LR, Hawkins PT (2002) The PX domain: a new phosphoinositide-binding module. J Cell Sci 115:1099–1105

Elwell CA, Czudnochowski N, von Dollen J, Johnson JR, Nakagawa R, Mirrashidi K, Krogan NJ, Engel JN, Rosenberg OS (2017) Chlamydia interfere with an interaction between the mannose-6-phosphate receptor and sorting nexins to counteract host restriction. elife 6: e22709

Gallop JL, Walrant A, Cantley LC, Kirschner MW (2013) Phosphoinositides and membrane curvature switch the mode of actin polymerization via selective recruitment of toca-1 and Snx9. Proc Natl Acad Sci U S A 110:7193–7198

Ghai R, Mobli M, Norwood SJ, Bugarcic A, Teasdale RD, King GF, Collins BM (2011) Phox homology band 4.1/ezrin/radixin/moesin-like proteins function as molecular scaffolds that interact with cargo receptors and Ras GTPases. Proc Natl Acad Sci U S A 108:7763–7768

Ghai R, Bugarcic A, Liu H, Norwood SJ, Skeldal S, Coulson EJ, Li SS, Teasdale RD, Collins BM (2013) Structural basis for endosomal trafficking of diverse transmembrane cargos by PX-FERM proteins. Proc Natl Acad Sci U S A 110:E643–E652

Ghai R, Tello-Lafoz M, Norwood SJ, Yang Z, Clairfeuille T, Teasdale RD, Merida I, Collins BM (2015) Phosphoinositide binding by the SNX27 FERM domain regulates its localization at the immune synapse of activated T-cells. J Cell Sci 128:553–565

Hammond GR, Balla T (2015) Polyphosphoinositide binding domains: key to inositol lipid biology. Biochim Biophys Acta 1851:746–758

Harrison MS, Hung CS, Liu TT, Christiano R, Walther TC, Burd CG (2014) A mechanism for retromer endosomal coat complex assembly with cargo. Proc Natl Acad Sci U S A 111:267–272

Harterink M, Port F, Lorenowicz MJ, McGough IJ, Silhankova M, Betist MC, van Weering JRT, van Heesbeen R, Middelkoop TC, Basler K et al (2011) A SNX3-dependent retromer pathway mediates retrograde transport of the Wnt sorting receptor Wntless and is required for Wnt secretion. Nat Cell Biol 13:914–923

Henne WM, Zhu L, Balogi Z, Stefan C, Pleiss JA, Emr SD (2015) Mdm1/Snx13 is a novel ER-endolysosomal interorganelle tethering protein. J Cell Biol 210:541–551

Hiroaki H, Ago T, Ito T, Sumimoto H, Kohda D (2001) Solution structure of the PX domain, a target of the SH3 domain. Nat Struct Biol 8:526–530

Hodgkin MN, Masson MR, Powner D, Saqib KM, Ponting CP, Wakelam MJ (2000) Phospholipase D regulation and localisation is dependent upon a phosphatidylinositol 4,5-biphosphate-specific PH domain. Current Biol 10:43–46

Honbou K, Minakami R, Yuzawa S, Takeya R, Suzuki NN, Kamakura S, Sumimoto H, Inagaki F (2007) Full-length p40phox structure suggests a basis for regulation mechanism of its membrane binding. EMBO J 26:1176–1186

Jang IH, Lee S, Park JB, Kim JH, Lee CS, Hur EM, Kim IS, Kim KT, Yagisawa H, Suh PG et al (2003) The direct interaction of phospholipase C-gamma 1 with phospholipase D2 is important for epidermal growth factor signaling. J Biol Chem 278:18184–18190

Kanai F, Liu H, Field SJ, Akbary H, Matsuo T, Brown GE, Cantley LC, Yaffe MB (2001) The PX domains of p47phox and p40phox bind to lipid products of PI(3) K. Nat Cell Biol 3:675–678

Karathanassis D, Stahelin RV, Bravo J, Perisic O, Pacold CM, Cho W, Williams RL (2002) Binding of the PX domain of p47(phox) to phosphatidylinositol 3,4-bisphosphate and phosphatidic acid is masked by an intramolecular interaction. EMBO J 21:5057–5068

Kim JH, Ohba M, Suh PG, Ryu SH (2005) Novel functions of the phospholipase D2-Phox homology domain in protein kinase Czeta activation. Mol Cell Biol 25:3194–3208

Koharudin LM, Furey W, Liu H, Liu YJ, Gronenborn AM (2009) The phox domain of sorting nexin 5 lacks PTDINS(3)P specificity and preferentially binds to PTDINS(4,5)P2. J Biol Chem 284:23697–23707

Kutateladze TG (2010) Translation of the phosphoinositide code by PI effectors. Nat Chem Biol 6:507–513

Lee HY, Park JB, Jang IH, Chae YC, Kim JH, Kim IS, Suh PG, Ryu SH (2004) Munc-18-1 inhibits phospholipase D activity by direct interaction in an epidermal growth factor-reversible manner. J Biol Chem 279:16339–16348

Lee JS, Kim JH, Jang IH, Kim HS, Han JM, Kazlauskas A, Yagisawa H, Suh PG, Ryu SH (2005) Phosphatidylinositol (3,4,5)-trisphosphate specifically interacts with the phox homology domain of phospholipase D1 and stimulates its activity. J Cell Sci 118:4405–4413

Lee CS, Kim IS, Park JB, Lee MN, Lee HY, Suh PG, Ryu SH (2006) The phox homology domain of phospholipase D activates dynamin GTPase activity and accelerates EGFR endocytosis. Nat Cell Biol 8:477–484

Lemmon MA (2008) Membrane recognition by phospholipid-binding domains. Nat Rev Mol Cell Biol 9:99–111

Lucas M, Gershlick DC, Vidaurrazaga A, Rojas AL, Bonifacino JS, Hierro A (2016) Structural mechanism for cargo recognition by the retromer complex. Cell 167(1623–1635):e1614

Mirrashidi KM, Elwell CA, Verschueren E, Johnson JR, Frando A, Von Dollen J, Rosenberg O, Gulbahce N, Jang G, Johnson T et al (2015) Global mapping of the Inc-human Interactome reveals that retromer restricts chlamydia infection. Cell Host Microbe 18:109–121

Narayan K, Lemmon MA (2006) Determining selectivity of phosphoinositide-binding domains. Methods 39:122–133

Niu Y, Dai Z, Liu W, Zhang C, Yang Y, Guo Z, Li X, Xu C, Huang X, Wang Y et al (2017) Ablation of SNX6 leads to defects in synaptic function of CA1 pyramidal neurons and spatial memory. eLife 6: e20991

Paul B, Kim HS, Kerr MC, Huston WM, Teasdale RD, Collins BM (2017) Structural basis for the hijacking of endosomal sorting nexin proteins by chlamydia trachomatis. eLife 6:pii: e22311

Ponting CP (1996) Novel domains in NADPH oxidase subunits, sorting nexins, and PtdIns 3-kinases: binding partners of SH3 domains? Protein Sci 5:2353–2357

Posor Y, Eichhorn-Gruenig M, Puchkov D, Schoneberg J, Ullrich A, Lampe A, Muller R, Zarbakhsh S, Gulluni F, Hirsch E et al (2013) Spatiotemporal control of endocytosis by phosphatidylinositol-3,4-bisphosphate. Nature 499:233–237

Pylypenko O, Lundmark R, Rasmuson E, Carlsson SR, Rak A (2007) The PX-BAR membrane-remodeling unit of sorting nexin 9. EMBO J 26:4788–4800

Qin B, He M, Chen X, Pei D (2006) Sorting nexin 10 induces giant vacuoles in mammalian cells. J Biol Chem 281:36891–36896

Sato TK, Overduin M, Emr SD (2001) Location, location, location: membrane targeting directed by PX domains. Science 294:1881–1885

Schaff UY, Shih HH, Lorenz M, Sako D, Kriz R, Milarski K, Bates B, Tchernychev B, Shaw GD, Simon SI (2008) SLIC-1/sorting nexin 20: a novel sorting nexin that directs subcellular distribution of PSGL-1. Eur J Immunol 38:550–564

Seet LF, Hong W (2006) The Phox (PX) domain proteins and membrane traffic. Biochim Biophys Acta 1761:878–896

Shmelzer Z, Karter M, Eisenstein M, Leto TL, Hadad N, Ben-Menahem D, Gitler D, Banani S, Wolach B, Rotem M et al (2008) Cytosolic phospholipase A2alpha is targeted to the p47phox-PX domain of the assembled NADPH oxidase via a novel binding site in its C2 domain. J Biol Chem 283:31898–31908

Song X, Xu W, Zhang A, Huang G, Liang X, Virbasius JV, Czech MP, Zhou GW (2001) Phox homology domains specifically bind phosphatidylinositol phosphates. Biochemistry 40:8940–8944

Stahelin RV, Ananthanarayanan B, Blatner NR, Singh S, Bruzik KS, Murray D, Cho W (2004) Mechanism of membrane binding of the phospholipase D1 PX domain. J Biol Chem 279:54918–54926

Stahelin RV, Karathanassis D, Bruzik KS, Waterfield MD, Bravo J, Williams RL, Cho W (2006) Structural and membrane binding analysis of the Phox homology domain of phosphoinositide 3-kinase-C2alpha. J Biol Chem 281:39396–39406

Stahelin RV, Karathanassis D, Murray D, Williams RL, Cho W (2007) Structural and membrane binding analysis of the Phox homology domain of Bem1p: basis of phosphatidylinositol 4-phosphate specificity. J Biol Chem 282:25737–25747

Strochlic TI, Setty TG, Sitaram A, Burd CG (2007) Grd19/Snx3p functions as a cargo-specific adapter for retromer-dependent endocytic recycling. J Cell Biol 177:115–125

Stroupe C, Collins KM, Fratti RA, Wickner W (2006) Purification of active HOPS complex reveals its affinities for phosphoinositides and the SNARE Vam7p. EMBO J 25:1579–1589

Sun Y, Hedman AC, Tan X, Schill NJ, Anderson RA (2013) Endosomal type Igamma PIP 5-kinase controls EGF receptor lysosomal sorting. Dev Cell 25:144–155

Teasdale RD, Collins BM (2012) Insights into the PX (phox-homology) domain and SNX (sorting nexin) protein families: structures, functions and roles in disease. Biochem J 441:39–59

Ueno H, Huang X, Tanaka Y, Hirokawa N (2011) KIF16B/Rab14 molecular motor complex is critical for early embryonic development by transporting FGF receptor. Dev Cell 20:60–71

Viaud J, Mansour R, Antkowiak A, Mujalli A, Valet C, Chicanne G, Xuereb JM, Terrisse AD, Severin S, Gratacap MP et al (2016) Phosphoinositides: important lipids in the coordination of cell dynamics. Biochimie 125:250–258

Virbasius JV, Song X, Pomerleau DP, Zhan Y, Zhou GW, Czech MP (2001) Activation of the Akt-related cytokine-independent survival kinase requires interaction of its phox domain with endosomal phosphatidylinositol 3-phosphate. Proc Natl Acad Sci U S A 98:12908–12913

Wang Q, Kaan HY, Hooda RN, Goh SL, Sondermann H (2008) Structure and plasticity of endophilin and sorting nexin 9. Structure 16:1574–1587

Worby CA, Dixon JE (2002) Sorting out the cellular functions of sorting nexins. Nat Rev Mol Cell Biol 3:919–931

Wu C, Ma MH, Brown KR, Geisler M, Li L, Tzeng E, Jia CY, Jurisica I, Li SS (2007) Systematic identification of SH3 domain-mediated human protein-protein interactions by peptide array target screening. Proteomics 7:1775–1785

Xu Y, Hortsman H, Seet L, Wong SH, Hong W (2001) SNX3 regulates endosomal function through its PX-domain-mediated interaction with PtdIns(3)P. Nat Cell Biol 3:658–666

Xu J, Xu T, Wu B, Ye Y, You X, Shu X, Pei D, Liu J (2013) Structure of sorting nexin 11 (SNX11) reveals a novel extended PX domain (PXe domain) critical for the inhibition of sorting nexin 10 (SNX10) induced vacuolation. J Biol Chem 288(23):16598–16605

Xu T, Xu J, Ye Y, Wang Q, Shu X, Pei D, Liu J (2014) Structure of human SNX10 reveals insights into its role in human autosomal recessive osteopetrosis. Proteins 82:3483–3489

Xu J, Zhang L, Ye Y, Shan Y, Wan C, Wang J, Pei D, Shu X, Liu J (2017) SNX16 regulates the recycling of E-cadherin through a unique mechanism of coordinated membrane and cargo binding. Structure 25 (1251–1263):e1255

Yu JW, Lemmon MA (2001) All phox homology (PX) domains from Saccharomyces cerevisiae specifically recognize phosphatidylinositol 3-phosphate. J Biol Chem 276:44179–44184

Zhan Y, He D, Newburger PE, Zhou GW (2004) p47 (phox) PX domain of NADPH oxidase targets cell membrane via moesin-mediated association with the actin cytoskeleton. J Cell Biochem 92:795–809

Zhou CZ, Li de La Sierra-Gallay I, Quevillon-Cheruel S, Collinet B, Minard P, Blondeau K, Henckes G, Aufrere R, Leulliot N, Graille M et al (2003) Crystal structure of the yeast Phox homology (PX) domain protein Grd19p complexed to phosphatidylinositol-3-phosphate. J Biol Chem 278:50371–50376

Adv Exp Med Biol - Protein Reviews (2019) 20: 19–32
https://doi.org/10.1007/5584_2018_236
© Springer Nature Singapore Pte Ltd. 2018
Published online: 2 August 2018

Modeling of PH Domains and Phosphoinositides Interactions and Beyond

Jiarong Feng, Lei He, Yuqian Li, Fei Xiao, and Guang Hu

Abstract

Pleckstrin homology (PH) domains form a large family of protein modules within membrane-targeting domains. PH domains can function as lipid-binding modules, and in particular bind with different specificities and affinities to phosphoinositides (PIs). Understanding the association of PH domains to PIs is critical for many aspects of cellular biology. Bioinformatics and computational modeling approaches have become standard tools to study the structure and dynamics of PH domains and PIs. In this review, recent advances in the binding specificity of PH domains and their interactions with PIs, using bioinformatics tools for the prediction of PIs binding sites, performing molecular dynamics simulations to study PH domains-PIs interactions, as well as the computational inhibitor design for PH domains guided signaling pathways have been discussed.

Keywords

Protein-lipid interactions · Drug design · Signaling pathway · Bioinformatics · Molecular dynamics

Jiarong Feng, Lei He and Yuqian Li contributed equally to this work.

J. Feng and F. Xiao (✉)
School of Biology and Basic Medical Sciences, Medical College, Soochow University, Suzhou, China
e-mail: xiaofei@suda.edu.cn

L. He
School of Radiation Medicine and Protection, Soochow University, Suzhou, China

Cambridge-Suda (Cam-SU) Genomic Resource Center, Soochow University, Suzhou, China
e-mail: lhe@suda.edu.cn

Y. Li
School of Electronic Engineering, University of Electronic Science and Technology of China, Chengdu, China
e-mail: yuqianli@uestc.edu.cn

G. Hu (✉)
Center for Systems Biology, Soochow University, Suzhou, China
e-mail: huguang@suda.edu.cn

Abbreviations

PtdIns(4)P	Phosphatidylinositol 4-phosphate
PtdIns(3,4)P$_2$	Phosphatidylinositol 3,4-bisphosphate
PtdIns(3,5)P$_2$	Phosphatidylinositol 3,5-bisphosphate
PtdIns(4,5)P$_2$	Phosphatidylinositol 4,5-bisphosphate
PtdIns(3,4,5)P$_3$	Phosphatidylinositol 3,4,5-triphosphate;
PtdIns(1,3,4,5)P$_4$	Phosphatidylinositol 1,3,4,5-trisphosphate
Ins (4, 5)P$_2$	Inositol 4,5-bisphosphate
Ins(1,4,5)P$_3$	Inositol 1,4,5-trisphosphate

Ins(1,3,4,5) P_4	Inositol tetrakisphosphate	1,3,4,5-

1   Introduction

Many globular domains of signaling and trafficking proteins bind to the surface of cellular membranes, or to specific lipids in these membranes, providing strict control and versatility of the protein binding, acting like lipid clamps (Lemmon 2008). Plecktrin homology (PH) domains comprise a major type of membrane binding domains in the human proteome (Maffucci and Falasca 2001). The structure of a PH domain belonging to the human protein, Rho guanine nucleotide exchange factor 11 (PDZ-RhoGEF) is shown in Fig. 1. On the other hand, the conserved domains database (Marchler-Bauer et al. 2015) is a resource for the annotation of conserved domains within proteins collected from the National Center for Biotechnology Information (NCBI), accessed at https://www.ncbi.nlm.nih.gov/cd. Using the simple modular architecture research tool (SMART) (Letunic et al. 2015) (http://smart.embl.de/), 55,759 PH domains in 48,568 proteins have been collected and annotated to date.

With the development of high-throughput methods has allowed more and more structures of PH domains to be identified in the human proteome (Cozier et al. 2004). A number of PH domains have been well characterized as molecular modules for binding to phosphoinositides (PIs) as well as other proteins with varying specificities to perform multiple biological functions (Lemmon and Ferguson 2000). Despite these progressions there remain a number of unanswered questions. These include: (i) if and how a PH domain interacts with PIs, (ii) what is the structural feature of a PH domain underlying its diverse binding capabilities to PIs, and (iii) whether this binding to PIs is membrane-binding and, if so, how? For these questions, computational methods including bioinformatics analyses, molecular dynamics (MD) simulation, along with various structure-based methods are needed. To

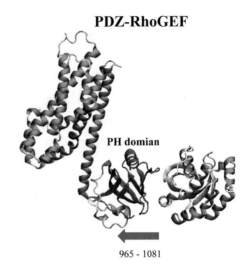

PDZ-RhoGEF

PH domian

965 - 1081

Fig. 1 The distribution of PH domain in the complex of PDZ-RhoGEF (PDB:3kz1). The PH domain (residue number: 965-1081) is displayed in red color

date, these methods have been applied to predict PIs binding sites, to investigate PH domain-PIs interactions, and to have computer-aided inhibitor design.

2   Sequence and Structural Conservation of PH Domains

The biological functions of PH domains can be predicted both from their sequences and structures. Within the family of PH domains there have been three conserved features noted. The sequence length of a typical PH domain is 100–120 amino acid residues (Rebecchi and Scarlata 1998). However, as seen in the sequence alignment (Fig. 2a), PH domains of different proteins lack primary sequence similarity, ranging from 10% to 30% (Lemmon et al. 1996). This is the first conserved feature of PH domains characterized to date. Despite this variability seen in PH domains from different classes, the sequence similarity remains relatively high within the same class.

Conservation of different properties of each PH domain can either be made within different protein families or for individual residues in one PH domain. To further understand the

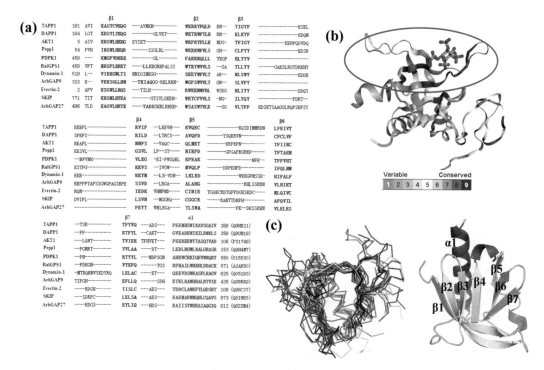

Fig. 2 Three conservation features of PH domains. (a) The sequence alignment of PH domains based on 3D structures (TAPP1: 1EAZ (PDB code), DAPP1: 1FAO, AKT1: 1UNQ, Pepp1: 1UPR, PDPK1: 1W1D, RalGPS1: 2DTC, Dynamin-1: 2DYN, ArhGAP9: 2P0H, Evectin-2: 3AJ4, SKIP: 3CXB and ArhGAP27: 3PP2) show low sequence similarity, while the SwissPort accession numbers are shown at the end of the sequences. (b) The 3D structure of Bruton's tyrosine kinase (PDB code: 1B55) colored with conservation grades. The red circle highlights the PH domain, which has relatively high conservation. (c) The structure alignment of PH domains as used in the sequence alignment, showing a quite high conservation for topological folds. The common folds of the PH domains from different proteins are colored in yellow (β-sandwich) and red (α-helix), respectively

conservation of a PH domain within a protein, a conservation analysis of the protein Bruton's tyrosine kinase (Btk, PDB code: 1B55), which contains a PH domain itself, was performed. This analysis of Btk was performed by using the ConSurf web server (Ashkenazy et al. 2016) (http://consurf.tau.ac.il), based on its structure. When the conservation scores are generated by generated by ConSurf, they are normalized and grouped into nine conservation grades (1–9), where 1 denotes the least conserved reside, 5 denotes positions of moderately conserved resides, and 9 denotes the most conserved residue. The PH domain of Btk is relatively conserved (Fig. 2b). Remarkably, the region with the highest conservation is located in the ligand binding pocket, with the most conserved residues being Gln 15, Gln 16, and Ser21 located in the β1–β2 loop. Moreover, two reported interacting residues (Baraldi et al. 1999), Tyr 39 and Gly53 located in the β3 and β4, respectively, were also highlighted to have high conservation, with a conservation grade of 8. This is the second conserved feature of all characterized PH domains.

To date, more than 30 3D structures of non-redundant PH domains have been determined (http://smart.embl.de/). Despite PH domains having low sequence identities, their 3D structures are quite similar due to their conserved folds. For example, when the same 11 PH domains that were shown to have low sequence similarity (Fig. 2a) are structurally aligned, a good level of conservation is seen

(Fig. 2c). This is the third conserved feature of PH domains. Previously, it has been shown that PH domains are comprised of seven β-strands that form two perpendicular antiparallel β-sheets and one C-terminal α-helix that packs against an edge of the β-sheet structure, stabilizing the entire fold (Rebecchi and Scarlata 1998). With the β-sheets of PH domains closely packed, there is no space for comparable hydrophobic ligands. Therefore, the loops connecting these β-strands play an important role in ligands binding. Another interesting feature of a PH domain is the distribution of charged residues, with generally three residues of Lys, Arg, and His forming a positively charged surface on one side, while the C-terminal α-helix located at the opposite face of the domain, is occupied by acidic residues. This distribution of charges also suggests possible ligand-binding sites, which are located at the center of the positively charged surface.

These three levels of conservation of PH domains of low sequence homology but high sequence and structural homology of the binding pockets allow these domains to carry out their required function. The low sequence homology of PH domains means that PH domains have the ability to have varying binding specificities and different mutational sites. However, the high conservation of both the amino acid residues and the structure of the binding pocket suggests that it may be a potential target for further drug discovery.

3 PH Domains Binds to the Lipid Phosphoinositie

PH domains can bind to different ligands, performing different biological functions (Lemmon 2007). Known ligands to bind to PH domains include structurally diverse proteins and lipids such as PIs and inositol (Ins) polyphosphates/pyrophosphates. PIs are the phosphorylated derivatives of phosphatidylinositol (PtdIns), where they are reversibly phosphorylated at three positions D3, D4, and D5. Among these different ligands, PIs are the most important lipids that bind to membrane-target domains, playing key roles in lipid signaling, cell signaling, and membrane trafficking (Riehle et al. 2013). PIs bind to their target proteins, including PH domains, by their negatively charged headgroups interacting with disordered regions harboring clusters of basic residues within the proteins (Krauss and Haucke 2007). Both the physiological roles (De Craene et al. 2017) and biological processes of PIs (Balla 2013), especially cellular physiology and disease regulation by different PIs have been discussed comprehensively.

There are seven variations of PIs, including three monophosphates (PtdIns3P, PtdIns4P, and PtdIns5P) (Stahelin et al. 2014), three bisphosphates (PtdIns(4,5)P_2, PtdIns(3,4)P_2, and PtdIns(3,5)P_2), and one triphosphate, PtdIns(3,4,5)P_3. It should be noted that all PIs except PtdIns5P can bind to a PH domain. However, high-affinity and specific recognition have only were reported for four PIs, PtdIns3P, PtdIns(3,4)P_2, PtdIns(3,5)P_2, and PtdIns(3,4,5)P_3, which are produced by activated phosphoinositide 3-kinase (PI3K). The binding specificities and additional membrane phosphoinosities for PH domains have been discussed previously (DiNitto and Lambright 2006). In addition to these PIs, PI derivatives have also been reported to bind to PH domains such as C4PtdIns (3, 4, 5) P_3, which has been shown in a complex with Btk PH domain was determined in a dimerization state (Murayama et al. 2008).

PH domains can be classified into four groups according to their PIs binding specificity (Hurley and Misra 2000). It has been shown that the triphosphate PIs, PtdIns(3,4,5)P_3, represents the most important PIs for binding with PH domains, as it is noted in two of the four PH domain groups. As the product of the class I PI3K activation, PtdIns(3,4,5)P_3 is capable of activating a wide range of proteins, including Akt (also known as kinase PKB) and Pyruvate dehydrogenase lipoamide kinase isozyme 1(PDK1) via their PH domains(Riehle et al. 2013). The PI3K pathway is one of the most activated proliferation pathways in human disease (Liu et al. 2009). Many studies have demonstrated that PtdIns(3,4,5)P_3 is a prime mediator of various metabolic diseases associated with the PI3K signaling pathway (Luo and Mondal 2015; Ghigo et al. 2017). For this reason, further study of interactions between PtdIns $(3,4,5)P_3$ and target proteins is of particular importance. An initial investigation into the conformational arrangements of the inositol rings, have been performed to help understand the interactions between phosphoinositide and proteins. This theoretical investigation of PtdIns $(3,4,5)P_3$ was performed using quantum mechanics calculations to describe the energetic and structural properties of eleven isomers of its head group (PtdIns$(1,3,4,5)P_4$) (Rosen et al. 2011). From this study, parameters for PtdIns $(3,4,5)P_3$ have been generated, and can now be included in the Amber force field, for MD simulations of different protein interactions with PtdIns$(1,3,4,5)P_4$.

4   Binding Specificities of PH Domains with PIs

So far in this review, PH domains have been demonstrated to serve as modules for PIs binding. However, only a minority of PH domains actually target PIs with high affinity, with the vast majority of PH domains not binding PIs at all (Lemmon 2007). Previously, electrostatic modeling of 33 PH domains encoded in the *Saccharomyces cerevisiae* genome revealed that only the PH domain of the Nuclear migration protein NUM1 had high binding affinity to PIs (Yu et al. 2004). Furthermore, the first PI-binding PH domain was reported in the substrate of Protein Kinase C (Yao et al. 1994); but now there has been more than 30 3D structures of PH domain-PIs complexes determined. When the affinities and specificities for each of the established functional subclasses of PH domains are observed, it can be seen that group 1, including the Btk PH domain and group 3 including Akt and PDK1 PH domains bind PIs with similar affinities, group 2 including phospholipase C-delta 1 (PLCδ1) PH domains have weaker affinities, and group 4 including dynamin PH domains have lower specificity as well as affinity.

Although PH domains bind PIs with varying degrees of specificity, there have been three proposed common features needed for PIs recognition, determined by the investigation of the structural properties of interactions between PH domains and PIs. All PH domains that bind reasonably strong to PIs have two conserved basic amino acids, they have positively charged variable loops serve as PIs binding site, and PtdIns$(3,4)P_2$, PtdIns$(3,5)P_2$, and PtdIns$(3,4,5)P_3$ are three PIs that always show binding specificity with PH domains. Furthermore, PH domains bind to PIs through two different binding pockets; with PH domains of proteins such as PLCδ, the protein kinase B, and Phosphatidylinositol 3,4,5-trisphosphate-dependent Rac exchanger 1 protein (P-Rex1) binding between β1-β2 and β3-β4 loops (Fig. 3a) (Cash et al. 2016), and PH domains of proteins such as Plasma membrane proteins Slm1, T-cell lymphoma invasion and metastasis-inducing protein 1 (Tiam1), Rho GTPase activating protein 9 (ArhGAP9) and β-spectrin binding between β1-β2 and β5-β6 loops (Fig. 3b) (Anand et al. 2012).

However, the structural basis underlying the diverse specificities of PIs to PH domains remains elusive. Recently, newly determined 3D structures of PH domain-PI complexes have provided more insights into the binding specificity. As previously stated, PI$(4,5)P_2$, PI$(3,4,5)P_3$, and PI$(3,4)P_2$ are three products of PI3K, which can bind to Tiam1 and ArhGAP9 PH domains in the non-canonical model. In addition, the structural analysis suggested a role for the three loops (β1-β2, β3-β4, and β6-β7 loops) in binding to the PIs (Ceccarelli et al. 2007). Each of the PH domains have different binding affinities with different types of PIs. For example, the Btk PH domain can bind to PtdIns $(3, 4, 5)P_3$ more specifically than any other PIs (Murayama et al. 2008). Additionally, it has also been shown that the Btk PH domain has a much higher affinity for binding to three derivatives of PtdIns$(3, 4, 5)P_3$; C4PtdIns$(3,4,5)P_3$, C8PtdIns$(3,4,5)P_3$, C16PtdIns $(3,4,5)P_3$.

This difference in specificity and affinity is even evident in PH domains belonging to the same family of proteins. For example, three isoforms, kindlin-1, kindlin-2, and kindlin-3, all

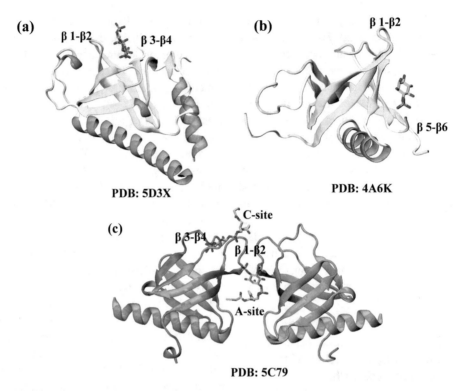

Fig. 3 Different types of binding specificities of PH domains with PIs. (**a**) P-Rex1 PH domain bind to PtdIns (3,4,5)P₃ with the pocket between β1-β2 and β3-β4 loops. (**b**) Slm1-PH domain bind to (PtdIns(4)P) with the pocket between β1-β2 and β5-β6 loops, (**c**) ASAP1 PH domains bind to PtdIns(4,5)P₂ with two different binding sites, which are located at "atypical" binding site at the dimer interface and the canonical site between β1-β2 and β3-β4 loops

contain PH domains, are important for integrin activation. The Kindlin-1 PH domain will only bind to Ins (3, 4, 5)P₃ with weak affinities (Yates et al. 2012). Whereas, Kindlin-2 PH domain bind with high affinity multiple PIs, including Ins (3, 4, 5)P₃, Ins (4, 5)P₂, and Ins (1,3, 4, 5)P₄ (Liu et al. 2011). Most recently the PH domain in Kindlin-3 was shown to only c bind to two PIs, Ins (3, 4, 5) P₃ with a high affinity and Ins (4, 5)P₂ with low affinity (Ni et al. 2017). Furthermore, PH domains may also bind to different ligands or identical ligands at different sites cooperatively (Roy et al. 2016). This so-called allosteric binding property suggests a new type of binding specificity of PH domains, potentially increasing binding specificity for a particular membrane. These allosteric PH domains have been found in 2 ADP- ribosylation factor exchange factors (Grp1 and Brag2) and the Arf GTPase-activating

protein (ASAP1). Specifically, the PH domain of the Arf GEF Grp1 bound structure was found to bind to two different ligands, PI(3,4,5)P3 and Arf6·GTP, at two different sites of PH domain (Jian et al. 2015). Additionally, the recently solved structure of PtdIns(4,5)P2 bound to the ASAP1 PH domain suggests that PtdIns(4,5)P2 binds to ASAP1 dimer through the canonical site and the "atypical" binding site cooperatively (Fig. 3c).

Lastly, some PH domains cannot bind to PIs at all. For example, the crystal structure of the PH domains of guanine nucleotide exchange factor (hGEF-H1) clearly highlights the close proximity of the β6/β7 loop to the β3/β4 loop, preventing the binding of PIs (Jiang et al. 2016). However, hGEF-H1 still has the ability to bind to other proteins to regulate many cellular signal processes, such as the RAS/MAPK pathway as well as the Nucleotide-Binding Oligomerization Domain 2 (NOD2) signaling pathway. Therefore, the ability of the PH domain of hGEF-H1 to interact with other proteins may be attributed to the conserved surface, comprised of the beta-sheets, β8, β9, and β10.

5 Prediction of the Binding Sites of PIs

In addition to the characterization of binding sites, the prediction of ligand-binding sites is also crucial in the understanding of protein function, and thus is helpful in drug design. Two recent sequence-based tools, MBPpred (Nastou et al. 2016) (http://bioinformatics.biol.uoa.gr/MBPpred) and HHpred (Fidler et al. 2016), have been utilized to begin a genome-wide identification of PH domains. Despite very low sequence similarities, the analyses of PH domains have found some common binding properties for PIs. Thus, these tools provide useful discriminators for sequence-based prediction of PIs-binding even in the absence of 3D structures.

Form the sequences of PH domains, specific physic chemical properties such as hydropathy and flexibility of the PH domains have been predicted (Shen et al. 2008). The hydropathy profile represents the hydrophobic or hydrophilic properties of amino acids, useful in indicating key residues for protein function and stability (Gallet et al. 2000). Flexibility is another indicator of protein functions, always required for ligand recognition. The hydropathy and flexibility profiles calculated for different PH domains have been shown to have very similar profiles; despite different in the primary sequences. Using these two properties, known X-linked agammaglobulinemia (XLA)-causing mutations in the Btk PH domain can be predicted (Mohamed et al. 2009). In addition to sequence-based properties, electronic charge distribution based on the structure can be used for the prediction of PIs binding sites (Ojteg et al. 1989). It is well established that electrostatic interactions are very important in the binding between PH domains and PIs.

Previously, electrostatic potentials of both PH domain structures and PIs have been calculated, showing that PH domain binding sites are generally hydrophilic, flexible, and charged, and that the binding specificities of PIs correlate well with the electrostatic properties of the bound inositol phosphate (Jiang et al. 2015). Taken together, the combined sequence and structure analysis of PH domains shows that charge concentration, hydrophilicity, and flexibility are the main factors in determining the binding affinity and specificity.

The effect of sequence mutations on the binding specificity of PH domains has also been studied. The combination of sequence and structure-based methods, including sequence alignments, in silico single point mutations, and the energy minimization calculation have been applied to three mutated structures of the PLCδ1 PH domain (Morales et al. 2017). The in silico modelling combined with experimental studies demonstrated that these mutated models show different binding specificities to PIs, and the presence of aromatic amino acids at the binding pockets were necessary. Furthermore, the Phyre server (www.sbg.bio.ic.ac.uk/phyre/) has been used to predict PIs binding sites of the PH domains of MyoIcs from its ab initio sequence prediction and secondary structural analysis (Hokanson et al. 2006). The results highlight that the sequence of the PH domain is conserved among other myosin-I isoforms, and the binding sites of identified anionic phospholipids have overlaps with predicted membrane binding sites located in the IQ motifs of the proteins. An IQ motif has been defined as a particular sequence of 21–25 amino acids, that PH domain use to bind to calmodulin and calmodulin-like proteins (Tang et al. 2002). Also based on the multiple sequence alignment, a recursive functional classification (RFC) score was defined for a given domain sequence S as:

$$\text{RFC score (S)} = \sum_{si \in S} \log\left(\frac{P_{si,\ i,+}}{P_{si,\ i,-}}\right)$$

where $P_{si,\ i,+}$ and $P_{si,\ i,-}$ denote the probability of the ith amino acid, with s and i denote binding and non-binding, respectively. The RFC score can be used to predict lipid binding (Kallberg et al. 2012).

Finally, for predicting the membrane binding sites of PH domains, the membrane optimal docking area (MODA) program was utilized on a genome-wide scale (Lenoir et al. 2015). The sequence and structural data suggests that basic and proximal hydrophobic residues, as well as the β1-β2 loop are important in bilayer interaction. Additionally, the Phyre server (Hokanson et al. 2006) also predicts membrane binding sites, and these correlated well with classic lipid binding sites. In the case the PH domain of the protein, Myosin 1G, two hydrophobic residues Lys and Arg, two basic residues for membrane association, were conserved in the PIs-binding pocket. Therefore, it seems that hydrophobic residues have high binding specificity to membranes.

6 MD Simulations of the Interactions Between PH Domains and PIs

The MD simulation is a computational simulation method for studying protein dynamics. The MD simulation is rapidly becoming a standard tool for modeling the molecular interactions between PH domains and PIs, including the binding profile of PIs and the prediction of key residues in the binding PIs to PH domains. To our knowledge, the first MD study was performed on the Akt PH domain, investigating the interactions between the protein and three PI3K-generated PIs (Rong et al. 2001). It was computationally predicted that three loops between β1-β2, β3-β4, and β6-β7, and five resides Lys 14, Arg 25, Tyr 38, Arg 48, and Arg 86 were the binding pocket and key residues interacting with PIs, which was in agreement with experiment results. The computational method has also been used to compare the PIs binding properties to β, γ, δ, and ε- PLC PH domains (Singh and Murray 2003). Due to very little sequence conservation, information from crystal structures and homology modelling needed to be considered for the different types of PH domains. A set of computational tools, including fold recognition, sequence-to-profile alignments, homology modelling, and the continuum electrostatic calculation were adopted to compare the binding patterns (Rosen et al. 2012).

Modeling protein flexibility constitutes another major challenge in the prediction of protein-ligand interactions and the target-based drug design. The MD simulation and free energy calculation of kindlin-3 PH domain demonstrated that loops connecting β-strands with high flexibility serve as PIs binding sites (Ni et al. 2017). It seems that PI-binding sites prefer to locate at the loop regions of the domain. Furthermore, a study of the SmCesA2 PH domain and PtdIns (3, 4, 5) P_3 suggested that the binding pocket was located at β1, β2 and the β1-β2 loop, and that Lys 88, Lys 100 and Arg 102 were the three residues that interact with the free phosphate groups, leading to a high binding affinity (Kuang et al. 2016). Also, MD simulations have been performed on 12 Btk-PH domains including one wild type and 11 mutants (Lu et al. 2013). The free energy landscapes were calculated to predict their binding affinities with Ins(1,3,4,5)P_4, and the mutant PH domains were able to be classified into two groups according to their binding affinity. In this study, relationship between the binding affinities of PIs and mutations of PH domains were explored. Additionally, the effect of mutations on the Akt1 PH domain and inhibitor binding have also been investigated using MD simulation, and the rapid conformational drift observed in the mutant structure was suggested as the reason for related pathological consequences (Kumar and Purohit 2013).

Moreover, there have been a number of significant contributions to this field (Kalli and Sansom 2014; Psachoulia and Sansom 2008; Lumb et al. 2011; Lumb and Sansom 2012; Yamamoto et al. 2016; Buyan et al. 2016). A set of MD simulations have been performed for the interactions between a PH domain and PIs in the environment of membrane bilayers. Additionally, MD simulations into the complex of the C-δ1 PH domain and PtdIns(4,5)P2, and the PH domain of the general receptor for phosphoinositides 1 (GRP1) PH domain and PI(3,4,5)P_3, allowed the "dual-recognition" binding model to be proposed. Taken together, the binding behavior of a

PH domain with PIs, affects the binding to the membrane bilayers, with the PIs binding enhancing the interaction of a PH domain with a membrane. Furthermore, Brownian dynamics combined with high-throughput MD comparative simulations, the molecular mechanism of such complex interactions both between PH domains and PIs, and between PH domains and membrane bilayers have been investigated. More recently, the fractional Brownian motion and the temporally fluctuating diffusivity of a PH domain bound to both PIs and a membrane have also been investigated (Yamamoto et al. 2017). And once again, the MD simulation proved that the PH domian -PIs interactions participate in the PH domain -membrane interactions.

In some cases, cooperation between the PH domain and the BAR domain can help ACAP1 (Arf gap with coiled coil, Ankyrin repeat, and PH domain protein 1) proteins in membrane binding. Again, the MD simulation was able to shed some insights into the role of the PH domain, where a positively changed patch and a loop contribute to membrane binding (Pang et al. 2014). Based on MD simulations, further comprehensive orientation and potential mean force analysis of PH domains in unbound and bound states, revealed the molecular basis of PIs binding to PH domains during the membrane remodeling process of ACAP1(Chan et al. 2017).

7 Computer–Aided Drug Design Based on PH Domains

7.1 PH Domains and Diseases

PH domains have emerged as critical regulators of various cellular processes, and thus have been implicated in various diseases, including Alzheimer's disease (Ogawa et al. 2013), leukemia (Miroshnychenko et al. 2010), obesity and diabetes (Manna and Jain 2015), and some cancers (Bar-Shavit et al. 2016). Diseases related to PH domains are caused by preventing their binding to PIs, influencing important signaling pathways. For example, a set of point mutations in the Btk PH domain have been reported to result in the severe human immunodeficiency, known as XLA (Hyvonen and Saraste 1997). Consequently, there has been the establishment of adatabase (Valiaho et al. 2015) (http://structure.bmc.lu.se/PON-BTK/) containing XLA-causing mutations in Btk PH domains for further analysis.

The PI3K pathway is one of the most activated proliferation pathways in cancer (Rodgers et al. 2017). A previous systems biology studies showed that the PI3K pathway plays a number of important roles in cancer. For example, PI3K mediates oncogenesis in various malignancies, and Akt has been proposed as a potential biomarkers as well as attractive target for cancer drug discovery. However, the molecular mechanism of PI3K pathway in cancer remains elusive (Wadhwa et al. 2017). PI3Ks can be classified into three groups (Hawkins and Stephens 2016); however, of most interest there is group of PI3Ks mostly mediated by the interaction of Akt PH domains with PtdIns(3,4,5)P_3 and PtdIns(3,4)P_2 (Li and Marshall 2015). Akt consists of different cellular isoforms, Akt1 (PKBα), Akt2 (PKBβ), and Akt3 (PKBγ). The first mutation related to the Akt1 PH domain (E17K) was found to correspond to breast, colorectal, and ovarian cancers (Carpten et al. 2007). In addition there have been other mutations such as the D412G mutation in the mSin1 PH domain (Liu et al. 2015) and the H349A mutation in PAR2 PH domain (Kancharla et al. 2015) identified in ovarian cancer and breast cancer, respectively. Strikingly, a systems analysis of Akt mutations revealed that up to 5% of human cancers are related to mutations in PH and kinase domains of proteins (Parikh et al. 2012). Besides Akt, the PI3K pathway can also be activated by its lipid product binding to other downstream signaling proteins through their PH domains, including PDK1, mSin1, Rac-GEFs, and TEC family kinase (Btk) (Lien et al. 2017). It is therefore of great interest to design drugs to suppress the interaction between Akt PH domain and PIs, as the exciting new therapeutic targets.

7.2   Inhibitors Design for PH Domains

Computational methods continue to play an important role in novel therapeutics discovery and development. The previously established PIs binding site prediction and the MD simulation of PH domain structures and interactions could help improve computer-aided PH domains inhibitor design. Furthermore, the Akt PH domain is a critical component in the PI3K signaling pathway involved in cancer, thus proving an exciting drug target for design inhibitors (Politz et al. 2017). There have been many attempts to develop small molecule inhibitors, especially several analogs of PtdIns(3,4,5)P$_3$, to inhibit the binding of the Akt PH domains to PIs (Meuillet 2011; Mak et al. 2011). In particular, the PHusis platform (http://www.phusistherapeutics.com/index.html) has been built to develop PH domain inhibitors through integrated computational approaches.

Some common computational-aided methods used in drug design include not only the computational modeling and MD simulations, but also the quantitative structure-activity relationship (QSAR) study, molecular docking, and virtual screening approaches. Similar to MD simulations, molecular docking can also be carried out to predict binding affinities. Using the searching algorithm and calculating the scoring function, the best ligand binding position can be obtained. Various docking programs have been developed, such as FlexX, GOLD, and Glide. The first molecules to target the PH domain of Akt have been identified using the silico library screening and interactive molecular docking (Mahadevan et al. 2008). Two lipid-based derivatives were optimization docked and proposed as molecules for inhibiting the Akt PH domain (Du-Cuny et al. 2009). Since then, more compounds including 4-dodecyl-*N*-(1,3,4-thiadiazol-2-yl) benzenesulfonamide (Moses et al. 2009), PITenins, quinoline-4-carboxamide and 2-[4-(cyclohexa-1,3-dien-1-yl)-1H-pyrazol-3-yl] phenol, MK2206, Lig1, and Lig2 have been computationally developed as antagonist inhibitors of PtdIns(3,4,5)P$_3$ signaling (Miao et al. 2010). It should be noted that the last five compounds and ARQ 092 (Lapierre et al. 2016) correspond to allosteric drugs of Akt (Yilmaz et al. 2014; Chen et al. 2014), which inhibit the activity of Akt via their binding in a pocket between the PH domain and the kinase domain. Molecular docking has also been used in the design of triazole-based compounds for the inhibition of the PLCδ1-PH domain (Gorai et al. 2015).

Grb2-associated binding protein 1 (GAB1) is another PH domain containing protein that is associated with the PI3K pathway. Since the GAB1 PH domain structure is not available, a pipeline of structure-based drug discovery for PH domains with no structural data was proposed (Chen et al. 2015). In this pipeline, the structure is predicted by generating the specific position-site matrixes based on sequence analysis, and by homology modeling based on all non-redundant PH domain structures. This is followed by performing a high-throughput virtual screening of drug compounds collected from various chemical databases. Finally, the MD simulation is used to refine the structure of the PH domain in a complex with inhibitors. This proposed pipeline has been applied to the GAB1 PH domain, and five potential drug molecules have been predicted successfully. In addition, there is evidence that the binding sites of PH domains undergo conformational changes, along with the binding of PIs, which should be considered in docking simulations. To address this, a normal mode-based docking method has been proposed (Tran and Zhang 2011). The basis behind this method is the integration of sampling structures, obtained by global modes of the Elastic network mode, into molecular docking which eliminates the limitation of traditional docking methods, which can only consider receptors as rigid objects. The application of this method has been validated by describing several inhibitors for the Akt PH domain complexes.

QSAR is a theoretical model that uses a set of predictors including physio-chemical properties and molecular descriptors of chemicals for predicting biological activity. The 4D QSAR model has been successfully used to find new potent Btk inhibitors, from 96 nicotinamide analogs (Santos-Garcia et al. 2016). Structural descriptors of both protein targets such as the Akt2 PH domain and small molecules, including two hydrogen bond acceptors, two hydrogen bond donors, and one hydrophobic feature at a certain distance from each other, are needed to calculate the Grid-Independent Molecular Descriptor (GRIND) (Akhtar and Jabeen 2016). Additionally, integrating 3D structural features of the PH domain of Akt with 2D-QSAR, calculating physicochemical parameters, would provide a more useful mode to develop. Furthermore, a group-based QSAR method has also been developed to analyze a variety of chemically diverse inhibitors for Akt1 PH domain (Ajmani et al. 2010). This study revealed that alignment-independent topological descriptors are quite sensitive for different domain targets such as the Kinase domain and PH domain, of Akt1. Based on this finding, a QSAR modeling approach including alignment-independent topological information along with other descriptors has also been used to discover novel inhibitors for Akt2 (Davis and Vasanthi 2015).

8   Conclusion

In summary, this review has summarized recent advances in the computational study of the interaction between PH domains and PIs. PH domains were shown to have conserved features in their sequence as well as their structure. Furthermore, it was highlighted that PH domains bind to PIs to perform many cellar functions, which means they have also been implicated in human diseases. Novel binding specificities have been discovered by newly determined structures of the complex formed between PH domains and PIs. Additionally, computational methods including bioinformatics tools and MD simulation methods have been used to predict PI binding sites and study the molecular mechanism of the interactions. Finally, integrating various computational strategies involving QASR, molecular docking, along with virtual screening has provided efficient tools in the design of novel drugs based on

structures of PH domains. However, the application of these computational methods to PH domains is still in its early stage, with many challenges in the field of bioinformatics and computational structural biology still remaining. However, to study other protein-protein interactions facilitated by PH domains, more sophisticated computational methods are needed.

Acknowledgements This work was supported by the National Natural Science Foundation of China (61401068), the China Postdoctoral Science Foundation (2016 M590495), Jiangsu College Natural Science Research Key Program (17KJA520004), and the Jiangsu Planned Projects for Postdoctoral Research Funds (1601168C).

References

Ajmani S, Agrawal A, Kulkarni SA (2010) A comprehensive structure-activity analysis of protein kinase B-alpha (Akt1) inhibitors. J Mol Graph Model 28 (7):683–694

Akhtar N, Jabeen I (2016) A 2D-QSAR and grid-independent molecular descriptor (GRIND) analysis of Quinoline-type inhibitors of Akt2: exploration of the binding mode in the Pleckstrin homology (PH) domain. PLoS One 11(12):e168806

Anand K, Maeda K, Gavin AC (2012) Structural analyses of the Slm1-PH domain demonstrate ligand binding in the non-canonical site. PLoS One 7(5):e36526

Ashkenazy H, Abadi S, Martz E et al (2016) ConSurf 2016: an improved methodology to estimate and visualize evolutionary conservation in macromolecules. Nucleic Acids Res 44(W1):W344–W350

Balla T (2013) Phosphoinositides: tiny lipids with giant impact on cell regulation. Physiol Rev 93 (3):1019–1137

Baraldi E, Djinovic CK, Hyvonen M et al (1999) Structure of the PH domain from Bruton's tyrosine kinase in complex with inositol 1,3,4,5-tetrakisphosphate. Structure 7(4):449–460

Bar-Shavit R, Maoz M, Kancharla A et al (2016) Protease-activated receptors (PARs) in cancer: Novel biased signaling and targets for therapy. Methods Cell Biol 132:341–358

Buyan A, Kalli AC, Sansom MS (2016) Multiscale simulations suggest a mechanism for the association of the Dok7 PH domain with PIP-containing membranes. PLoS Comput Biol 12(7):e1005028

Carpten JD, Faber AL, Horn C et al (2007) A transforming mutation in the pleckstrin homology domain of AKT1 in cancer. Nature 448(7152):439–444

Cash JN, Davis EM, Tesmer JJ (2016) Structural and Biochemical Characterization of the Catalytic Core of the Metastatic Factor P-Rex1 and Its Regulation by PtdIns(3,4,5)P3. Structure 24(5):730–740

Ceccarelli DF, Blasutig IM, Goudreault M et al (2007) Non-canonical interaction of phosphoinositides with pleckstrin homology domains of Tiam1 and ArhGAP9. J Biol Chem 282(18):13864–13874

Chan KC, Lu L, Sun F, Fan J (2017) Molecular details of the PH domain of ACAP1BAR-PH protein binding to PIP-containing membrane. J Phys Chem B 121 (15):3586–3596

Chen SF, Cao Y, Han S, Chen JZ (2014) Insight into the structural mechanism for PKBalpha allosteric inhibition by molecular dynamics simulations and free energy calculations. J Mol Graph Model 48:36–46

Chen L, Du-Cuny L, Moses S et al (2015) Novel inhibitors induce large conformational changes of GAB1 Pleckstrin homology domain and kill breast Cancer cells. PLoS Comput Biol 11(e10040211):e1004021

Cozier GE, Carlton J, Bouyoucef D, Cullen PJ (2004) Membrane targeting by pleckstrin homology domains. Curr Top Microbiol Immunol 282:49–88

Davis GD, Vasanthi AH (2015) QSAR based docking studies of marine algal anticancer compounds as inhibitors of protein kinase B (PKBbeta). Eur J Pharm Sci 76:110–118

De Craene JO, Bertazzi DL, Bar S, Friant S (2017) Phosphoinositides, major actors in membrane trafficking and lipid signaling pathways. Int J Mol Sci 18 (3):634

DiNitto JP, Lambright DG (2006) Membrane and juxtamembrane targeting by PH and PTB domains. Biochim Biophys Acta 1761(8):850–867

Du-Cuny L, Song Z, Moses S et al (2009) Computational modeling of novel inhibitors targeting the Akt pleckstrin homology domain. Bioorg Med Chem 17 (19):6983–6992

Fidler DR, Murphy SE, Courtis K et al (2016) Using HHsearch to tackle proteins of unknown function: a pilot study with PH domains. Traffic 17 (11):1214–1226

Gallet X, Charloteaux B, Thomas A, Brasseur R (2000) A fast method to predict protein interaction sites from sequences. J Mol Biol 302(4):917–926

Ghigo A, Laffargue M, Li M, Hirsch E (2017) PI3K and calcium signaling in cardiovascular disease. Circ Res 121(3):282–292

Gorai S, Bagdi PR, Borah R et al (2015) Insights into the inhibitory mechanism of triazole-based small molecules on phosphatidylinositol-4,5-bisphosphate binding pleckstrin homology domain. Biochem Biophys Rep 2:75–86

Hawkins PT, Stephens LR (2016) Emerging evidence of signalling roles for PI(3,4)P2 in class I and II PI3K-regulated pathways. Biochem Soc Trans 44(1):307–314

Hokanson DE, Laakso JM, Lin T et al (2006) Myo1c binds phosphoinositides through a putative pleckstrin homology domain. Mol Biol Cell 17(11):4856–4865

Hurley JH, Misra S (2000) Signaling and subcellular targeting by membrane-binding domains. Annu Rev Biophys Biomol Struct 29:49–79

Hyvonen M, Saraste M (1997) Structure of the PH domain and Btk motif from Bruton's tyrosine kinase: molecular explanations for X-linked agammaglobulinaemia. EMBO J 16(12):3396–3404

Jian X, Tang WK, Zhai P et al (2015) Molecular basis for cooperative binding of anionic phospholipids to the PH domain of the Arf GAP ASAP1. Structure 23 (11):1977–1988

Jiang Z, Liang Z, Shen B, Hu G (2015) Computational analysis of the binding specificities of PH domains. Biomed Res Int 2015:792904

Jiang Y, Jiang H, Zhou S et al (2016) Crystal structure of hGEF-H1 PH domain provides insight into incapability in phosphoinositide binding. Biochem Biophys Res Commun 471(4):621–627

Kallberg M, Bhardwaj N, Langlois R, Lu H (2012) A structure-based protocol for learning the family-specific mechanisms of membrane-binding domains. Bioinformatics 28(18):i431–i437

Kalli AC, Sansom MS (2014) Interactions of peripheral proteins with model membranes as viewed by molecular dynamics simulations. Biochem Soc Trans 42 (5):1418–1424

Kancharla A, Maoz M, Jaber M et al (2015) PH motifs in PAR1&2 endow breast cancer growth. Nat Commun 6:8853

Krauss M, Haucke V (2007) Phosphoinositides: regulators of membrane traffic and protein function. FEBS Lett 581(11):2105–2111

Kuang G, Bulone V, Tu Y (2016) Computational studies of the binding profile of phosphoinositide PtdIns (3,4,5) P(3) with the pleckstrin homology domain of an oomycete cellulose synthase. Sci Rep 6:20555

Kumar A, Purohit R (2013) Cancer associated E17K mutation causes rapid conformational drift in AKT1 pleckstrin homology (PH) domain. PLoS One 8(5): e64364

Lapierre JM, Eathiraj S, Vensel D et al (2016) Discovery of 3-(3-(4-(1-Aminocyclobutyl)phenyl)-5-phenyl-3H-imidazo[4,5-b]pyridin-2-yl)pyridin −2-amine (ARQ 092): an orally bioavailable, selective, and potent allosteric AKT inhibitor. J Med Chem 59(13):6455–6469

Lemmon MA (2007) Pleckstrin homology (PH) domains and phosphoinositides. Biochem Soc Symp 74:81–93

Lemmon MA (2008) Membrane recognition by phospholipid-binding domains. Nat Rev Mol Cell Biol 9(2):99–111

Lemmon MA, Ferguson KM (2000) Signal-dependent membrane targeting by pleckstrin homology (PH) domains. Biochem J 350(Pt 1):1–18

Lemmon MA, Ferguson KM, Schlessinger J (1996) PH domains: diverse sequences with a common fold recruit signaling molecules to the cell surface. Cell 85 (5):621–624

Lenoir M, Kufareva I, Abagyan R, Overduin M (2015) Membrane and protein interactions of the Pleckstrin homology domain superfamily. Membranes (Basel) 5 (4):646–663

Letunic I, Doerks T, Bork P (2015) SMART: recent updates, new developments and status in 2015. Nucleic Acids Res 43(Database issue):D257–D260

Li H, Marshall AJ (2015) Phosphatidylinositol (3,4) bisphosphate-specific phosphatases and effector proteins: a distinct branch of PI3K signaling. Cell Signal 27(9):1789–1798

Lien EC, Dibble CC, Toker A (2017) PI3K signaling in cancer: beyond AKT. Curr Opin Cell Biol 45:62–71

Liu P, Cheng H, Roberts TM, Zhao JJ (2009) Targeting the phosphoinositide 3-kinase pathway in cancer. Nat Rev Drug Discov 8(8):627–644

Liu J, Fukuda K, Xu Z et al (2011) Structural basis of phosphoinositide binding to kindlin-2 protein pleckstrin homology domain in regulating integrin activation. J Biol Chem 286(50):43334–43342

Liu P, Gan W, Chin YR et al (2015) PtdIns(3,4,5)P3-dependent activation of the mTORC2 kinase complex. Cancer Discov 5(11):1194–1209

Lu D, Jiang J, Liang Z et al (2013) Molecular dynamic simulation to explore the molecular basis of Btk-PH domain interaction with Ins(1,3,4,5)P4. Sci World J 2013:580456

Lumb CN, Sansom MS (2012) Finding a needle in a haystack: the role of electrostatics in target lipid recognition by PH domains. PLoS Comput Biol 8(7): e1002617

Lumb CN, He J, Xue Y et al (2011) Biophysical and computational studies of membrane penetration by the GRP1 pleckstrin homology domain. Structure 19 (9):1338–1346

Luo HR, Mondal S (2015) Molecular control of PtdIns (3,4,5)P3 signaling in neutrophils. EMBO Rep 16 (2):149–163

Maffucci T, Falasca M (2001) Specificity in pleckstrin homology (PH) domain membrane targeting: a role for a phosphoinositide-protein co-operative mechanism. FEBS Lett 506(3):173–179

Mahadevan D, Powis G, Mash EA et al (2008) Discovery of a novel class of AKT pleckstrin homology domain inhibitors. Mol Cancer Ther 7(9):2621–2632

Mak LH, Georgiades SN, Rosivatz E et al (2011) A small molecule mimicking a phosphatidylinositol (4,5)-bisphosphate binding pleckstrin homology domain. ACS Chem Biol 6(12):1382–1390

Manna P, Jain SK (2015) Phosphatidylinositol-3,4,5-triphosphate and cellular signaling: implications for obesity and diabetes. Cell Physiol Biochem 35 (4):1253–1275

Marchler-Bauer A, Derbyshire MK, Gonzales NR et al (2015) CDD: NCBI's conserved domain database. Nucleic Acids Res 43(Database issue):D222–D226

Meuillet EJ (2011) Novel inhibitors of AKT: assessment of a different approach targeting the pleckstrin homology domain. Curr Med Chem 18(18):2727–2742

Miao B, Skidan I, Yang J et al (2010) Small molecule inhibition of phosphatidylinositol-3,4,5-triphosphate (PIP3) binding to pleckstrin homology domains. Proc Natl Acad Sci U S A 107(46):20126–20131

Miroshnychenko D, Dubrovska A, Maliuta S et al (2010) Novel role of pleckstrin homology domain of the Bcr-Abl protein: analysis of protein-protein and protein-lipid interactions. Exp Cell Res 316(4):530–542

Mohamed AJ, Yu L, Backesjo CM et al (2009) Bruton's tyrosine kinase (Btk): function, regulation, and transformation with special emphasis on the PH domain. Immunol Rev 228(1):58–73

Morales J, Sobol M, Rodriguez-Zapata LC et al (2017) Aromatic amino acids and their relevance in the specificity of the PH domain. J Mol Recognit 30:e2649

Moses SA, Ali MA, Zuohe S et al (2009) In vitro and in vivo activity of novel small-molecule inhibitors targeting the pleckstrin homology domain of protein kinase B/AKT. Cancer Res 69(12):5073–5081

Murayama K, Kato-Murayama M, Mishima C et al (2008) Crystal structure of the Bruton's tyrosine kinase PH domain with phosphatidylinositol. Biochem Biophys Res Commun 377(1):23–28

Nastou KC, Tsaousis GN, Papandreou NC, Hamodrakas SJ (2016) MBPpred: proteome-wide detection of membrane lipid-binding proteins using profile hidden Markov models. Biochim Biophys Acta 1864(7):747–754

Ni T, Kalli AC, Naughton FB et al (2017) Structure and lipid-binding properties of the kindlin-3 pleckstrin homology domain. Biochem J 474(4):539–556

Ogawa A, Yamazaki Y, Nakamori M et al (2013) Characterization and distribution of adaptor protein containing a PH domain, PTB domain and leucine zipper motif (APPL1) in Alzheimer's disease hippocampus: an immunohistochemical study. Brain Res 1494:118–124

Ojteg G, Lundahl P, Wolgast M (1989) The net electric charge of proteins. A comparison of determinations by Donnan potential measurements and by gel electrophoresis. Biochim Biophys Acta 991(2):317–323

Pang X, Fan J, Zhang Y et al (2014) A PH domain in ACAP1 possesses key features of the BAR domain in promoting membrane curvature. Dev Cell 31(1):73–86

Parikh C, Janakiraman V, Wu WI et al (2012) Disruption of PH-kinase domain interactions leads to oncogenic activation of AKT in human cancers. Proc Natl Acad Sci U S A 109(47):19368–19373

Politz O, Siegel F, Barfacker L et al (2017) BAY 1125976, a selective allosteric AKT1/2 inhibitor, exhibits high efficacy on AKT signaling-dependent tumor growth in mouse models. Int J Cancer 140(2):449–459

Psachoulia E, Sansom MS (2008) Interactions of the pleckstrin homology domain with phosphatidylinositol phosphate and membranes: characterization via molecular dynamics simulations. Biochemistry 47 (14):4211–4220

Rebecchi MJ, Scarlata S (1998) Pleckstrin homology domains: a common fold with diverse functions. Annu Rev Biophys Biomol Struct 27:503–528

Riehle RD, Cornea S, Degterev A (2013) Role of phosphatidylinositol 3,4,5-trisphosphate in cell signaling. Adv Exp Med Biol 991:105–139

Rodgers SJ, Ferguson DT, Mitchell CA, Ooms LM (2017) Regulation of PI3K effector signalling in cancer by the phosphoinositide phosphatases. Biosci Rep 37(1): BSR20160432

Rong SB, Hu Y, Enyedy I et al (2001) Molecular modeling studies of the Akt PH domain and its interaction with phosphoinositides. J Med Chem 44(6):898–908

Rosen SA, Gaffney PR, Gould IR (2011) A theoretical investigation of inositol 1,3,4,5-tetrakisphosphate. Phys Chem Chem Phys 13(3):1070–1081

Rosen SA, Gaffney PR, Spiess B, Gould IR (2012) Understanding the relative affinity and specificity of the pleckstrin homology domain of protein kinase B for inositol phosphates. Phys Chem Chem Phys 14 (2):929–936

Roy NS, Yohe ME, Randazzo PA, Gruschus JM (2016) Allosteric properties of PH domains in Arf regulatory proteins. Cell Logist 6(2):e1181700

Santos-Garcia L, Assis LC, Silva DR et al (2016) QSAR analysis of nicotinamidic compounds and design of potential Bruton's tyrosine kinase (Btk) inhibitors. J Biomol Struct Dyn 34(7):1421–1440

Shen B, Bai J, Vihinen M (2008) Physicochemical feature-based classification of amino acid mutations. Protein Eng Des Sel 21(1):37–44

Singh SM, Murray D (2003) Molecular modeling of the membrane targeting of phospholipase C pleckstrin homology domains. Protein Sci 12(9):1934–1953

Stahelin RV, Scott JL, Frick CT (2014) Cellular and molecular interactions of phosphoinositides and peripheral proteins. Chem Phys Lipids 182:3–18

Tang N, Lin T, Ostap EM (2002) Dynamics of myo1c (myosin-ibeta) lipid binding and dissociation. J Biol Chem 277(45):42763–42768

Tran HT, Zhang S (2011) Accurate prediction of the bound form of the Akt pleckstrin homology domain using normal mode analysis to explore structural flexibility. J Chem Inf Model 51(9):2352–2360

Valiaho J, Faisal I, Ortutay C et al (2015) Characterization of all possible single-nucleotide change caused amino acid substitutions in the kinase domain of Bruton tyrosine kinase. Hum Mutat 36(6):638–647

Wadhwa B, Makhdoomi U, Vishwakarma R, Malik F (2017) Protein kinase B: emerging mechanisms of isoform-specific regulation of cellular signaling in cancer. Anti-Cancer Drugs 28(6):569–580

Yamamoto E, Kalli AC, Yasuoka K, Sansom MS (2016) Interactions of Pleckstrin homology domains with membranes: adding back the bilayer via high-throughput molecular dynamics. Structure 24 (8):1421–1431

Yamamoto E, Akimoto T, Kalli AC et al (2017) Dynamic interactions between a membrane binding protein and lipids induce fluctuating diffusivity. Sci Adv 3(1): e1601871

Yao L, Kawakami Y, Kawakami T (1994) The pleckstrin homology domain of Bruton tyrosine kinase interacts with protein kinase C. Proc Natl Acad Sci U S A 91 (19):9175–9179

Yates LA, Lumb CN, Brahme NN et al (2012) Structural and functional characterization of the kindlin-1 pleckstrin homology domain. J Biol Chem 287 (52):43246–43261

Yilmaz OG, Olmez EO, Ulgen KO (2014) Targeting the Akt1 allosteric site to identify novel scaffolds through virtual screening. Comput Biol Chem 48:1–13

Yu JW, Mendrola JM, Audhya A et al (2004) Genome-wide analysis of membrane targeting by S. cerevisiae pleckstrin homology domains. Mol Cell 13(5):677–688

Adv Exp Med Biol - Protein Reviews (2019) 20: 33–53
https://doi.org/10.1007/5584_2018_259
© Springer Nature Switzerland AG 2018
Published online: 28 August 2018

BAR Domain Proteins Regulate Rho GTPase Signaling

Pontus Aspenström

Abstract

The Bin–Amphiphysin–Rvs (BAR) domain is a membrane lipid binding domain present in a wide variety of proteins, often proteins with a role in Rho-regulated signaling pathways. BAR domains do not only confer binding to lipid bilayers, they also possess a membrane sculpturing ability and thereby directly control the topology of biomembranes. BAR domain-containing proteins participate in a plethora of physiological processes but the common denominator is their capacity to link membrane dynamics to actin dynamics and thereby integrate processes such as endocytosis, exocytosis, vesicle trafficking, cell morphogenesis and cell migration. The Rho family of small GTPases constitutes an important bridging theme for many BAR domain-containing proteins. This review article will focus predominantly on the role of BAR proteins as regulators or effectors of Rho GTPases and it will only briefly discuss the structural and biophysical function of the BAR domains.

Keywords

BAR domain · Rho GTPase · RhoGAP · RhoGEF

P. Aspenström (✉)
Department of Microbiology, and Tumor and Cell Biology, Karolinska Institutet, Stockholm, Sweden
e-mail: pontus.aspenstrom@ki.se

1 Introduction

BAR domain-containing proteins, or BAR proteins for short, comprise a heterogeneous group of multi-domain proteins with diverse biological functions; the common denominator being the Bin–Amphiphysin–Rvs (BAR) domain that not only confers targeting to lipid bilayers, but also provides the supporting scaffolding to shape lipid membranes into concave or convex surfaces (Peter et al. 2004; Qualmann et al. 2011). This function of BAR proteins is an important determinant in the dynamic reconstruction of membrane vesicles, as well as of the plasma membrane. Membrane deformation is for instance required for membrane fission and fusion events during endocytosis, exocytosis and vesicularization of endomembranes, such as in the endoplasmic reticulum and the Golgi complex (Qualmann et al. 2011).

Several BAR proteins fulfil specific roles in linking actin cytoskeleton dynamics and membrane trafficking, two vital cellular processes that need to be precisely coordinated. Bridges between these processes are provided by direct interactions between BAR proteins and actin nucleation-promoting factors (NPFs) of the Wiskott-Aldrich syndrome protein (WASP) and the Diaphanous-related formin (DRFs) families (Chesarone et al. 2010; Rotty et al. 2013). The Rho GTPases are key factors in this intricate interplay, BAR proteins can function upstream as well as downstream of the Rho GTPases.

The first member of the Rho GTPase, named Rho for Ras homolog, was discovered in 1985 (Madaule and Axel 1985). We now know that the human genome harbors 20 members of the Rho branch of the Ras superfamily of small GTPases and many of them function in a similar mode as the Ras oncoproteins (Jaffe and Hall 2005; Aspenström 2018). The Rho GTPases are, as one can deduct from the name, GTP hydrolyzing enzymes and they function as binary switches based on their intrinsic ability to cycle between inactive GDP-bound conformations and active GTP-bound conformations (Jaffe and Hall 2005). However, the enzyme activity of the Rho GTPases on their own is low; instead they require auxiliary proteins, so-called GTPase activating proteins (GAP), which increase the hydrolysis activity manifold (Tcherkezian and Lamarche-Vane 2007). In order to get into the active conformation, the Rho GTPases need to be activated by so called guanine nucleotide exchange factors (GEFs) that facilitate the exchange of GDP for GTP (Cook et al. 2014). In addition, many Rho GTPases are regulated by so called GDP dissociation inhibitors (GDIs), which sequester the Rho GTPase in the inactive GDP-bound conformation (DerMardirossian and Bokoch 2005). Intriguingly, it is only half of the Rho GTPases that follow this classical scheme of regulation. These members are referred to as the classical Rho GTPases and comprise RhoA, RhoB, RhoC, Rac1, Rac2, Rac3, RhoG, Cdc42, TCL(RhoJ) and TC10(RhoQ). The other half are referred to as atypical Rho GTPases and can be further divided into GTPase deficient Rho GTPases (Rnd1, Rnd2, Rnd3(RhoE), RhoH, RhoBTB1 and RhoBTB2) and fast-cycling Rho GTPases (Wrch-1(RhoU), Chp(RhoV), RhoD and Rif (RhoF) (Table 1) (Aspenström 2018).

BAR proteins collaborate with Rho GTPases at many levels and, as already mentioned, they can function both upstream and downstream of the Rho GTPases. The over-all picture is complex, because BAR proteins and Rho GTPases tend to interact in multi-protein complexes that involve several additional subunits. These

Table 1 The Rho GTPases

A *Classical Rho GTPases*
Rho subfamily: RhoA, RhoB, RhoC
Rac subfamily: Rac1, Rac2. Rac3, RhoG
Cdc42 subfamily: Cdc42, TC10(RhoQ), TCL(RhoJ)
B *Atypical Rho GTPases*
GTPase-deficient Rho GTPases
Rnd1, Rnd2, Rnd3(RhoE), RhoH, RhoBTB1, RhoBTB2
Fast-cycling Rho GTPases
RhoD, Rif(RhoF), Wrch-1(RhoU), Chp(RhoV)

components can for instance be proteins that regulate actin polymerization, most notably the WASP and DRFs families of actin NPFs (Chesarone et al. 2010; de Kreuk et al. 2011; Rotty et al. 2013). Both WASP and DRFs contain binding sites for Rho GTPases, which means that it is difficult to determine if signaling occurs directly from the BAR proteins or from the actin regulators. It is worth noticing that, so far, the BAR proteins seems to be linked predominantly to signaling involving the classical Rho GTPases and not the atypical ones.

This review will focus on the interplay between Rho GTPases and BAR proteins and describe how defects in this interplay can cause human diseases such as neuropathological conditions and cancer. The mutual collaboration between Rho GTPases and BAR protein is a central factor in the regulation of vital cellular processes, such as cell migration, cytokinesis, intracellular transport, endocytosis and exocytosis. This overview over BAR proteins will be restricted with the BAR proteins with a known function in Rho signaling (Table 2). There are several excellent reviews focusing on the function of the various types of BAR domains and can provide entrance points for further reading on the structural and biophysical aspects of BAR proteins (Qualmann et al. 2011; Salzer et al. 2017).

2   BAR Domains

The concept of BAR domain was established in a study by Sakamuro et al. (1996) on the Myc-binding protein Bin1. The authors of this

Table 2 BAR proteins involved in Rho regulation

A *BAR proteins with RhoGEF domains*
The Tuba subfamily: Tuba (DMNBP, KIAA1010), ARHGEF38(Tuba2), ARGEF37(Tuba3)
B *BAR proteins with RhoGAP domains*
The GRAF subfamily: GRAF1(ARHGAP26), ARHGAP10, ARHGAP42, Oligophrenin
The RICH/Nadrin subfamily: SH3BP1, RICH1(Nadrin), RICH2
The HMHA subfamily: HMHA-1, Gmip, PARG(ARHGAP29)
The srGAP subfamily: SrGAP1, srGAP2, srGAP3, RhoGAP4
C *BAR proteins serving as Rho effector proteins*
F-BAR proteins: CIP4, FBP17, Toca-1, PACSIN2/syndapin-2
I-BAR proteins: IRSp53, IRTKS

study noticed an area of similarity between Bin1, Amphiphysin and the yeast protein Rvs167, hence the acronym Bin/Amphiphysin/Rvs (BAR) domain. The functional role of BAR domains was further identified in a study on the Amphiphysin BAR domain (Takei et al. 1999). The authors showed that Amphiphysin possessed the ability to catalyze the formation of membrane tubules in a cell-free liposome-binding assay. It was furthermore found that the tubulating activity resided in the N-terminal fragment encompassing amino acid residues 1–276, i.e. the fragment that contains the BAR domain (Takei et al. 1999). BAR domains confer binding to phospholipids and the phospholipid composition is important for their lipid-binding capacity. By using liposomes with defined ratios between posphatidylcholine (PC), phosphatidylethanolamine (PE) and phosphatidylserine (PS), Itoh et al. found that the optimal binding of the FBP17 and Syndapin F-BAR domains to liposomes occurred with 70% PC, 20% PE and 10% PS. The lipid binding was highly dependent on the presence of PS (Itoh et al. 2005).

The capacity to induce various forms of membrane tubulation constitutes the common denominator of the majority of BAR domains. Tubulation can be triggered *in vitro*, by incubation of isolated BAR domains with liposomes with defined composition of phospholipids. This phenomenon can be studied by ectopic expression of BAR domains and can be visualized as the formation of extended invaginations and tubulations of the plasma membrane and of endomembranes. The extensive membrane tubulation is considered to be an over-activity of

a natural process of extension of membrane tubules that occurs during endocytosis. However, under normal conditions the tubulation is counteracted and regulated by proteins with membrane-fission activity, such as Dynamin (Sweitzer and Hinshaw 1998).

The similarity between the different types of BAR domains is not that apparent if one looks at the amino-acid sequence level. Importantly, structural analysis of the three-dimensional fold has revealed a high degree of similarity between the different types of BAR domains (Peter et al. 2004). The BAR proteins can be subdivided into six classes: N-BAR, F-BAR, I-BAR, BAR-PH, PX-BAR and BAR-PDZ (Salzer et al. 2017). The family of BAR proteins has been constantly growing and there are probably still proteins with BAR domain function to be identified. Commonly, BAR domains are formed by three extended α-helices that fold into crescent-shaped antiparallel dimers (Peter et al. 2004). However, the curvature of this crescent-shaped structure varies between the different classes of BAR domains. N-BAR and BAR-PH domains have a higher degree of curvature compared to the F-BAR and PX-BAR domains, which form a more shallow-curved 'banana' shape (Peter et al. 2004; Shimada et al. 2007; Salzer et al. 2017) (Fig. 1). In contrast, the I-BAR proteins fold into extended, zeppelin-shaped, structures (Millard et al. 2005). The concave surfaces of the crescent-shaped BAR domains have surplus of positively charged residues that can bind to negatively charged head groups on lipid membranes (Fig. 2). This structural organization is true for the N-BAR, BAR-PH and PX-BAR

Fig. 1 Three dimensional structures of three representative BAR domains. (**a**) endophilin, (**b**) CIP4, (**c**) IRSp53. The three dimensional structures were visualized by the Protein Data Bank

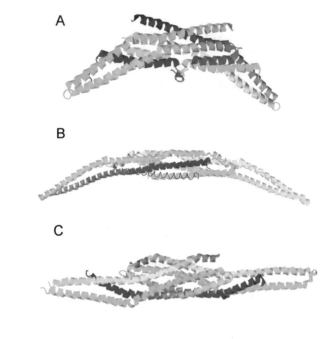

Fig. 2 Schematic representation of the lipid binding surfaces and their effect on membrane topology. (**a**) BAR domains with concave binding surfaces induce membrane invaginations. (**b**) BAR domains with convex binding surface induces membrane tubulation. (**c**) I-BAR domains with convex binding surfaces induce membrane tubulation

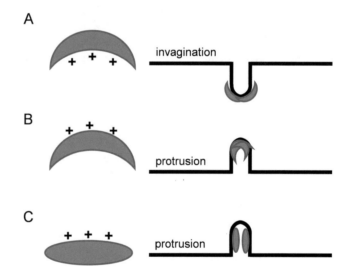

domains. A majority of the F-BAR domains also bind lipid bilayers through concave surfaces but there are exceptions to this rule since, some of them bind lipids through their convex surfaces and thereby give rise to an inverse curvature (Fig. 2). This mechanism of interaction is also true for I-BAR proteins; all of them, with the exception of a protein known as Pinkbar, interact through convex surfaces and trigger the formation of lipid membranes with inverse curvature (Millard et al. 2005; Pykäläinen et al. 2011).

The mechanisms by which BAR-domain-induced membrane deformation drives membrane tubulation are only partially known. The membrane binding and membrane bending abilities of F-BAR domains appears to occur independently of each other. According to a model put forward by Frost et al., F-BAR domains do not bind primarily to curved membranes, instead, F-BAR domains bind along their side to flat membranes. Subsequently, oligomerization of F-BAR domains results in the relocation of the curved surface to the face of the membrane, and hence drives membrane deformation (Frost et al. 2008). This mode of action to coordinate oligomerization and tabulation is not restricted to F-BAR proteins, since N-BAR proteins, like endophilin, can also assemble into regular helical scaffolds to promote membrane curvature, however, in the latter case the oligomerization was found to be stabilized through interactions between the N-terminal helices of endophilin (Mim et al. 2012). BAR domains do not only induce membrane tubulation, they can also promote the formation of stable phosphoinositide-containing microdomains on lipid bilayers (Zhao et al. 2013).

3   BAR Proteins with RhoGEF Activity

There is just one group of BAR domain proteins which also possess a RhoGEF domain. The founding member of this group is the dynamin-binding protein Tuba (named after the musical instrument in accordance with a tradition of naming large endocytic proteins after musical instruments) (Salazar et al. 2003). Tuba is quite unique in comparison to other RhoGEFs, because it does not have a membrane targeting pleckstrin homology (PH) domain adjacent to the Dbl Homology (DH domain-the domain that possesses the GEF activity). In the case of Tuba, the BAR domain appears to fulfil this role, since the Tuba DH domain alone has a very low activity in the absence of the BAR domain. Tuba is a Cdc42 GEF and has a very limited ability to catalyze the activation of Rac1 and RhoA. In addition to the DH and BAR domains, Tuba has six SH3 domains, four in the N-terminus and two in the C-terminus (Fig. 3). Tuba (it is also known as KIAA1010 or Dynamin binding protein-DMNBP) is ubiquitously expressed in vertebrate tissue and several splice-variants have been described (Salazar et al. 2003). There are two additional Tuba-like proteins; ARHGEF37 (or Tuba3) and ARHGEF38 (or Tuba2) (Salazar et al. 2003; Cook et al. 2014). ARHGEF37 appears to be ubiquitously expressed in many cell-types but ARHGEF38 has a more restricted expression pattern. Both proteins lack the N-terminal SH3 domains and very little is known about their biological function.

Tuba binds Dynamin through the N-terminal SH3 domains, whereas the most C-terminally positioned SH3 domain binds N-WASP, as well as other known actin regulators. The original observation implicated a role of Tuba in the coordination of endocytosis and actin dynamic at the presynaptic compartment (Salazar et al. 2003). Overexpression of Tuba in B16 melanoma cells stimulated the formation of dorsal ruffles by the coordinated localization of lipid micro-domains and the actin regulating machinery (Kovacs et al. 2006). In epithelial Caco-2 cells, Tuba was found to localize to the apical part of cell-cell junctions and bound directly to the tight junction protein ZO-1 (Otani et al. 2006). Knock-down of Tuba resulted in a perturbed distribution of the junctional protein E-cadherin and filamentous actin. Epithelial organs consist of a monolayer of polarized epithelial cells that surround a central lumen. To produce an apical surface and a lumen, this polarization requires interactions between the membrane sorting machinery and signaling pathways that define cortical domains (Bryant et al. 2010; Kovacs et al. 2011). Cdc42 is a master regulator in this process, predominantly through its interaction with the Par6–Par3–atypical PKC polarity complex. The role of Tuba during lumenogenesis is to provide activated Cdc42 at specific sites at the apical side of epithelial cells, thereby influencing the cell junction integrity and spindle orientation (Bryant et al. 2010; Kovacs et al. 2011; Qin et al. 2010). To summarize, the proposed role of Tuba in the studies described above is to coordinate Cdc42 activation and localization of the N-WASP nucleation promotion factor. Tuba is also required for the formation of primary cilia and knock-down of Tuba in MDCK cells results in loss of the primary cilia and inhibition of HGF-induced tubulogenesis (Baek et al. 2016). Dysfunction of renal primary cilia is

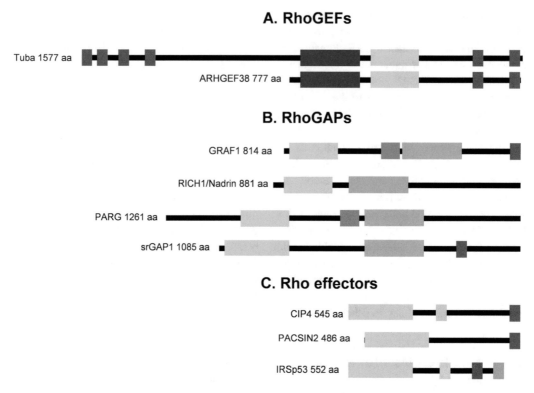

Fig. 3 Schematic representations of the BAR protein involved in Rho signaling. (**a**) BAR proteins with a RhoGEF domains: Tuba, ARHGEF38. (ARHGEF37 has the same domain organization as ARHGEF38). (**b**) BAR proteins with a RhoGAP domains: GRAF1 (ARHGAP10, ARHGAP42 and oligophrenin have similar domain organizations), RICH1/Nadrin (RICH2 and SH3BP1 have similar domain organizations), PARG (HMHA-1 and Gmip have similar domain organizations), srGAP1 (srGAP2, srGAP3 and ARHGAP4 have similar domain organizations). (**c**) Effectors for Rho GTPases with BAR domains: CIP4 (Toca1 and FBP17 have similar domain organizations), PACSIN2, IRSp53 (IRTKS has similar domain organization). Brown denotes RhoGEF domain, Green denotes RhoGAP domain, Yellow denotes BAR domain, Red denotes SH3 domain, blue denotes PH domain, Orange denotes C1 (cysteine-rich phorbol ester binding) domain, Light Blue denotes HR1 (protein kinase C-related kinase homology region 1) domain, Light Green denotes GBD (GTPase-binding) domain, purple denotes WH2 (WASP homology 2) domain

implicated in polycystic kidney disease. Interestingly, ablation of Tuba in zebrafish by the injection morpholinos directed to Tuba led to abnormal kidney development and defective cilia formation (Baek et al. 2016).

A role for Tuba in cell polarization was suggested by the observation that the protein was localized to the Golgi complex in neuronal cells (Salazar et al. 2003). In HeLa cells, Tuba was found to bind directly to GM130, through the N-terminal part of the protein, to form a trimeric complex consisting of Tuba, Cdc42 and GM130 (Kodani et al. 2009). The role of Tuba has

furthermore been shown to collaborate with the BAR-domain-containing RhoGAP protein ARHGAP10 in the control of Cdc42 activation at the Golgi cisternae during centrosome regulation (Herrington et al. 2017).

4 BAR Proteins with RhoGAP Activity

The human genome harbors genes encoding around 70 RhoGAP proteins and at least 17 of them have BAR domains in addition to the

GTPase-activating domain (Csépányi-Kömi et al. 2013) (Fig. 3). They will be described family by family below, starting with the four members of the GRAF family of RhoGAPs and continuing with the RICH, HMHA-1 and srGAP families of proteins.

4.1 GRAF1

GTPase regulator associated with focal adhesion kinase-1 (GRAF1) (also known as ARHGAP26) is a BAR-domain and PH domain-containing RhoGAP with a C-terminal SH3 domain. GRAF1 has been characterized as a Cdc42- and RhoA-specific RhoGAP. GRAF1 binds Focal adhesion kinase (FAK) and was originally ascribed a role in the regulation of cell adhesion (Hildebrand et al. 1996). More recent research has demonstrated that GRAF1 has a critical role in Clathrin-independent endocytosis through the so-called CLIC/GEEC pathway (Clathrin-Independent Carriers/GPI-Enriched Endocytic Pathway) (Lundmark et al. 2008). There is a communication between Clathrin-independent endocytosis and cell spreading and GRAF1 is an important messenger in this communication, since knock-down of GRAF1 inhibits CLIC generation and results in defective cell spreading and decreased cell migration (Doherty et al. 2011). Clearly, there is a mutual dependency between actin dynamics and plasma membrane turnover in order to control cell morphology and cell migration. GRAF1is involved in this finely tuned cooperation by transient interactions with Cdc42 and GRAF1, which regulate the endocytic membrane turnover at the leading edge of migrating cells (Francis et al. 2015). GRAF1 was also found to be involved in cell polarization, since depletion of GRAF1 in MDCK cells resulted in impaired epithelial morphogenesis seen as decreased lumen formation and spindle orientation when the cells were cultured in 3D cell cultures, a response that also involved deregulation of Rab8 (Vidal-Quadras et al. 2017). Another role of the GRAF1 BAR domain was identified in studies on lipid droplet formation in non-adipose cells. Overexpression of GRAF1 resulted in a clustering of lipid droplets and the GRAF1 protein accumulated at the interphase between the droplets, suggesting a role of the membrane sculpturing ability of the BAR domain in the regulation of lipid droplets homeostasis (Lucken-Ardjomande Häsler et al. 2014).

Intriguingly, GRAF1 has been implicated in muscle differentiation. The skeletal muscles consist of fibers formed by the fusion of precursor cells, so called myoblasts. As a result of this fusion, muscle fibers consist of large multinucleated entities. Studies in Xenopus frogs and in mice have shown that animals lacking GRAF1 have a severely impaired myoblast fusion capacity resulting in muscle with decreased force generation abilities (Doherty et al. 2011; Lenhart et al. 2014). GRAF1 dysfunction has also been linked to a neuropathological condition called subacute cerebellar ataxia, i.e. patients suffer from unsteady gait, lack of limb coordination and dizziness (Wallwitz et al. 2017). One cause of the disease has been shown to be an autoimmunological response and the appearance of an anti-neuronal autoantibody called anti-Ca. Interestingly, protein screening experiments revealed that GRAF1 was the target antigen (Jarius et al. 2015; Wallwitz et al. 2017). The condition is very rare and so far only a limited number of patients with anti-Ca antibodies have been identified but it demonstrates that inactivation or ablation of GRAF1 can cause severe neurological disorders.

GRAF1 has been suggested to function as tumor suppressor. Recurrent fusions between the GRAF1 gene (gene name ARHGAP26) and the gene encoding the cell junction protein claudin 18 (*CLDN18*), *CLDN18-ARHGAP26*, have been identified in gastric cancer in Southeast Asian patients (Yao et al. 2015). Ectopic expression of *CLDN18-ARHGAP26* in epithelial cells resulted in reduced cell-cell and cell-extracellular adhesion, as well as loss of the epithelial phenotype by activation of pathways regulating epithelial-mesenchymal transition (EMT). These characteristics were furthermore accompanied with an increased invasive capacity in 3D Matrigel chamber assays (Yao et al. 2015). The mechanisms underlying the potential involvement of GRAF1 in cancer progression are not clear but a recent study by Holst et al. (2017) has shed some light on the matter. Key to the topic is the aforementioned role of GRAF1 to coordinate actin dynamics and plasma membrane turnover through clathrin-independent endocytosis (Francis et al. 2015). Cells lacking GRAF1 have defective clathrin-independent endocytosis because the GRAF1-dependent control of surface tension is lost (Holst et al. 2017). As a result, the membrane blebbing increases a response that is at least partially caused by a lost ability of GRAF1 to act as a RhoGAP for RhoA. Importantly, the increased membrane blebbing is accompanied with an increased invasion as seen in the Matrigel 3D invasion assay (Holst et al. 2017).

4.2   The GRAF1-Like RhoGAPs ARHGAP10 and ARHGAP42

ARHGAP10 (also known as GRAF2 or PSGAP) has the same domain organization as GRAF1 (Fig. 3). It was cloned in a yeast two-hybrid screen for proteins interacting with the C-terminus of the non-receptor tyrosine kinase Pyk2. The interaction involved the SH3 domain of ARGAP10 binding to a proline-rich stretch in the C-terminal portion of Pyk2 (Ren et al. 2001). ARHGAP10 is predominantly a RhoA-GAP but it also exhibits mild GTPase enhancing activity on Cdc42, but not Rac1. Pyk2 can phosphorylate ARHGAP10 and this phosphorylation result in a decreased Cdc42-GAP activity; however, it is not known if also the RhoA-GAP activity is under negative control by Pyk2-dependent phosphorylation (Ren et al. 2001). In contrast, another study argues that ARHGAP10 is predominantly a Cdc42-GAP (Koeppel et al. 2004). ARHGAP10 was found to be involved in the regulation of PAK2-induced apoptosis. PAK2 can be constitutively activated by caspase cleavage and ARHGAP10 interacts selectively with the activated fragment, PAK2p34. This leads to a decreased kinase activity of the PAK2 fragment which, in turn, leads to a decreased apoptotic response (Koeppel et al. 2004). ARHGAP10 was, moreover, found to interact with the cell adhesion protein α-catenin at cell-cell junctions. ARHGAP10 knock-down resulted in derailed recruitment of α-catenin to the junctions (Sousa et al. 2005). Similar to GRAF1, ARHGAP10 has been implicated to function as a tumor suppressor. Decreased ARHGAP10 expression was found in 58 out of 75 ovarian cancer tissues analyzed, in contrast to non-tumorous counterparts. Increased expression of ARHGAP10 in low expression ovarian cancer cell lines effectively decreased cell proliferation in tissue culture and tumorigenicity in xenograft models in nude mice (Luo et al. 2016).

ARHGAP42 (also known as GRAF3) is also likely to be a RhoA-GAP, since ARHGAP42 depletion results in increased RhoA activity (Bai et al. 2013). ARHGAP42 is highly and selectively expressed in smooth muscle tissue. Studies of ARHGAP42 hypomorphic mice, in which the ARHGAP42 message was significantly, but not completely, depleted showed that ARHGAP42 is involved in the regulation of blood pressure homeostasis (Bai et al. 2013). ARHGAP42 was shown to be under control of the non-receptor tyrosine kinase Src. The BAR domain is a negative regulator of the GAP activity of ARHGAP42. If this occurs through an autoinhibitory interaction is not known, but this is a plausible mechanism. Src phosphorylation at tyrosine residue 376 counteracted the inhibitory influence of the BAR domain, resulting in a decreased pool of GTP-bound RhoA, which, in turn, led to increased focal adhesion dynamics, and more active lamellopodia protrusions (Luo et al. 2017).

4.3   Oligophrenin

Oligophrenin-1(OPHN1) was identified in a hunt for the gene causing the rare condition nonspecific X-linked mental retardation (Billuart et al. 1998). It was discovered that mutations in the *OPHN1* gene, which encodes a BAR domain- and RhoGAP-domain-containing protein was the cause of the disease. It is similar to GRAF1, ARHGAP10 and ARHGAP42 in its over-all domain organization. OPHN1 is required for dendritic spine morphogenesis, and its RhoGAP activity appears to be important in the modulation of the length of the dendritic spines (Govek et al. 2004). Inactivation of OPHN1 results in increased RhoA activity that is associated with significantly increased length of the dendritic spines. This phenotype can be suppressed by inhibition of ROCK, which is acting downstream of RhoA. OPHN1 is a clathrin-mediated endocytosis regulator, and moreover, it binds the N-BAR protein Endophilin A1 and regulates the endocytosis of synaptic vesicles (Nakano-Kobayashi et al. 2009). Inhibition of OPHN1 results in reduced endocytosis, and more particularly, in reduced internalization of the AMPA receptor, which is a receptor for glutamate and is an established regulator of synaptic plasticity (Khelfaoui et al. 2009).

A recent study in chromaffin cells (neuroendochrine cells) from *OPHN1* knock out mice has implicated a double function of OPHN1 in a process called compensatory endocytosis. This is a process that follows exocytosis of secretory granules from chromaffin cells in order to retain the net cell surface area as well as to recover components of the exocytosis machinery. Interestingly OPHN1 was shown to have roles in both these processes through distinct mechanisms; membrane fusion during the exocytosis of secretory granules required a functional RhoGAP domain, whereas the BAR domain was needed for the endocytosis process (Houy et al. 2015).

4.4 RICH/SH3BP1

This group of RhoGAP domain proteins consist of three paralogs; RICH1/Nadrin, RICH2 and SH3BP1. SH3BP1 was the first of the three to be identified in a screen for substrate for the SH3 domain of the non-receptor tyrosine kinase Abl (Cicchetti et al. 1992; Ren et al. 1993). These studies are interesting from a historical point of view, since they actually helped to establish the nowadays well-known concept of SH3 domains interacting with proline-rich motifs (Ren et al. 1993).

Rac1 needs to be under a dynamic regime of cycling between its GTP-bound and GDP-bound conformations. SH3BP1 has been implicated to participate in the regulation of cell migration by specifically and locally down-regulating Rac1 at the leading edge. Down-regulation of SH3BP1 leads to decreased cell migration and disorganized peripheral protrusions (Parrini et al. 2011). SH3BP1 also has a role in junctional assembly during epithelial morphogenesis. Knock-down of SH3BP1 expression results in aberrant membrane remodeling and cell: cell junction formation (Elbediwy et al. 2012).

RICH1 (RhoGAP Interacting with Cdc42-interacting protein 4) was found in a yeast two-hybrid screen for proteins binding to the SH3 domain of the F-BAR protein CIP4 (Richnau and Aspenström 2001). A rat ortholog of RICH1, called Nadrin, was described roughly at the same time, as a result, there is an ambiguity regarding the naming of the protein in the literature (Harada et al. 2000). Moreover, the protein is sometimes referred to as ARHGAP17, which is the gene name. The initial molecular characterization of RICH demonstrated that it contained several proline-rich motives, a C-terminal PDZ domain-binding motif and an N-terminal "endophilin-like domain". This domain was later found to fall into the category of BAR domains (in this case an N-BAR domain). As for the specificity of RICH1, it is a Cdc42- and Rac1-GAP and it shows only a limited activity on RhoA (Richnau and Aspenström 2001).

RICH1 has been ascribed a role in the regulation of calcium-dependent exocytosis in neuronal cells (Harada et al. 2000). In addition, RICH1 was found to be involved in the organization of apical polarity in cells. RICH1 binding to the scaffolding protein angiomotin targets it to a protein complex at tight junctions that contains additional polarity proteins, such as Pals1, Patj, and PAR-3. Cdc42 is an important factor for apical cell polarity in epithelial cells, and regulation of Cdc42 by RICH1 is necessary for the maintenance of tight junctions (Wells et al. 2006). There is a connection between RICH1 and the merlin tumor suppressor protein, since RICH1 is mediating the function of merlin by downregulating Rac1 activity after being released from the RICH1:angiomotin complex at tight junctions (Yi et al. 2011). Tight junctions are essential for a proper barrier function of epithelial cell layers in the body. Mice lacking RICH1 are viable and fertile but there is an aberrant localization of adherence junction proteins in the luminal epithelium. Defects in the integrity of the epithelial cell layer can cause disease conditions, such as colitis. The $RICH1^{-/-}$ mice do not develop spontaneous colitis, however, if the mice are challenged with dextran sulfate sodium, which is a well-established model for colitis induction, this results in destruction of the inner mucus layer and a much severe disease condition compared to the wild-type mice (Lee et al. 2016). These data suggest an important function of RICH1 in the maintenance of the integrity of intestinal barriers.

A RICH1-like protein found in the genomic data-bases was given the name RICH2 (Richnau and Aspenström 2001). RICH2 appears to have a role in cell polarity, through interactions with a lipid-raft-associated integral membrane protein known as CD317 (or tetherin). This protein is expressed at the apical surface of polarized epithelial cells, where it has been shown to interact indirectly with the underlying actin cytoskeleton through RICH2, EBP50, and ezrin. Knock-down of CD317 resulted in increased Rac1 activity, which was accompanied by a loss of apical microvilli (Rollason et al. 2009). RICH2 has also been shown to have a role in the regulation of AMPA receptor internalization, via an interaction with a postsynaptic scaffolding protein called Shank3. Knock-down of RICH2 using RNA interference, or disruption of the RICH2–Shank3 complex, interfered with the recycling of the AMPA receptor, and thereby the control of synaptic plasticity (Raynaud et al. 2013). RICH2 seems to have a prominent role in the brain, since the protein was found to be enriched in dendritic spines of hippocampal pyramidal neurons in mice during early development (Raynaud et al. 2014). RICH2 was shown to regulate the size and density of dendritic spines in a Rac1-dependent manner. Studies on RICH2 knockout mice have confirmed a role of RICH2 in the regulation of dendritic spine density (Sarowar et al. 2016). Mice lacking RICH2 displayed an increased stereotypic behavior, impaired motor

learning as well as a highly significant fear of novel objects, so called neophobia. This is an concrete example of that a defect in the fine-tuning of activity of a Rho GTPases, in this case Rac1, has a direct implication on the psychology of living beings.

5   HMHA-1, Gmip and ARHGAP29

The human minor histocompatibility antigen1 (HMHA-1) is an F-BAR protein with a RhoGAP domain. In this case, HMHA-1 was found to stimulate GTP hydrolysis of Rac1 and RhoA (de Kreuk et al. 2013). There are two additional RhoGAP proteins which are related to HMHA-1, however it is not entirely clear if they contain fully functional BAR domains (Csépányi-Kömi et al. 2013). There is, however, a crystal structure at 2.4 Å resolution of the Gmip BAR domain in the protein data bank (PBD ID:3QWE) demonstrating the presence of a completely folded F-BAR domain. The common denominator for all three members of this group, HMHA-1, Gmip and ARHGAP29, is that they all catalyze GTP hydrolysis on RhoA. Gmip, is an effector for the Ras-like small GTPase Gem and has a role in the regulation of stress fiber dissolution and focal adhesion disassembly (Hatzoglou et al. 2007). This effect on actin reorganization appears to be of particular importance during exocytosis, as down-regulation of the RhoA pathway is a critical step during exocytosis of secretory granules in neutrophils (Johnson et al. 2012).

The third member of this group of RhoGAP proteins ARHGAP29 is also known PTLP1-associated RhoGAP (PARG). It was originally cloned as a protein binding to one of the PDZ domains of the intracellular tyrosine phosphatase PTLP1 (Saras et al. 1997). ARHGAP29 appears to be under control of the small GTPase Rap1 in the regulation of the integrity of endothelial junctions, which controls endothelial barrier function (Post et al. 2013, 2015). Rap1 does not bind directly to ARHGAP29, instead the interaction is mediated through two Rap1 effectors, Rasip1 and Radil (Post et al. 2015). Rap1 has been shown to regulate the localization of a multiprotein complex involving these proteins which is required for Rap1-induced inhibition of Rho signaling and correct endothelial barrier function. ARHGAP29 has, similar to several other BAR domain-containing RhoGAP proteins, been linked to cancer. A recent study by Qiao et al. has specifically linked ARHGAP29 to gastric cancer metastasis, and patients with increased ARHGAP29 expression exhibit shortened survival (Qiao et al. 2017). Moreover, mutations in ARHGAP29 has been linked a rare condition called non-syndromic cleft lip with or without cleft palate, which is a complex congenital malformation (Savastano et al. 2017).

5.1 srGAP and ARHGAP4

The Slit-Robo RhoGAPs (srGAP1-3) were originally identified as binding partners for repulsion receptor roundabout 1 (Robo1). The Slit–Robo ligand–receptor system is expressed in the developing nervous system and has crucial roles, for instance, during axon guidance (Ballard and Hinck 2012). The srGAPs play important roles in the regulation of neuronal morphogenesis and synaptic plasticity. Slit binding to the Robo1 receptor is an important determinant for the regulation of neuronal migration, and srGAP1, which is a Cdc42-specific and RhoA-specific RhoGAP, is an important signaling molecule downstream of Robo1(Wong et al. 2001; Madura et al. 2004). The srGAPs directly bind to components in the actin polymerization machinery and could therefore bridge signals from the Slit-Robo signaling system to the proteins regulating the organization and dynamics of the actin filament system. The catalytic specificity and interaction spectrum differs a bit between the three srGAPs; srGAP3, for instance, is a Rac1-specific and Cdc42-specific GAP and a binding partner of WAVE1 (Endris et al. 2002; Soderling et al. 2002, 2007). Genetic analysis of patients with severe types of mental retardation has implicated *SrGAP3*, which is also known as *MEGAP* or *WRP*, as one of the defective genes, and as causative of this neuropathological condition (Endris et al. 2002; Waltereit et al. 2012). Ablation of the *WAVE1*

gene results in sensomotor slowing in mice, which support the notion of the existence of a srGAP–WAVE1-dependent pathway in mental retardation (Soderling et al. 2002). SrGAP2 was found to regulate neuronal cell migration and to induce neurite outgrowth and branching (Guerrier et al. 2009). Interestingly, srGAP2 induces filopodia-like membrane protrusions that resemble those induced by I-BAR domains, despite having an F-BAR domain (Coutinho-Budd et al. 2012). This I-BAR–like activity appears unique for srGAP2 and srGAP3, while the srGAP1 F-BAR domain functions more in congruence with other know F-BAR domains. There is also a possible hierarchical relationship between the srGAPs, as srGAP2 acts through srGAP3 in neurite outgrowth in neuroblastoma cells (Ma et al. 2013).

ARHGAP4 is a srGAP-related protein but it has not been linked to Slit-Robo signaling. It was found to be expressed in the nervous system of rats and localize the Golgi apparatus (Foletta et al. 2002). It is a RhoGAP predominantly for Rac1 and Cdc42 and to a lesser extent for RhoA. ARHGAP4 was also found to localize to the leading edge of migrating cells and in growth cones, implicating a role in cell migration (Vogt et al. 2007). Similar to srGAP3, mutations in ARHGAP4 has been linked to mental retardation (Liu et al. 2016).

6 BAR Proteins as Effectors for Rho GTPases

Bar proteins that serve as effectors for Rho GTPases, i.e. proteins that bind to the active conformation of Rho GTPases, are restricted to the F-BAR and I-BAR subfamilies of BAR proteins.

6.1 CIP4 Family of F-BAR Proteins

The F-BAR protein were previously known as *Pombe* Cdc15 homology (PCH) proteins, which can be important to know for readers wanting to access the early literature on the subject (Aspenström 2009). The first described F-BAR protein, Cdc42-interacting protein 4 (CIP4), was identified in a yeast two-hybrid screen for proteins binding to the active conformation of Cdc42 (Aspenström 1997). The study identified a region in the N-terminal domain of CIP4 with sequence similarity to the FER and Fes non-receptor tyrosine kinases, as well as to a number of additional eukaryotic proteins. The domain was therefore named FER-CIP4 homology (FCH) domain (unfortunately often erroneously referred to as Fes-CIP4 domain). The FCH domain was furthermore found to be proximal to a α-helical region, which was named Cdc15-like coiled-coil (Aspenström 1997). Subsequent analysis showed that the Cdc15-like coiled-coil had significant similarities to BAR domain and the term F-BAR (FCH and BAR) was coined, initially also referred to as EFC (extended FCH) domain (Itoh et al. 2005; Tsujita et al. 2006). CIP4 was shown to function as a Cdc42 effector and coexpression of CIP4 and Cdc42 in Swiss 3 T3 fibroblasts resulted in a relocalization of CIP4 to the cell edges and to tubular structures in the cytoplasm (Aspenström 1997). The effect on CIP4 localization was specific for Cdc42, since neither Rac1 nor RhoA caused a similar relocalization of CIP4.

Two additional CIP4-like proteins have been identified; FBP17 and Toca-1. The main function of the CIP4-like proteins is at the interface between cytoskeletal dynamics and membrane trafficking. For instance, CIP4 and FBP17 have important, but complex, roles in the regulation of clathrin-mediated endocytosis (Taylor et al. 2011). One role of Toca-1 is to bridge Cdc42 to the actin regulator N-WASP and Arp2/3-driven actin polymerization (Ho et al. 2004). According to a model put forward in this study, the WASP-interacting protein (WIP) binds N-WASP in an autoinhibited conformation in non-stimulated cells. Cell activation results in the dissociation of the N-WASP:WIP complex and a release of the autoinhibited N-WASP. The role of Cdc42 in this context is presumably to alter the affinity of Toca-1 for N-WASP and/or for WIP. In addition, translocation of Toca-1 to the plasma membrane, or to internal membranes via the F-BAR domain,

might alter the affinity of Toca-1 for the N-WASP:WIP complex (Ho et al. 2004). Another study showed that N-WASP:WIP-mediated actin polymerization is regulated by PS-containing membranes and Toca-1 or FBP17 (Takano et al. 2008). Cdc42 was found to bind simultaneously to Toca-1 and N-WASP, demonstrating that Cdc42, Toca-1 and N-WASP can form a trimeric complex (Bu et al. 2010).

Cdc42 is not the only Rho GTPase member that binds F-BAR proteins and different Rho GTPases can promote a redistribution of F-BAR proteins to distinct cellular locations, and this redistribution is likely to contribute to the regulation of the CIP4-like proteins. As already mentioned, Cdc42 can target CIP4 to the plasma membrane, as well as to tubular cytoplasmic structures. TCL (also known as RhoJ) targets CIP4 predominantly to the plasma membrane, where CIP4 localizes to membrane protrusions and lamellopodia (Toguchi et al. 2010). The subcellular localization of CIP4 differs between different cell types. In fibroblasts and epithelial cells, CIP4 is localized to cytoplasmic vesicles and tubules, whereas in cortical neurons, CIP4 localized to lamellopodia and peripheral protrusions (Saengsawang et al. 2012). This variation in the targeting of CIP4 mediated by selective binding to different Rho GTPases might constitute a mechanism for the regulation of CIP4 function. Moreover, there are differences between the three CIP4-like proteins in their binding capacity to different Rho GTPases. It is debatable if FBP17 can bind Cdc42 despite the sequence similarity to CIP4, but it has at least been shown to bind the Cdc42-like GTPase TCL in the yeast two-hybrid assay (Toguchi et al. 2010). In addition, FBP17 interacts and cooperates with the atypical Rho member Rnd2 in the regulation of dendritic spines in hippocampal neurons, however, CIP4 and Toca-1 does not bind to this atypical Rho GTPase (Wakita et al. 2011). Interestingly, FBP17 have been found to coordinate the actin polymerization machinery, through N-WASP, during cell migration (Tsujita et al. 2015). Plasma membrane tension at the leading edge causes a polarized accumulation of FBP17 to small invaginations, which triggers actin polymerization at the cell forefront. The actin reorganization, in turn, results in an increase in the local plasma membrane tension, causing a feed-forward loop, again involving FBP17, that aids in directed cell migration. One study has implicated CIP4 in cell polarization through an interaction to the centrosome protein AKAP350 (Tonucci et al. 2015). CIP4 was found to localize to centromeres and depletion of CIP4 or loss of the CIP4:AKAP350 interaction resulted in derailed centrosome positioning and defective cell polarization.

Several recent reports have demonstrated that the CIP4-like proteins have important roles in endocytosis and trafficking of cell-surface receptors, such as the epidermal growth factor receptor, the platelet-derived growth factor receptor, and the transferrin receptor (Tsujita et al. 2006; Hu et al. 2009; Toguchi et al. 2010). Another example where CIP4 has a role in regulating the subcellular localization of other proteins has been provided by studies on the trafficking of the glucose transporter GLUT4. GLUT4 is the major insulin-responsive glucose transporter for the removal of glucose from the bloodstream (Hou and Pessin 2007). In resting cells, GLUT4 is predominantly localized to the cytoplasm, however, insulin stimulation results in a rapid shift in the balance between endocytosis and of exocytosis GLUT4 to favor the latter. This results in that about 50% of GLUT4 is localized to the plasma membrane. The critical step in this pathway is the activation of the Rho family member TC10 (also known as RhoQ). This activation occurs via C3G, which is a GEF for another small GTPase, Rap1, as well as for TC10 (also known as RhoQ) (Chiang et al. 2001). Interestingly, TC10, in turn, bind CIP4, or rather, to a splice variant of CIP4 called CIP4/2, and CIP4/2 is involved in insulin-dependent translocation of GLUT4 (Chang et al. 2002). Importantly, CIP4 does not bind GLUT4 directly; instead, this CIP4-dependet translocation of GLUT4 occurs via another protein, Gapex-5. Gapex-5 is a GEF for yet another small GTPase, Rab31. In the absence of insulin, Gapex-5 maintains Rab31 in its active,

GTP-bound state, which in turn maintains a significant pool of GLUT4 storage vesicles (Lodhi et al. 2007). Upon insulin stimulation, TC10 is activated and relocates CIP4–Gapex-5 to the plasma membrane. As a result, the amount of active Rab31 is decreased, and GLUT4 is transported to the plasma membrane through the process of exocytosis.

CIP4-like proteins also participate in DRF-dependent actin polymerization, however, the link is not as clear as the link between BAR proteins and the WASP family of NPFs. The most thorough study comes from the fission yeast *Schizosaccharomyces pombe*, where the CIP4 ortholog Cdc15 was shown to bind the formin Cdc12, and to have a role in the organization of the contractile actomyosin ring during cytokinesis (Carnahan and Gould 2003). In *Drospohila* CIP4 and Diaphanous (a fly DRF) was shown to interact and CIP4 was found to antagonize actin polymerization triggered by Diaphanous (Yan et al. 2013). FBP17 was actually originally identified in a screen to identify proteins binding to Formin, a protein involved in limb and renal development (Bedford et al. 1997). CIP4-like proteins was found to bind the DRF members Dia1, Dia2 and Disheveled associated activator of morphogenesis (DAAM1), and CIP4 participates with Src, Rho GTPases and DAAM1 in the formation of filopodia (Aspenström et al. 2006). Nervous wreck (Nwk), a protein first identified in Drosophila, has also been found to work in concert with Cdc42 to coordinate actin polymerization and endocytosis during synaptic growth in the fly larvae, however it is not clear if Nwk and Cdc42 interacts through a direct interaction (Rodal et al. 2008).

A model of a complex between Cdc42 and the Toca-1 Cdc42-binding domain was recently published showing that the interaction between Toca-1 and Cdc42 is weaker that the interaction between N-WASP and Cdc42. Based on these observations, a model was put forward suggesting that N-WASP is able to displace Cdc42 from Toca-1 and that the Toca-1:Cdc42 interaction most likely is an early step in the Cdc42-regulated pathway that regulates actin polymerization (Watson et al. 2016). Interestingly, the F-BAR protein Pacsin2 was shown to bind Rac1 through an unconventional type of interaction. In this case, the hypervariable C-terminal region of Rac1 was shown to mediate the Rac1 interaction with the SH3 domain of Pacsin2 at cytoplasmic membrane tubules and early endosomes (de Kreuk et al. 2011). There is a mutual regulatory function in this interaction, as Rac1 inhibition results in the accumulation of Pacsin2 at tubular structures. Moreover, knockdown of Pacsin2 results in increased GTP-bound Rac1, and in increased Rac1-dependent cellular responses, such as cell spreading and cell migration. This demonstrates that a possible additional function of BAR proteins is to sequester Rho GTPases, and thereby to modulate their functions.

Interestingly, there are indications that increased expression of CIP4 can be found in human cancer. For instance, CIP4 expression has been found to promote metastasis of several cancer types including lung adenocarcinoma, triple-negative breast cancer and nasopharyngeal cancer (Truesdell et al. 2015; Cerqueira et al. 2015; Meng et al. 2017).

6.2   I-BAR Proteins

The domain type we now name I-BAR domain was initially called IMD (IRSp53-MIM homology domain) and was found in the N-termini of insulin receptor tyrosine kinase substrate p53 (IRSp53) and missing-in-metastasis (MIM) (Yamagishi et al. 2004). Overexpression of the IMD domains alone isolated from IRSp53 or MIM triggered the formation of filopodia. Analysis of the three-dimensional structure of the IRSp53 IMD domain showed that this potential coiled-coil domain folded into 180-Å-long zeppelin-shaped dimers (Millard et al. 2005). Similar to BAR domains, the IMD domain was shown to induce membrane tubulation; however, in contrast to the other BAR domains, the IRSp53 I-BAR domain caused membrane protrusions rather than invaginations, hence the name inverse BAR (I-BAR) (Suetsugu et al. 2006; Mattila et al. 2007). The I-BAR proteins subfamily consists of

IRSp53, MIM, insulin receptor tyrosine kinase substrate (IRTKS), and actin-bundling protein with BAIAP2 homology (ABBA) (Zhao et al. 2011). The BAR domain of a related protein, called Pinkbar, has strong structural similarity to the I-BAR domain, but has been shown to induce flat sheets of lipids rather than membrane protrusions, and can therefore be considered to constitute a unique type of BAR domain (Pykäläinen et al. 2011).

IRSp53 have the potential to bind both Cdc42 and Rac1, however, the two Rho GTPases do not share binding sites. Cdc42 interacts through a CRIB-like motif, whereas Rac1 binds to the I-BAR domain itself (Miki et al. 2000; Govind et al. 2001; Krugmann et al. 2001). Importantly, the Cdc42 binds IRSp53 in its active, GTP-bound, conformation, whereas the Rac1: IRSp53 interaction appears to be independent of nucleotide bound state of Rac1(Miki et al. 2000; Millard et al. 2007). IRSp53 has been suggested to form a link between WAVE2 and Rac1. None of the other I-BAR proteins bind Cdc42, but at least MIM and IRSTK bind both GTP-bound and GDP-bound Rac1 via their I-BAR domains (Bompard et al. 2005; Millard et al. 2007). ABBA has also been shown to be involved in signaling via Rac1, and although it is not clear if the cellular effects are mediated by Rac1, or if ABBA binds Rac1 directly (Zheng et al. 2010).

There have been several observations that demonstrate increased Rac1 activation in response to overexpression of I-BAR proteins (i.e., MIM and ABBA); however, the mechanism involved here is not clear, since I-BAR proteins have no RhoGEF domains nor do they possess any GEF activity (Millard et al. 2007; Zheng et al. 2010; Dawson et al. 2012). It is possible that the activation occurs though interacting RhoGEFs, and one such protein, Tiam1, has been shown to promote the relocalization of IRSp53 to Rac1-induced lamellipodia, rather than to Cdc42-induced filopodia (Connolly et al. 2005). Tiam1 was also found to direct IRSp53 to Rac1 signaling by enhancing IRSp53 binding to both active Rac1 and the WAVE2 complex. That there exists an important link between Tiam1 and IRSp53 was further substatiated by the observation that IRSp53 depletion in cells prevents the lamellipodia formation induced by Tiam1 overexpression or PDGF stimulation (Connolly et al. 2005). IRSp53-dependent Rac1 activation can be mediated by Eps8 via a complex that includes WAVE2, Abi1 and the GEF Sos-1 (Funato et al. 2004; Suetsugu et al. 2006). Since Cdc42-induced filopodia are abrogated by depletion of IRSp53 or Eps8, this pathway is likely to be involved in filopodia formation (Disanza et al. 2006). Furthermore, IRSp53 can interact with mDia1 and WAVE2 in the production of filopodia (Goh et al. 2012). These findings suggest a model in which activated GTP-bound Cdc42 binds IRSp53 and promotes the localization and activation of Rac via RhoGEFs. Intriguingly, the Eps8–Abi-1–Sos-1 tri-complex has an important role in metastasis of ovarian tumor cells, presumably in association with IRSp53 (Funato et al. 2004; Chen et al. 2010). It is not completely true that F-BAR and I-BAR proteins only bind classical Rho GTPases. One I-BAR protein, IRTKS has been found to bind the fast-cycling Rho member Rif (also known as RhoF) (Sudhaharan et al. 2016). In this study Rif was shown to induce filopodia via a mechanism that, in addition to IRTKS, also involved Eps8 and the WASP family protein WAVE2. Rif was also found to bind PinkBAR, however the biological function of this interaction is not clear (Sudhaharan et al. 2016).

6.3 Bin

Other BAR domain-containing proteins affect Rho signaling, but it is not clear if the effect is direct or indirect but for completeness sake they are mentioned here. One example is the N-BAR protein Bin3 that is involved in the regulation of myofiber size during early myogenesis (Simionescu-Bankston et al. 2013). Bin3-knockout cells migrate over shorter distances than wild-type cells, and have reduced levels of activated Rac1 and Cdc42. This suggests a functional link between Bin3 and Rho GTPases, although the underlying mechanism is currently not known.

Another example is Bin1 since RhoB has been linked to Bin1 in apoptosis induced by farnesyl-transferase inhibitors (DuHadaway et al. 2003).

7 Concluding Remarks

It is clear from the studies presented in this review article that a majority of BAR proteins appears to possess overlapping functions. Most of them are involved in coordinating membrane trafficking and cytoskeletal dynamics. The obvious question that comes into mind is; where is the specificity? There is no complete answer to this question, however, there is a cell-type specific distribution of BAR proteins and they are not all expressed at the same level in every cell in the body. Several of the BAR proteins, in particular the RhoGAP domain-containing proteins, are mutated or dysfunctional neurological pathologies. Importantly, they are not involved in the same neuropathology, demonstrating that they have unique biological functions.

The cytoskeletal reorganization that controls cell migration is mainly orchestrated by dynamic interactions between the Rho GTPases, the WASP family of NPFs, and the DRFs. During recent years, the concept of membrane dynamics has reappeared in the field of cell migration. Cell migration clearly requires the very intricate coordination of cytoskeletal dynamics and the machineries that control endocytosis, exocytosis and intracellular transport of endomembranes. The cooperation between the BAR proteins and the actin regulatory NPFs and DRFs that is brought about through the Rho GTPases provides the central players in this control. It is worth noting that BAR proteins are not only implicated in actin dynamics and endosome function, there is an emerging awareness that they regulate a great variety of vital cellular processes such as cell polarity, gene transcription, stress signaling and tumor promotion.

Acknowledgments PA has been supported by grants from the Swedish Cancer Society and Karolinska Institutet.

References

Aspenström P (1997) A Cdc42 target protein with homology to the non-kinase domain of FER has a potential role in regulating the actin cytoskeleton. Curr Biol 7 (7):479–487

Aspenström P (2009) Roles of F-BAR/PCH proteins in the regulation of membrane dynamics and actin reorganization. Int Rev Cell Mol Biol 272:1–31. https://doi.org/10.1016/S1937-6448(08)01601-8

Aspenström P (2018) Fast-cycling Rho GTPases. Small GTPases 29:1–8. https://doi.org/10.1080/21541248.2017.1391365

Aspenström P, Richnau N, Johansson AS (2006) The diaphanous-related formin DAAM1 collaborates with the Rho GTPases RhoA and Cdc42, CIP4 and Src in regulating cell morphogenesis and actin dynamics. Exp Cell Res 312(12):2180–2194

Baek JI, Kwon SH, Zuo X, Choi SY, Kim SH, Lipschutz JH (2016) Dynamin binding protein (Tuba) deficiency inhibits ciliogenesis and nephrogenesis in vitro and in vivo. J Biol Chem 291(16):8632–8643. https://doi.org/10.1074/jbc.M115.688663

Bai X, Lenhart KC, Bird KE, Suen AA, Rojas M, Kakoki M, Li F, Smithies O, Mack CP, Taylor JM (2013) The smooth muscle-selective RhoGAP GRAF3 is a critical regulator of vascular tone and hypertension. Nat Commun 4:2910. https://doi.org/10.1038/ncomms3910

Ballard MS, Hinck L (2012) A roundabout way to cancer. Adv Cancer Res 114:187–235. https://doi.org/10.1016/B978-0-12-386503-8.00005-3

Bedford MT, Chan DC, Leder P (1997) FBP WW domains and the Abl SH3 domain bind to a specific class of proline-rich ligands. EMBO J 16(9):2376–2383

Billuart P, Bienvenu T, Ronce N, des Portes V, Vinet MC, Zemni R, Roest Crollius H, Carrié A, Fauchereau F, Cherry M, Briault S, Hamel B, Fryns JP, Beldjord C, Kahn A, Moraine C, Chelly J (1998) Oligophrenin-1 encodes a rhoGAP protein involved in X-linked mental retardation. Nature 392(6679):923–926

Bompard G, Sharp SJ, Freiss G, Machesky LM (2005) Involvement of Rac in actin cytoskeleton rearrangements induced by MIM-B. J Cell Sci 118 (Pt 22):5393–5403

Bryant DM, Datta A, Rodríguez-Fraticelli AE, Peränen J, Martín-Belmonte F, Mostov KE (2010) A molecular network for de novo generation of the apical surface and lumen. Nat Cell Biol 12(11):1035–1045. https://doi.org/10.1038/ncb2106

Bu W, Lim KB, Yu YH, Chou AM, Sudhaharan T, Ahmed S (2010) Cdc42 interaction with N-WASP and Toca-1 regulates membrane tubulation, vesicle formation and vesicle motility: implications for endocytosis. PLoS One 5(8):e12153. https://doi.org/10.1371/journal.pone

Carnahan RH, Gould KL (2003) The PCH family protein, Cdc15p, recruits two F-actin nucleation pathways to coordinate cytokinetic actin ring formation in Schizosaccharomyces pombe. J Cell Biol 162(5):851–862

Cerqueira OL, Truesdell P, Baldassarre T, Vilella-Arias SA, Watt K, Meens J, Chander H, Osório CA, Soares FA, Reis EM, Craig AW (2015) CIP4 promotes metastasis in triple-negative breast cancer and is associated with poor patient prognosis. Oncotarget 6 (11):9397–9408

Chang L, Adams RD, Saltiel AR (2002) The TC10-interacting protein CIP4/2 is required for insulin-stimulated Glut4 translocation in 3T3L1 adipocytes. Proc Natl Acad Sci U S A 99(20):12835–12840

Chen H, Wu X, Pan ZK, Huang S (2010) Integrity of SOS1/EPS8/ABI1 tri-complex determines ovarian cancer metastasis. Cancer Res 70(23):9979–9990. https://doi.org/10.1158/0008-5472.CAN-10-2394

Chesarone MA, DuPage AG, Goode BL (2010) Unleashing formins to remodel the actin and microtubule cytoskeleton. Nat Rev Mol Cell Biol 11(1):62–74. https://doi.org/10.1038/nrm2816

Chiang SH, Baumann CA, Kanzaki M, Thurmond DC, Watson RT, Neudauer CL, Macara IG, Pessin JE, Saltiel AR (2001) Insulin-stimulated GLUT4 translocation requires the CAP-dependent activation of TC10. Nature 410(6831):944–948

Cicchetti P, Mayer BJ, Thiel G, Baltimore D (1992) Identification of a protein that binds to the SH3 region of Abl and is similar to Bcr and GAP-rho. Science 257 (5071):803–806. https://doi.org/10.1126/science.1379745

Connolly BA, Rice J, Feig LA, Buchsbaum RJ (2005) Tiam1-IRSp53 complex formation directs specificity of rac-mediated actin cytoskeleton regulation. Mol Cell Biol 25(11):4602–4614. Erratum in: Mol Cell Biol 25(17):7928

Cook DR, Rossman KL, Der CJ (2014) Rho guanine nucleotide exchange factors: regulators of Rho GTPase activity in development and disease. Oncogene 33 (31):4021–4035. https://doi.org/10.1038/onc.2013.362

Coutinho-Budd J, Ghukasyan V, Zylka MJ, Polleux F (2012) The F-BAR domains from srGAP1, srGAP2 and srGAP3 regulate membrane deformation differently. J Cell Sci 125(Pt 14):3390–3401. https://doi.org/10.1242/jcs.098962

Csépányi-Kömi R, Sáfár D, Grósz V, Tarján ZL, Ligeti E (2013) In silico tissue-distribution of human Rho family GTPase activating proteins. Small GTPases 4 (2):90–101. https://doi.org/10.4161/sgtp.23708

Dawson JC, Bruche S, Spence HJ, Braga VM, Machesky LM (2012) Mtss1 promotes cell-cell junction assembly and stability through the small GTPase Rac1. PLoS One 7(3):e31141. https://doi.org/10.1371/journal.pone.0031141

de Kreuk BJ, Nethe M, Fernandez-Borja M, Anthony EC, Hensbergen PJ, Deelder AM, Plomann M, Hordijk PL (2011) The F-BAR domain protein PACSIN2 associates with Rac1 and regulates cell spreading and migration. J Cell Sci 124(Pt 14):2375–2388. https://doi.org/10.1242/jcs.080630

de Kreuk BJ, Schaefer A, Anthony EC, Tol S, Fernandez-Borja M, Geerts D, Pool J, Hambach L, Goulmy E,

Hordijk PL (2013) The human minor histocompatibility antigen1 is a RhoGAP. PLoS One 8(9):e73962. https://doi.org/10.1371/journal.pone.0073962 eCollection 2013

DerMardirossian C, Bokoch GM (2005) GDIs: central regulatory molecules in Rho GTPase activation. Trends Cell Biol 15(7):356–363

Disanza A, Mantoani S, Hertzog M, Gerboth S, Frittoli E, Steffen A, Berhoerster K, Kreienkamp HJ, Milanesi F, Di Fiore PP, Ciliberto A, Stradal TE, Scita G (2006) Regulation of cell shape by Cdc42 is mediated by the synergic actin-bundling activity of the Eps8-IRSp53 complex. Nat Cell Biol 8(12):1337–1347

Doherty GJ, Åhlund MK, Howes MT, Morén B, Parton RG, McMahon HT, Lundmark R (2011) The endocytic protein GRAF1 is directed to cell-matrix adhesion sites and regulates cell spreading. Mol Biol Cell 22 (22):4380–4389. https://doi.org/10.1091/mbc.E10-12-0936

DuHadaway JB, Du W, Donover S, Baker J, Liu AX, Sharp DM, Muller AJ, Prendergast GC (2003) Transformation-selective apoptotic program triggered by farnesyltransferase inhibitors requires Bin1. Oncogene 22(23):3578–3588

Elbediwy A, Zihni C, Terry SJ, Clark P, Matter K, Balda MS (2012) Epithelial junction formation requires confinement of Cdc42 activity by a novel SH3BP1 complex. J Cell Biol 198(4):677–693. https://doi.org/10.1083/jcb.201202094

Endris V, Wogatzky B, Leimer U, Bartsch D, Zatyka M, Latif F, Maher ER, Tariverdian G, Kirsch S, Karch D, Rappold GA (2002) The novel Rho-GTPase activating gene MEGAP/srGAP3 has a putative role in severe mental retardation. Proc Natl Acad Sci U S A 99 (18):11754–11759

Foletta VC, Brown FD, Young WS 3rd (2002) Cloning of rat ARHGAP4/C1, a RhoGAP family member expressed in the nervous system that colocalizes with the Golgi complex and microtubules. Brain Res Mol Brain Res 107(1):65–79

Francis MK, Holst MR, Vidal-Quadras M, Henriksson S, Santarella-Mellwig R, Sandblad L, Lundmark R (2015) Endocytic membrane turnover at the leading edge is driven by a transient interaction between Cdc42 and GRAF1. J Cell Sci 128(22):4183–4195. https://doi.org/10.1242/jcs.174417

Frost A, Perera R, Roux A, Spasov K, Destaing O, Egelman EH, De Camilli P, Unger VM (2008) Structural basis of membrane invagination by F-BAR domains. Cell 132(5):807–817. https://doi.org/10.1016/j.cell.2007.12.041

Funato Y, Terabayashi T, Suenaga N, Seiki M, Takenawa T, Miki H (2004) IRSp53/Eps8 complex is important for positive regulation of Rac and cancer cell motility/invasiveness. Cancer Res 64(15):5237–5244

Goh WI, Lim KB, Sudhaharan T, Sem KP, Bu W, Chou AM, Ahmed S (2012) mDia1 and WAVE2 proteins interact directly with IRSp53 in filopodia and are involved in filopodium formation. J Biol Chem 287

(7):4702–4714. https://doi.org/10.1074/jbc.M111.305102

Govek EE, Newey SE, Akerman CJ, Cross JR, Van der Veken L, Van Aelst L (2004) The X-linked mental retardation protein oligophrenin-1 is required for dendritic spine morphogenesis. Nat Neurosci 7 (4):364–372

Govind S, Kozma R, Monfries C, Lim L, Ahmed S (2001) Cdc42Hs facilitates cytoskeletal reorganization and neurite outgrowth by localizing the 58-kD insulin receptor substrate to filamentous actin. J Cell Biol 152(3):579–594

Guerrier S, Coutinho-Budd J, Sassa T, Gresset A, Jordan NV, Chen K, Jin WL, Frost A, Polleux F (2009) The F-BAR domain of srGAP2 induces membrane protrusions required for neuronal migration and morphogenesis. Cell 138(5):990–1004. https://doi.org/10.1016/j.cell.2009.06.047

Harada A, Furuta B, Takeuchi K, Itakura M, Takahashi M, Umeda M (2000) Nadrin, a novel neuron-specific GTPase-activating protein involved in regulated exocytosis. J Biol Chem 275(47):36885–36891

Hatzoglou A, Ader I, Splingard A, Flanders J, Saade E, Leroy I, Traver S, Aresta S, de Gunzburg J (2007) Gem associates with Ezrin and acts via the Rho-GAP protein Gmip to down-regulate the Rho pathway. Mol Biol Cell 18(4):1242–1252

Herrington KA, Trinh AL, Dang C, O'Shaughnessy E, Hahn KM, Gratton E, Digman MA, Sütterlin C (2017) Spatial analysis of Cdc42 activity reveals a role for plasma membrane-associated Cdc42 in centrosome regulation. Mol Biol Cell 28(15):2135–2145. https://doi.org/10.1091/mbc

Hildebrand JD, Taylor JM, Parsons JT (1996) An SH3 domain-containing GTPase-activating protein for Rho and Cdc42 associates with focal adhesion kinase. Mol Cell Biol 16(6):3169–3178

Ho HY, Rohatgi R, Lebensohn AM, Le M, Li J, Gygi SP, Kirschner MW (2004) Toca-1 mediates Cdc42-dependent actin nucleation by activating the N-WASP–WIP complex. Cell 118(2):203–216

Holst MR, Vidal-Quadras M, Larsson E, Song J, Hubert M, Blomberg J, Lundborg M, Landström M, Lundmark R (2017) Clathrin-independent endocytosis suppresses cancer cell blebbing and invasion. Cell Rep 20(8):1893–1905. https://doi.org/10.1016/j.celrep.2017.08.006

Hou JC, Pessin JE (2007) Ins (endocytosis) and outs (exocytosis) of GLUT4 trafficking. Curr Opin Cell Biol 19(4):466–473

Houy S, Estay-Ahumada C, Croisé P, Calco V, Haeberlé AM, Bailly Y, Billuart P, Vitale N, Bader MF, Ory S, Gasman S (2015) Oligophrenin-1 connects Exocytotic fusion to compensatory endocytosis in neuroendocrine cells. J Neurosci 35(31):11045–11055. https://doi.org/10.1523/JNEUROSCI.4048-14.2015

Hu J, Troglio F, Mukhopadhyay A, Everingham S, Kwok E, Scita G, Craig AW (2009) F-BAR-containing adaptor CIP4 localizes to early endosomes and regulates epidermal growth factor receptor trafficking and downregulation. Cell Signal 21(11):1686–1697. https://doi.org/10.1016/j.cellsig.2009.07.007

Itoh T, Erdmann KS, Roux A, Habermann B, Werner H, De Camilli P (2005) Dynamin and the actin cytoskeleton cooperatively regulate plasma membrane invagination by BAR and F-BAR proteins. Dev Cell 9 (6):791–804

Jaffe AB, Hall A (2005) Rho GTPases: biochemistry and biology. Annu Rev Cell Dev Biol 21:247–269

Jarius S, Wildemann B, Stöcker W, Moser A, Wandinger KP (2015) Psychotic syndrome associated with anti-Ca/ARHGAP26 and voltage-gated potassium channel antibodies. J Neuroimmunol 286:79–82. https://doi.org/10.1016/j.jneuroim.2015.07.009

Johnson JL, Monfregola J, Napolitano G, Kiosses WB, Catz SD (2012) Vesicular trafficking through cortical actin during exocytosis is regulated by the Rab27a effector JFC1/Slp1 and the RhoA-GTPase-activating protein gem-interacting protein. Mol Biol Cell 23 (10):1902–1916. https://doi.org/10.1091/mbc.E11-12-1001

Khelfaoui M, Pavlowsky A, Powell AD, Valnegri P, Cheong KW, Blandin Y, Passafaro M, Jefferys JG, Chelly J, Billuart P (2009) Inhibition of RhoA pathway rescues the endocytosis defects in Oligophrenin1 mouse model of mental retardation. Hum Mol Genet 18(14):2575–2583. https://doi.org/10.1093/hmg/ddp189

Kodani A, Kristensen I, Huang L, Sütterlin C (2009) GM130-dependent control of Cdc42 activity at the Golgi regulates centrosome organization. Mol Biol Cell 20(4):1192–1200. https://doi.org/10.1091/mbc.E08-08-0834

Koeppel MA, McCarthy CC, Moertl E, Jakobi R (2004) Identification and characterization of PS-GAP as a novel regulator of caspase-activated PAK-2. J Biol Chem 279(51):53653–53664

Kovacs EM, Makar RS, Gertler FB (2006) Tuba stimulates intracellular N-WASP-dependent actin assembly. J Cell Sci 119(Pt 13):2715–2726

Kovacs EM, Verma S, Thomas SG, Yap AS (2011) Tuba and N-WASP function cooperatively to position the central lumen during epithelial cyst morphogenesis. Cell Adhes Migr 5(4):344–350

Krugmann S, Jordens I, Gevaert K, Driessens M, Vandekerckhove J, Hall A (2001) Cdc42 induces filopodia by promoting the formation of an IRSp53: Mena complex. Curr Biol 11(21):1645–1655

Lee SY, Kim H, Kim K, Lee H, Lee S, Lee D (2016) Arhgap17, a RhoGTPase activating protein, regulates mucosal and epithelial barrier function in the mouse colon. Sci Rep 6:26923. https://doi.org/10.1038/srep26923

Lenhart KC, Becherer AL, Li J, Xiao X, McNally EM, Mack CP, Taylor JM (2014) GRAF1 promotes ferlin-dependent myoblast fusion. Dev Biol 393(2):298–311. https://doi.org/10.1016/j.ydbio.2014.06.025

Liu F, Guo H, Ou M, Hou X, Sun G, Gong W, Jing H, Tan Q, Xue W, Dai Y, Sui W (2016) ARHGAP4 mutated in a Chinese intellectually challenged family. Gene 578(2):205–209. https://doi.org/10.1016/j.gene.2015.12.035

Lodhi IJ, Chiang SH, Chang L, Vollenweider D, Watson RT, Inoue M, Pessin JE, Saltiel AR (2007) Gapex-5, a Rab31 guanine nucleotide exchange factor that regulates Glut4 trafficking in adipocytes. Cell Metab 5(1):59–72

Lucken-Ardjomande Häsler S, Vallis Y, Jolin HE, McKenzie AN, McMahon HT (2014) GRAF1a is a brain-specific protein that promotes lipid droplet clustering and growth, and is enriched at lipid droplet junctions. J Cell Sci 127(Pt 21):4602–4619. https://doi.org/10.1242/jcs.147694

Lundmark R, Doherty GJ, Howes MT, Cortese K, Vallis Y, Parton RG, McMahon HT (2008) The GTPase-activating protein GRAF1 regulates the CLIC/GEEC endocytic pathway. Curr Biol 18 (22):1802–1808. https://doi.org/10.1016/j.cub.2008.10.044

Luo N, Guo J, Chen L, Yang W, Qu X, Cheng Z (2016) ARHGAP10, downregulated in ovarian cancer, suppresses tumorigenicity of ovarian cancer cells. Cell Death Dis 7:e2157. https://doi.org/10.1038/cddis.2015.401

Luo W, Janoštiak R, Tolde O, Ryzhova LM, Koudelková L, Dibus M, Brábek J, Hanks SK, Rosel D (2017) ARHGAP42 is activated by Src-mediated tyrosine phosphorylation to promote cell motility. J Cell Sci 130(14):2382–2393. https://doi.org/10.1242/jcs.197434

Ma Y, Mi YJ, Dai YK, Fu HL, Cui DX, Jin WL (2013) The inverse F-BAR domain protein srGAP2 acts through srGAP3 to modulate neuronal differentiation and neurite outgrowth of mouse neuroblastoma cells. PLoS One 8(3):e57865. https://doi.org/10.1371/journal.pone.0057865

Madaule P, Axel R (1985) A novel Ras-related gene family. Cell 41(1):31–40. https://doi.org/10.1016/0092-8674(85)90058-3

Madura T, Yamashita T, Kubo T, Tsuji L, Hosokawa K, Tohyama M (2004) Changes in mRNA of Slit-Robo GTPase-activating protein 2 following facial nerve transection. Brain Res Mol Brain Res 123(1–2):76–80

Mattila PK, Pykäläinen A, Saarikangas J, Paavilainen VO, Vihinen H, Jokitalo E, Lappalainen P (2007) Missing-in-metastasis and IRSp53 deform PI(4,5)P2-rich membranes by an inverse BAR domain-like mechanism. J Cell Biol 176(7):953–964

Meng DF, Xie P, Peng LX, Sun R, Luo DH, Chen QY, Lv X, Wang L, Chen MY, Mai HQ, Guo L, Guo X, Zheng LS, Cao L, Yang JP, Wang MY, Mei Y, Qiang YY, Zhang ZM, Yun JP, Huang BJ, Qian CN (2017) CDC42-interacting protein 4 promotes metastasis of nasopharyngeal carcinoma by mediating invadopodia formation and activating EGFR signaling. Exp Clin Cancer Res 36(1):21. https://doi.org/10.1186/s13046-

016-0483-z. Erratum in: J Exp Clin Cancer Res 36 (1):33

Miki H, Yamaguchi H, Suetsugu S, Takenawa T (2000) IRSp53 is an essential intermediate between Rac and WAVE in the regulation of membrane ruffling. Nature 408(6813):732–735

Millard TH, Bompard G, Heung MY, Dafforn TR, Scott DJ, Machesky LM, Fütterer K (2005) Structural basis of filopodia formation induced by the IRSp53/MIM homology domain of human IRSp53. EMBO J 24 (2):240–250

Millard TH, Dawson J, Machesky LM (2007) Characterisation of IRTKS, a novel IRSp53/MIM family actin regulator with distinct filament bundling properties. J Cell Sci 120(Pt 9):1663–1672

Mim C, Cui H, Gawronski-Salerno JA, Frost A, Lyman E, Voth GA, Unger VM (2012) Structural basis of membrane bending by the N-BAR protein endophilin. Cell 149 (1):137–145. https://doi.org/10.1016/j.cell.2012.01.048

Nakano-Kobayashi A, Kasri NN, Newey SE, Van Aelst L (2009) The Rho-linked mental retardation protein OPHN1 controls synaptic vesicle endocytosis via endophilin A1. Curr Biol 19(13):1133–1139. https://doi.org/10.1016/j.cub.2009.05.022

Otani T, Ichii T, Aono S, Takeichi M (2006) Cdc42 GEF Tuba regulates the junctional configuration of simple epithelial cells. J Cell Biol 175(1):135–146

Parrini MC, Sadou-Dubourgnoux A, Aoki K, Kunida K, Biondini M, Hatzoglou A, Poullet P, Formstecher E, Yeaman C, Matsuda M, Rossé C, Camonis J (2011) SH3BP1, an exocyst-associated RhoGAP, inactivates Rac1 at the front to drive cell motility. Mol Cell 42 (5):650–661. https://doi.org/10.1016/j.molcel.2011.03.032

Peter BJ, Kent HM, Mills IG, Vallis Y, Butler PJ, Evans PR, McMahon HT (2004) BAR domains as sensors of membrane curvature: the amphiphysin BAR structure. Science 303(5657):495–499

Post A, Pannekoek WJ, Ross SH, Verlaan I, Brouwer PM, Bos JL (2013) Rasip1 mediates Rap1 regulation of Rho in endothelial barrier function through ArhGAP29. Proc Natl Acad Sci U S A 110(28):11427–121432. https://doi.org/10.1073/pnas.1306595110

Post A, Pannekoek WJ, Ponsioen B, Vliem MJ, Bos JL (2015) Rap1 spatially controls ArhGAP29 to inhibit Rho signaling during endothelial barrier regulation. Mol Cell Biol 35(14):2495–2502. https://doi.org/10.1128/MCB.01453-14

Pykäläinen A, Boczkowska M, Zhao H, Saarikangas J, Rebowski G, Jansen M, Hakanen J, Koskela EV, Peränen J, Vihinen H, Jokitalo E, Salminen M, Ikonen E, Dominguez R, Lappalainen P (2011) Pinkbar is an epithelial-specific BAR domain protein that generates planar membrane structures. Nat Struct Mol Biol 18(8):902–907. https://doi.org/10.1038/nsmb.2079

Qiao Y, Chen J, Lim YB, Finch-Edmondson ML, Seshachalam VP, Qin L, Jiang T, Low BC, Singh H, Lim CT, Sudol M (2017) YAP regulates actin

dynamics through ARHGAP29 and promotes metastasis. Cell Rep 19(8):1495–1502. https://doi.org/10.1016/j.celrep.2017.04.075

Qin Y, Meisen WH, Hao Y, Macara IG (2010) Tuba, a Cdc42 GEF, is required for polarized spindle orientation during epithelial cyst formation. J Cell Biol 189 (4):661–669. https://doi.org/10.1083/jcb.201002097

Qualmann B, Koch D, Kessels MM (2011) Let's go bananas: revisiting the endocytic BAR code. EMBO J 30(17):3501–3515. https://doi.org/10.1038/emboj.2011.266

Raynaud F, Janossy A, Dahl J, Bertaso F, Perroy J, Varrault A, Vidal M, Worley PF, Boeckers TM, Bockaert J, Marin P, Fagni L, Homburger V (2013) Shank3-Rich2 interaction regulates AMPA receptor recycling and synaptic long-term potentiation. J Neurosci 33(23):9699–9715. https://doi.org/10.1523/JNEUROSCI.2725-12.2013

Raynaud F, Moutin E, Schmidt S, Dahl J, Bertaso F, Boeckers TM, Homburger V, Fagni L (2014) Rho-GTPase-activating protein interacting with Cdc-42-interacting protein 4 homolog 2 (Rich2): a new Ras-related C3 botulinum toxin substrate 1 (Rac1) GTPase-activating protein that controls dendritic spine morphogenesis. J Biol Chem 289 (5):2600–2609. https://doi.org/10.1074/jbc.M113.534636

Ren R, Mayer BJ, Cicchetti P, Baltimore D (1993) Identification of a ten-amino acid proline-rich SH3 binding site. Science 259(5098):1157–1161. https://doi.org/10.1126/science.8438166

Ren XR, Du QS, Huang YZ, Ao SZ, Mei L, Xiong WC (2001) Regulation of CDC42 GTPase by proline-rich tyrosine kinase 2 interacting with PSGAP, a novel pleckstrin homology and Src homology 3 domain containing rhoGAP protein. J Cell Biol 152 (5):971–984

Richnau N, Aspenström P (2001) Rich, a rho GTPase-activating protein domain-containing protein involved in signaling by Cdc42 and Rac1. J Biol Chem 276 (37):35060–35070

Rodal AA, Motola-Barnes RN, Littleton JT (2008) Nervous wreck and Cdc42 cooperate to regulate endocytic actin assembly during synaptic growth. J Neurosci 28 (33):8316–8325. https://doi.org/10.1523/JNEUROSCI.2304-08.2008

Rollason R, Korolchuk V, Hamilton C, Jepson M, Banting G (2009) A CD317/tetherin-RICH2 complex plays a critical role in the organization of the subapical actin cytoskeleton in polarized epithelial cells. J Cell Biol 184(5):721–736. https://doi.org/10.1083/jcb.200804154

Rotty JD, Wu C, Bear JE (2013) New insights into the regulation and cellular functions of the ARP2/3 complex. Nat Rev Mol Cell Biol 14(1):7–12. https://doi.org/10.1038/nrm3492

Saengsawang W, Mitok K, Viesselmann C, Pietila L, Lumbard DC, Corey SJ, Dent EW (2012) The F-BAR protein CIP4 inhibits neurite formation by producing lamellipodial protrusions. Curr Biol 22 (6):494–501. https://doi.org/10.1016/j.cub.2012.01. 038

Sakamuro D, Elliott KJ, Wechsler-Reya R, Prendergast GC (1996) BIN1 is a novel MYC-interacting protein with features of a tumour suppressor. Nat Genet 14:69–77. https://doi.org/10.1038/ng0996-69

Salazar MA, Kwiatkowski AV, Pellegrini L, Cestra G, Butler MH, Rossman KL, Serna DM, Sondek J, Gertler FB, De Camilli P (2003) Tuba, a novel protein containing bin/amphiphysin/Rvs and Dbl homology domains, links dynamin to regulation of the actin cytoskeleton. J Biol Chem 278(49):49031–49043

Salzer U, Kostan J, Djinović-Carugo K (2017) Deciphering the BAR code of membrane modulators. Cell Mol Life Sci 74(13):2413–2438. https://doi.org/ 10.1007/s00018-017-2478-0

Saras J, Franzén P, Aspenström P, Hellman U, Gonez LJ, Heldin CH (1997) A novel GTPase-activating protein for Rho interacts with a PDZ domain of the protein-tyrosine phosphatase PTPL1. J Biol Chem 272 (39):24333–24338

Sarowar T, Grabrucker S, Föhr K, Mangus K, Eckert M, Bockmann J, Boeckers TM, Grabrucker AM (2016) Enlarged dendritic spines and pronounced neophobia in mice lacking the PSD protein RICH2. Mol Brain 9:28. https://doi.org/10.1186/s13041-016-0206-6

Savastano CP, Brito LA, Faria ÁC, Setó-Salvia N, Peskett E, Musso CM, Alvizi L, Ezquina SA, James C, GOSgene, Beales P, Lees M, Moore GE, Stanier P, Passos-Bueno MR (2017) Impact of rare variants in ARHGAP29 to the etiology of oral clefts: role of loss-of-function vs missense variants. Clin Genet 91(5):683–689. https://doi.org/10.1111/cge. 12823

Shimada A, Niwa H, Tsujita K, Suetsugu S, Nitta K, Hanawa-Suetsugu K, Akasaka R, Nishino Y, Toyama M, Chen L, Liu ZJ, Wang BC, Yamamoto M, Terada T, Miyazawa A, Tanaka A, Sugano S, Shirouzu M, Nagayama K, Takenawa T, Yokoyama S (2007) Curved EFC/F-BAR-domain dimers are joined end to end into a filament for membrane invagination in endocytosis. Cell 129 (4):761–772

Simionescu-Bankston A, Leoni G, Wang Y, Pham PP, Ramalingam A, DuHadaway JB, Faundez V, Nusrat A, Prendergast GC, Pavlath GK (2013) The N-BAR domain protein, Bin3, regulates Rac1- and Cdc42-dependent processes in myogenesis. Dev Biol 382(1):160–171. https://doi.org/10.1016/j.ydbio.2013. 07.004

Soderling SH, Binns KL, Wayman GA, Davee SM, Ong SH, Pawson T, Scott JD (2002) The WRP component of the WAVE-1 complex attenuates Rac-mediated signalling. Nat Cell Biol 4(12):970–975

Soderling SH, Guire ES, Kaech S, White J, Zhang F, Schutz K, Langeberg LK, Banker G, Raber J, Scott JD (2007) A WAVE-1 and WRP signaling complex

regulates spine density, synaptic plasticity, and memory. J Neurosci 27(2):355–365

Sousa S, Cabanes D, Archambaud C, Colland F, Lemichez E, Popoff M, Boisson-Dupuis S, Gouin E, Lecuit M, Legrain P, Cossart P (2005) ARHGAP10 is necessary for alpha-catenin recruitment at adherens junctions and for Listeria invasion. Nat Cell Biol 7 (10):954–960

Sudhaharan T, Sem KP, Liew HF, Yu YH, Goh WI, Chou AM, Ahmed S (2016) The Rho GTPase Rif signals through IRTKS, Eps8 and WAVE2 to generate dorsal membrane ruffles and filopodia. J Cell Sci 129 (14):2829–2840. https://doi.org/10.1242/jcs.179655

Suetsugu S, Murayama K, Sakamoto A, Hanawa-Suetsugu K, Seto A, Oikawa T, Mishima C, Shirouzu M, Takenawa T, Yokoyama S (2006) The RAC binding domain/IRSp53-MIM homology domain of IRSp53 induces RAC-dependent membrane deformation. J Biol Chem 281(46):35347–35358

Sweitzer SM, Hinshaw JE (1998) Dynamin undergoes a GTP-dependent conformational change causing vesiculation. Cell 93(6):1021–1029

Takano K, Toyooka K, Suetsugu S (2008) EFC/F-BAR proteins and the N-WASP-WIP complex induce membrane curvature-dependent actin polymerization. EMBO J 27(21):2817–2828. https://doi.org/10.1038/ emboj.2008.216

Takei K, Slepnev VI, Haucke V, De Camilli P (1999) Functional partnership between amphiphysin and dynamin in clathrin-mediated endocytosis. Nat Cell Biol 1(1):33–39

Taylor MJ, Perrais D, Merrifield CJ (2011) A high precision survey of the molecular dynamics of mammalian clathrin-mediated endocytosis. PLoS Biol 9(3): e1000604. https://doi.org/10.1371/journal.pbio. 1000604

Tcherkezian J, Lamarche-Vane N (2007) Current knowledge of the large RhoGAP family of proteins. Biol Cell 99(2):67–86

Toguchi M, Richnau N, Ruusala A, Aspenström P (2010) Members of the CIP4 family of proteins participate in the regulation of platelet-derived growth factor receptor-beta-dependent actin reorganization and migration. Biol Cell 102(4):215–230. https://doi.org/ 10.1042/BC20090033

Tonucci FM, Hidalgo F, Ferretti A, Almada E, Favre C, Goldenring JR, Kaverina I, Kierbel A, Larocca MC (2015) Centrosomal AKAP350 and CIP4 act in concert to define the polarized localization of the centrosome and Golgi in migratory cells. J Cell Sci 128 (17):3277–3289. https://doi.org/10.1242/jcs.170878

Truesdell P, Ahn J, Chander H, Meens J, Watt K, Yang X, Craig AW (2015) CIP4 promotes lung adenocarcinoma metastasis and is associated with poor prognosis. Oncogene 34(27):3527–3535. https://doi.org/10.1038/ onc.2014.280

Tsujita K, Suetsugu S, Sasaki N, Furutani M, Oikawa T, Takenawa T (2006) Coordination between the actin cytoskeleton and membrane deformation by a novel membrane tubulation domain of PCH proteins is involved in endocytosis. J Cell Biol 172(2):269–279

Tsujita K, Takenawa T, Itoh T (2015) Feedback regulation between plasma membrane tension and membrane-bending proteins organizes cell polarity during leading edge formation. Nat Cell Biol 17(6):749–758. https://doi.org/10.1038/ncb3162

Vidal-Quadras M, Holst MR, Francis MK, Larsson E, Hachimi M, Yau WL, Peränen J, Martín-Belmonte F, Lundmark R (2017) Endocytic turnover of Rab8 controls cell polarization. J Cell Sci 130 (6):1147–1157. https://doi.org/10.1242/jcs.195420

Vogt DL, Gray CD, Young WS 3rd, Orellana SA, Malouf AT (2007) ARHGAP4 is a novel RhoGAP that mediates inhibition of cell motility and axon outgrowth. Mol Cell Neurosci 36(3):332–342

Wakita Y, Kakimoto T, Katoh H, Negishi M (2011) The F-BAR protein Rapostlin regulates dendritic spine formation in hippocampal neurons. J Biol Chem 286 (37):32672–32683. https://doi.org/10.1074/jbc.M111.236265

Wallwitz U, Brock S, Schunck A, Wildemann B, Jarius S, Hoffmann F (2017) From dizziness to severe ataxia and dysarthria: new cases of anti-Ca/ARHGAP26 autoantibody-associated cerebellar ataxia suggest a broad clinical spectrum. J Neuroimmunol 309:77–81. https://doi.org/10.1016/j.jneuroim.2017.05.011

Waltereit R, Leimer U, von Bohlen Und Halbach O, Panke J, Hölter SM, Garrett L, Wittig K, Schneider M, Schmitt C, Calzada-Wack J, Neff F, Becker L, Prehn C, Kutscherjawy S, Endris V, Bacon C, Fuchs H, Gailus-Durner V, Berger S, Schönig K, Adamski J, Klopstock T, Esposito I, Wurst W, de Angelis MH, Rappold G, Wieland T, Bartsch D (2012) Srgap3-/- mice present a neurodevelopmental disorder with schizophrenia-related intermediate phenotypes. FASEB J 26(11):4418–4428. https://doi.org/10.1096/fj.11-202317

Watson JR, Fox HM, Nietlispach D, Gallop JL, Owen D, Mott HR (2016) Investigation of the interaction between Cdc42 and its effector TOCA1: handover of cdc42 to the actin regulator N-WASP is facilitated by differential binding affinities. J Biol Chem 291 (26):13875–13890. https://doi.org/10.1074/jbc.M116.724294

Wells CD, Fawcett JP, Traweger A, Yamanaka Y, Goudreault M, Elder K, Kulkarni S, Gish G, Virag C, Lim C, Colwill K, Starostine A, Metalnikov P, Pawson T (2006) A Rich1/Amot complex regulates the Cdc42 GTPase and apical-polarity proteins in epithelial cells. Cell 125(3):535–548

Wong K, Ren XR, Huang YZ, Xie Y, Liu G, Saito H, Tang H, Wen L, Brady-Kalnay SM, Mei L, Wu JY, Xiong WC, Rao Y (2001) Signal transduction in neuronal migration: roles of GTPase activating proteins and the small GTPase Cdc42 in the Slit-Robo pathway. Cell 107(2):209–221

Yamagishi A, Masuda M, Ohki T, Onishi H, Mochizuki N (2004) A novel actin bundling/filopodium-forming domain conserved in insulin receptor tyrosine kinase substrate p53 and missing in metastasis protein. J Biol Chem 279(15):14929–14936

Yan S, Lv Z, Winterhoff M, Wenzl C, Zobel T, Faix J, Bogdan S, Grosshans J (2013) The F-BAR protein Cip4/Toca-1 antagonizes the formin diaphanous in membrane stabilization and compartmentalization. J Cell Sci 126(Pt 8):1796–1805. https://doi.org/10.1242/jcs.118422

Yao F, Kausalya JP, Sia YY, Teo AS, Lee WH, Ong AG, Zhang Z, Tan JH, Li G, Bertrand D, Liu X, Poh HM, Guan P, Zhu F, Pathiraja TN, Ariyaratne PN, Rao J, Woo XY, Cai S, Mulawadi FH, Poh WT, Veeravalli L, Chan CS, Lim SS, Leong ST, Neo SC, Choi PS, Chew EG, Nagarajan N, Jacques PÉ, So JB, Ruan X, Yeoh KG, Tan P, Sung WK, Hunziker W, Ruan Y, Hillmer AM (2015) Recurrent fusion genes in gastric cancer: CLDN18-ARHGAP26 induces loss of epithelial integrity. Cell Rep 12(2):272–285. https://doi.org/10.1016/j.celrep.2015.06.020

Yi C, Troutman S, Fera D, Stemmer-Rachamimov A, Avila JL, Christian N, Persson NL, Shimono A, Speicher DW, Marmorstein R, Holmgren L, Kissil JL (2011) A tight junction-associated Merlin-angiomotin complex mediates Merlin's regulation of mitogenic signaling and tumor suppressive functions. Cancer Cell 19(4):527–540. https://doi.org/10.1016/j.ccr.2011.02.017

Zhao H, Pykäläinen A, Lappalainen P (2011) I-BAR domain proteins: linking actin and plasma membrane dynamics. Curr Opin Cell Biol 23(1):14–21. https://doi.org/10.1016/j.ceb.2010.10.005

Zhao H, Michelot A, Koskela EV, Tkach V, Stamou D, Drubin DG, Lappalainen P (2013) Membrane-sculpting BAR domains generate stable lipid microdomains. Cell Rep 4(6):1213–1223. https://doi.org/10.1016/j.celrep.2013.08.024

Zheng D, Niu S, Yu D, Zhan XH, Zeng X, Cui B, Chen Y, Yoon J, Martin SS, Lu X, Zhan X (2010) Abba promotes PDGF-mediated membrane ruffling through activation of the small GTPase Rac1. Biochem Biophys Res Commun 401(4):527–532. https://doi.org/10.1016/j.bbrc.2010.09.087

Adv Exp Med Biol - Protein Reviews (2019) 20: 55–76
https://doi.org/10.1007/5584_2018_218
© Springer Nature Singapore Pte Ltd. 2018
Published online: 18 May 2018

AP180 N-Terminal Homology (ANTH) and Epsin N-Terminal Homology (ENTH) Domains: Physiological Functions and Involvement in Disease

Sho Takatori and Taisuke Tomita

Abstract

The AP180 N-terminal homology (ANTH) and Epsin N-terminal homology (ENTH) domains are crucially involved in membrane budding processes. All the ANTH/ENTH-containing proteins share the phosphoinositide-binding activity and can interact with clathrin or its related proteins via multiple binding motifs. Their function also include promotion of clathrin assembly, induction of membrane curvature, and recruitment of various effector proteins, such as those involved in membrane fission. Furthermore, they play a role in the sorting of specific cargo proteins, thereby enabling the cargos to be accurately transported and function at their appropriate locations. As the structural bases underlying these functions are clarified, contrary to their apparent similarity, the mechanisms by which these proteins recognize lipids and proteins have unexpectedly been found to differ from each other. In addition, studies using knockout mice have suggested that their physiological roles may be more complicated than merely supporting membrane budding processes. In this chapter, we review the current knowledge on the biochemical features of ANTH/ENTH domains, their functions predicted from the phenotypes of animals deficient in these domain-containing proteins, and recent findings on the structural basis enabling specific recognition of their ligands. We also discuss the association of these domains with human diseases. Here we focus on CALM, a protein containing an ANTH domain, which is implicated in the pathogenesis of blood cancers and Alzheimer disease, and discuss how alteration of CALM function is involved in these diseases.

Keywords

Endocytosis · Clathrin · Phosphoinositide · ANTH · ENTH · Alzheimer disease

1 Introduction

The AP180 N-terminal homology (ANTH) and Epsin N-terminal homology (ENTH) domains represent an evolutionarily conserved protein module of which most members have been identified to play a role in membrane budding processes. Crystallographic studies have demonstrated that the two domains share poor sequence similarity, yet have a similar structure, and therefore it is appropriate to classify them as the "ENTH superfamily". In addition, from a

S. Takatori and T. Tomita (✉)
Laboratory of Neuropathology and Neuroscience,
Graduate School of Pharmaceutical Sciences,
The University of Tokyo, Tokyo, Japan
e-mail: taisuke@mol.f.u-tokyo.ac.jp

biochemical viewpoint, both domains share the ability to bind phosphoinositides; they are recruited to specific sites within the membrane through their lipid-binding activity, and interact with clathrin as well as clathrin-associated proteins by means of various binding motifs mainly residing in their carboxy termini. Therefore, ANTH and ENTH domains also have conserved functions in modulating the clathrin-dependent membrane budding process.

In this chapter, we will first summarize our current knowledge regarding the mechanism of clathrin-mediated endocytosis. We will then review the functions of individual members of the ANTH and ENTH proteins, particularly focusing on the knowledge regarding their mammalian homologues. Next, we will discuss the structural bases underlying their recognition of specific lipid and protein binding partners. In the last section, we will explain the importance of clathrin assembly lymphoid myeloid leukemia protein (CALM) in human diseases, especially focusing on Alzheimer disease.

2   Clathrin-Mediated Endocytosis

Eukaryotic cells are characterized by the presence of various membrane-bound organelles. They are maintained and replicated by budding, fusion, and expansion of pre-existing organelles. In particular, it is well known that invagination of the plasma membrane (PM) produces endosomes with diverse properties, and various research on endocytosis, which is the earliest event in this process, has been performed. Multiple, mechanistically distinct forms of endocytosis have been identified so far, but in simplest terms, endocytosis can be classified into two categories based on its clathrin dependency. Of these, we will focus only on clathrin-mediated endocytosis (CME) in this chapter. For a comprehensive review of endocytosis, the reader is referred to specific articles (Doherty and McMahon 2009).

Regarding the molecular mechanisms of membrane budding, CME is the best studied example among all forms of endocytosis. Formation of coated vesicles involves several morphologically

and molecularly distinct steps (Kaksonen and Roux 2018) (Fig. 1): In the earliest step, clathrin is recruited to the membrane, although the recruitment is a stochastic event, which often fails to produce a stable assembly (Loerke et al. 2009); once its polymerization is initiated, the membrane begins to invaginate and forms a clathrin-coated pit (CCP). In parallel, transmembrane cargoes are sorted into the pit; as the invagination grows deeper, a segment connecting the bud and planar membrane becomes narrow to form a tubular stalk, which is finally cleaved to yield a clathrin-coated vesicle (CCV). Previous studies have identified the core machinery working in these processes, and their functions have been extensively analyzed. We will briefly outline our knowledge about them below.

Clathrin, which is also called the clathrin triskelion, provides a structural scaffold for membrane deformation, and is a protein complex composed of three heavy chains and three light chains (Robinson 2015). Both types of chains lack the ability to bind to membranes, and therefore, they require the help of other proteins that can simultaneously bind to clathrin and the membrane. Whereas various factors that regulate this process have been identified, adaptor protein complex 2 (AP2) plays an indispensable role in CME. AP2 directly recognizes phosphatidylinositol (4,5)-bisphosphate (PtdIns(4,5)P_2), and can be recruited to the PM. This interaction is, however, so weak that AP2 easily dissociates from the PM, as demonstrated by total internal reflection fluorescence microscopy with single-molecule sensitivity (Cocucci et al. 2012). Interestingly, AP2 can be stably tethered to the membrane when at least two AP2 proteins and one clathrin triskelion coordinately arrive at the same place, where the assembly of CCPs is initiated. This suggests that clathrin can crosslink multiple AP2 molecules and strengthen their interaction with the membrane through an avidity effect. It is, however, controversial as to whether PtdIns(4,5)P_2 is solely sufficient for the recruitment of clathrin and AP2. Indeed, muniscin family proteins (FES-CIP4 homology domain only (FCHo)-1 and -2) and epidermal growth factor receptor substrate (EPS15) homology domain-

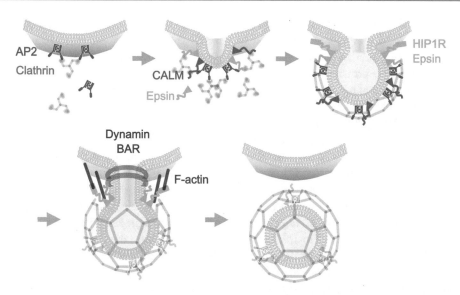

Fig. 1 Clathrin-mediated endocytosis and its sub-processes. The AP2 complex (red) anchors the clathrin triskelion (brown) to the membrane through the interaction with PtdIns(4,5)P$_2$, as well as with the "nucleator" complex containing FCHo-1/2, EPS15 and intersectin (not shown). CALM (magenta) and Epsin (blue), which are incorporated through interacting with PtdIns(4,5)P$_2$, clathrin and AP2, facilitate the clathrin assembly and membrane bending. They also function in the sorting of specific cargoes into the pit. As a pit grows deeply, the region connecting the bud and planner membrane becomes narrower to form a tubular stalk, which is generated and/or stabilized by several BAR domain-containing proteins. The "neck" is finally cleaved by dynamin, yielding a coated vesicle. HIP1R, which is recruited by the interaction with Epsin and PtdIns(3,4)P$_2$, is involved in a relatively late stage of CME possibly via F-actin-dependent mechanisms

containing proteins (EPS15 and intersectin) have been identified as components of a "nucleator" complex, which is targeted to the membrane independently of clathrin and AP2. Because the complex has multiple binding sites for AP2, it functions in the recruitment of clathrin at this site (Henne et al. 2010). Similarly to the above cases, the orchestrated actions of multiple proteins are important for the relocation of clathrin. As discussed later, many ANTH/ENTH proteins are also involved in this process through their direct binding to PtdIns(4,5)P$_2$, clathrin, and AP2.

Besides recruiting clathrin, AP2 is also important for incorporating transmembrane proteins for their internalization (Traub 2009). The μ2 subunit of AP2 directly interacts with the YxxΦ motif (where x is any amino acid and Φ is a bulky hydrophobic amino acid), which is found in the cytoplasmic tail of some cargoes, such as the transferrin receptor. The σ2 subunit is also involved in the recognition of another type of sorting signal [D/E]xxxL[L/I], also known as the acidic di-leucine motif. Of note is that the interaction with cargoes is assumed to take place only after AP2 arrives at the membrane, because its cargo binding sites are blocked by the β2 subunit when it is in the cytosol (Jackson et al. 2010). This means that a conformational change, possibly triggered by association with the nucleator complex (Ma et al. 2016), needs to occur for AP2 to interact with cargoes. This is considered as favorable to prevent the formation of ectopic pits, even if there are plenty of putative cargoes in organelles other than the PM. Besides AP2, some ANTH and ENTH proteins have cargo-binding sites with distinct specificities. All of these are believed to contribute for cargo proteins to be incorporated into the CCPs at their early stage of formation.

A number of CME-associated proteins are involved in the formation and/or recognition of membrane curvature (Itoh and De Camilli 2006). Among ANTH/ENTH proteins, Epsin is known to have a membrane-bending activity (Ford et al. 2002). It has an amphipathic helix and inserts it in the cytosolic leaflet of the lipid bilayer like a wedge, thereby inducing a membrane curvature in the membrane. Additionally, a large and flexible stretch of polypeptide at its C terminus contributes in some way to this membrane-bending activity (Stachowiak et al. 2012). Furthermore, Bin/Amphiphysin/Rvs (BAR) domain-containing proteins, such as FCHo-1/2, participate in the formation/recognition of the shallow pit at an early stage of CME (Henne et al. 2010). These are thought to act advantageously in terms of energy in forming a CCP with a high curvature from a planar cell surface membrane.

As a CCP grows, the region connecting the bud and planar membrane is constricted into a tubule, and the bud is finally pinched to yield a vesicle. A class of BAR domain-containing proteins have the activity to stabilize and/or constrict the tubular membrane (Itoh and De Camilli 2006). The guanosine triphosphate phosphatase dynamin forms a helical tube around the neck and promotes scission by tightening the "collar" through a structural change accompanying guanosine triphosphate hydrolysis (Sweitzer and Hinshaw 1998). Several BAR domain proteins such as Amphiphysin function to help recruiting dynamin to the tubular region of CCPs (Takei et al. 1999; Meinecke et al. 2013). Interestingly, it is proposed that the ENTH protein Epsin is also involved in both the formation and scission of membrane tubules, possibly through a mechanism that uses the wedge-like membrane insertion described above (Boucrot et al. 2012).

In parallel with these processes, it is known that fibrillar actin (F-actin) plays accessory roles, particularly in the regulation of late stages of CME. This was found because inhibitors of actin polymerization are inhibitory to some but not all types of CME. It was originally thought that, because F-actin forms a dense meshwork beneath the PM, it is necessary to rearrange this mesh in order for the newly formed CCV to move

inside the cytoplasm. It is also possible that the force generated by the movement of motor proteins along F-actin may facilitate constriction of the neck of CCPs or scission of CCVs (Buss et al. 2001). Furthermore, Boulant and colleagues noticed that the formation of CCP with high membrane curvature is energetically disadvantageous in tensioned cell membranes, and proposed an alternative model that F-actin counteracts the membrane tension through anchoring the membrane and pulling it toward the pit (Boulant et al. 2011). In this scenario, it is thought that the ANTH domain protein HIP1R plays an important role in anchoring the CCP to F-actin (Engqvist-Goldstein et al. 1999).

3 Identification and Functional Roles of ANTH/ENTH Proteins in Clathrin-Mediated Endocytosis

3.1 AP180 and Clathrin Assembly Lymphoid Myeloid Leukemia Protein (CALM)

AP180 was initially isolated as a component of CCVs from neuronal tissue. Its homologs have been identified in *Saccharomyces cerevisiae* (*Yap1801* and *Yap1802*), *Caenorhabditis elegans* (*Unc11*), and *Drosophila melanogaster* (*Lap1*). In mammals, AP180 has a close paralog named phosphatidylinositol-binding clathrin assembly protein (PICALM) or CALM. Whereas CALM is ubiquitously expressed, the expression of AP180 is restricted to neurons, particularly to synaptic terminals (Dreyling et al. 1996).

Both AP180 and CALM have an approximately 300 amino-acid conserved domain at their N terminus (Fig. 2). The crystal structure of this domain of CALM was first reported in 2001, in which the domain was demonstrated to contain a binding site for PtdIns(4,5)P_2 (Ford et al. 2001), in line with earlier findings on the phosphoinositide-binding activity of AP180 (Norris et al. 1995; Ye et al. 1995; Hao et al. 1997). Interestingly, the structure of the N-terminal half of the domain was found to

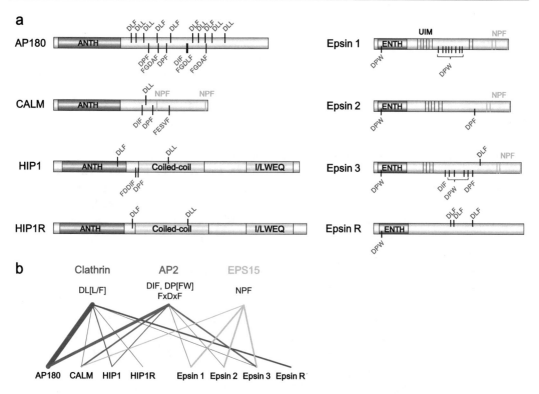

Fig. 2 Domain architectures of the ANTH and ENTH proteins. The ANTH/ENTH-containing proteins from human are illustrated with their domains, of which lengths are depicted proportionally to their amino-acid lengths (**a**). Each of these proteins possesses a phosphoinositide-binding ANTH (magenta) or ENTH (purple) domain at its N terminus. Binding motifs for clathrin (blue), AP2 (red) and EPS15 homology domain (green) are denoted with their amino acid sequences. Individual members possess various combinations of these motifs as summarized in (**b**). Both HIP1 and HIP1R have a coiled-coil domain that is implicated in their homo- and hetero-dimerization, and an I/LWEQ domain that binds to F-actin. Epsin 1–3 have a cluster of ubiquitin-interacting motifs (UIMs), which recognizes the Lys63-linked polyubiquitin chain

closely resemble that of another CME-associated protein, called Epsin (Hyman et al. 2000). The domain had been known to be conserved throughout the Epsin family, and was thus named the Epsin N-terminal homology domain (Kay et al. 1999). AP180 was also found to share a consensus motif of ENTH, although the function of this domain had remained unknown. Itoh and colleagues showed that the ENTH domain specifically recognizes PtdIns(4,5)P_2 (Itoh et al. 2001), suggesting that the N-terminal domain of AP180/CALM shares not only its structure but its function with the ENTH domain. Therefore, the N-terminal domain was named AP180 N-terminal homology domain.

In contrast to the N-terminal region, the remaining region does not adopt a particular secondary structure, but shows properties of an intrinsically disordered polypeptide (Kalthoff et al. 2002). It contains several binding motifs for AP2 (DxF and FxDxF) (Brett et al. 2006) and clathrin (L[L/I][D/E/N][L/F][D/E]). In addition, CALM, but not AP180, has Asn-Pro-Phe motifs, which bind to EPS15 homology domain-containing proteins (Morgan et al. 2003). The presence of these multiple binding motifs enables the enrichment of AP180/CALM in clathrin-coated structures. Indeed, AP180/CALM is a major component of CCVs (Prasad and Lippoldt 1988; Borner et al. 2012). It is equally abundant as the AP2 complex, which exists at a ratio of one adaptor per clathrin triskelion.

Biochemical studies have shown the functional significance of AP180 and CALM in the process of clathrin assembly. In a reconstitution experiment using purified proteins, AP180 in combination with AP2 was found to stimulate clathrin assembly more efficiently than each protein on its own (Hao et al. 1999). Interestingly, in the presence of PtdIns(4,5)P$_2$-containing lipid monolayers, AP180 on its own is able to induce clathrin assembly, although the formed lattice adopts a characteristic planar shape (Ford et al. 2001). A typical, invaginated pit is formed only in the coexistence of AP2 and AP180. Furthermore, the C-terminal region of AP180 (residues 530–915) shows an inhibitory effect on the assembly of clathrin when overexpressed in mammalian cells (Ford et al. 2001). This is possibly through competitive inhibition of the interaction between clathrin/AP2 and their binding proteins.

Studies using knockout animals have clarified the *in vivo* role of AP180/CALM. *Picalm*-deficient mice showed intrauterine growth retardation, dwarfism, and decreased lifespan as well as a change in fur color (Suzuki et al. 2012; Ishikawa et al. 2015). These knockout mice also had severe anemia due to defects in the development and maturation of erythrocytes. It is known that erythrocyte development depends highly on the cellular uptake of transferrin via its receptor. In fact, *Picalm* knockout cells showed defects in transferrin receptor endocytosis, concomitant with impaired clathrin-coat maturation. Detailed analyses using electron microscopy demonstrated that CALM depletion decreases the proportion of CCPs with a constricted neck, and increases the volume of both CCPs and CCVs (Meyerholz et al. 2005; Miller et al. 2015). These results suggest that CALM is important in the maturation of CCPs and the regulation of their size.

AP180/CALM have another important function in the recognition of certain specific cargoes. This function was first suggested by observations in budding yeast null mutants (Huang et al. 1999). The phenotypes of yeast deficient for both *yap1801* and *yap1802* were quite different from those observed in clathrin deficiency; on the contrary, the mutant phenocopied the deficiency of *snc1*, a yeast homolog of soluble N-ethylmaleimide-sensitive factor attachment

protein receptors (SNAREs) (Burston et al. 2009). This unexpected result has also been replicated in higher organisms. For example, the *C. elegans unc11* mutant showed synaptic defects caused by mislocalization of the SNARE synaptobrevin (Nonet et al. 1999; Dittman and Kaplan 2006). A study using mammalian cells demonstrated the underlying mechanism linking AP180/CALM and SNARE proteins (Miller et al. 2011). The authors found that in CALM-depleted cells, vesicle-associated membrane protein (VAMP) 2, VAMP3, and VAMP8, which act in the fusion of endosomes, were accumulated at the PM. CALM was found to physically bind to these SNARE proteins, thereby ensuring that they are internalized from the PM.

3.2 Other ANTH Proteins – Huntingtin-Interacting Protein 1 and Its Related Protein

Huntingtin-interacting protein 1 (HIP1) was originally identified as a binding protein of huntingtin, which is the product of the gene affected in Huntington disease, using a yeast-two hybrid assay (Wanker et al. 1997). Subsequently, a closely related paralog named HIP1-related protein (HIP1R) was identified, and they were found to be components of CCVs (Engqvist-Goldstein et al. 1999; Waelter et al. 2001). Additionally, The *HIP1/HIP1R* gene is evolutionarily conserved from budding yeast, which has a single homolog named *SLA2*, to humans.

HIP1/HIP1R have an ANTH domain at their N terminus (Fig. 2). In addition, HIP1, but not HIP1R, has multiple AP2-binding sites. Unlike AP180/CALM, both HIP1 and HIP1R have an actin-binding domain at their C terminus, called the I/LWEQ domain, or Talin-HIP1/R/Sla2p actin-tethering C-terminal homology domain. The I/LWEQ domain is approximately 250 amino-acid residues and specifically binds to F-actin. It is hypothesized that through F-actin, HIP1/HIP1R might transmit a force required for the invagination of CCPs (Boulant et al. 2011). In agreement with this idea, stalled CCPs were observed in HIP1/HIP1R-depleted cells, and bundles of F-actin were abnormally accumulated near the CCPs as demonstrated by freeze-etch electron microscopy (Engqvist-Goldstein et al. 2004).

Detailed biochemical analyses have identified additional features unique to HIP1/HIP1R. Firstly, their ANTH domains preferentially bind to phosphatidylinositol (3,4)-bisphosphate (PtdIns(3,4)P_2) and phosphatidylinositol (3,5)-bisphosphate, but not to PtdIns(4,5)P_2 (Hyun et al. 2004). PtdIns(3,4)P_2 is generated at the CCPs by phosphatidylinositol-3-kinase C2α, and possibly functions in the scission of the pit through recruiting sorting nexin 9, a BAR domain-containing protein (Posor et al. 2013). Moreover, PtdIns(3,4)P_2 is essential for a class of clathrin-independent endocytosis, termed fast endophilin-mediated endocytosis (Boucrot et al. 2015), which also depends on F-actin function. Although a functional role of HIP1/HIP1R has not been documented, it is possible that HIP1/HIP1R might regulate this clathrin-independent process through their actin-binding ability.

The second feature of HIP1/HIP1R is that they can interact with Epsin, an ENTH-domain containing protein. In the presence of liposomes, SLA2p and ENT1p (yeast homologs of HIP1R and Epsin, respectively) physically bind each other and deform the membrane into a tubular structure (Skruzny et al. 2015). The residues crucial for this interaction are conserved also in their mammalian counterparts. Importantly, HIP1R is not targeted to the PM when cells are depleted of all isoforms of Epsin (Messa et al. 2014). Therefore, it is possible that the phosphoinositide-binding activity of HIP1R is not involved in its targeting to the PM. Consistent with this, the phosphoinositide-binding interface of SLA2p was positioned far from the membrane in its complex with ENT1p (Skruzny et al. 2015). This study rather found that the complex formation occurred only when the phosphoinositide-binding capacities of both SLA2p and ENT1p were intact. These results suggest that the role of phosphoinositide binding is related to a conformational change of SLA2p, rather than its tethering at the membrane.

To understand the physiological roles of HIP1/HIP1R, several groups have reported the phenotypes of knockout mice. For HIP1, although multiple knockout mice have been established to date, their phenotypes are quite different from each other, making it difficult to interpret the results. The first reported knockout mice showed no obvious phenotype except for testicular degeneration (Rao et al. 2001). However, several independent mouse lines demonstrated severe kyphosis (abnormal convex curvature of the spine) in an age-dependent manner (Metzler et al. 2003; Oravecz-Wilson et al. 2004). The severity of kyphosis and other phenotypes were different for each mouse line. Since the murine *Hip1* gene spans 220 kilobase pairs in length and has more than 30 exons, complete gene deletion is difficult. In fact, each of these knockout mice was created by a different strategy, and in some mice it is possible that a hypomorphic allele might have been created unintentionally.

On the other hand, *Hip1r* knockout mice were viable and fertile without any morphological abnormalities. However, double-knockout mice, which were created by crossing them with *Hip1* knockout mice showing kyphosis, showed phenotypic exacerbations, such as the earlier onset of kyphosis (Hyun et al. 2004). These results suggest that HIP1 and HIP1R have redundant functions *in vivo*. Furthermore, spinal defects in double-knockout mice were completely suppressed by crossing them with human *HIP1* transgenic mice (Bradley et al. 2007). This strongly suggests that the kyphosis phenotype is indeed the result of a functional deletion of *Hip1*.

Currently, the molecular mechanism underlying this spinal degeneration remains unknown. Importantly, cells derived from double-knockout mice did not show any defects in endocytosis (Bradley et al. 2007), although decreased endocytosis of the glutamate receptor was observed in *Hip1* knockout neurons (Metzler et al. 2003). These results conflict with the importance of HIP1/HIP1 in the endocytic pathway, which has been suggested by studies using cultured cells. However, it is possible that there are other proteins that compensate for the loss of HIP1/

HIP1R. Further studies are required to clarify the mechanism underlying the phenotypic difference between knockout animals and cultured cells.

3.3 ENTH Domain-Containing Proteins

Epsin was initially identified as an EPS15-interacting protein (Chen et al. 1998; Rosenthal et al. 1999). In mammals, four paralogs have been found: Epsin 1, 2, and 3, and Epsin-related (Epsin R). Whereas the expression of Epsin 1 and 2 are ubiquitous with higher levels in the brain, Epsin 3 is expressed primarily in the stomach. Epsin 1–3 have an ENTH domain at the N terminus, through which they can specifically recognize $PtdIns(4,5)P_2$ (Itoh et al. 2001) (Fig. 2). In addition, Epsins have multiple binding motifs for clathrin, AP2, and EPS15 at the C terminus, which enable their interaction with CME machinery, such as AP180/CALM. Similar to AP180, Epsin can stimulate clathrin assembly in the presence of $PtdIns(4,5)P_2$-containing lipid monolayers (Ford et al. 2002). However, Epsin can also induce membrane invagination. This is in contrast to the fact that AP180 forms only a planar lattice of clathrin (Ford et al. 2001). Furthermore, Epsin can deform liposomes into tubules and vesicles (Ford et al. 2002). Such membrane-bending activities can be attributed primarily to the N-terminal helix of the ENTH domain, called helix 0. This helix has an amphipathic structure, which can be inserted in the membrane like a wedge, and thereby endows membrane with a positive curvature. Epsin-deficient cells showed impaired transferrin uptake and the accumulation of shallow and unconstricted (U-shaped) pits with markedly reduced turnover rates (Boucrot et al. 2012; Messa et al. 2014). These "immature" pits are devoid of dynamin and HIP1R, which are known to be recruited to CCPs at the timing of scission (Taylor et al. 2011), suggesting that endocytosis is arrested at the early or middle stage. Therefore, it is likely that Epsin regulates the maturation of CCP through its membrane-bending activity and recruitment of other effector proteins that function at later stages.

Besides the ENTH domain, Epsin 1–3 have ubiquitin-interacting motifs (UIMs) that recognize a polyubiquitin chain (Barriere et al. 2006; Hawryluk et al. 2006). Epsin has multiple UIMs, each of which adopts an α-helical structure and is separated at regular intervals by unstructured inter-UIM polypeptides. The interaction with a polyubiquitin chain induces a conformational change of the inter-UIM into an α-helix, rendering the multiple UIMs into a long continuous helix (Sims and Cohen 2009; Sato et al. 2009). Due to the appropriate length of its inter-UIM sequence, Epsin can specifically interact with the lysine 63-linked type of polyubiquitin chain, a post-translational modification found in transmembrane proteins that are internalized upon ligand stimulation (Hawryluk et al. 2006).

The phenotypes of Epsin-deficient animals highlight the importance of the cargo-sorting function of Epsin: *Liquid facets* (Lqf), the sole *Drosophila* Epsin, is essential for the patterning of imaginal discs, which are the larval body parts that give rise to adult organs, such as eyes and wings (Overstreet et al. 2003). In addition, double-knockout mice of Epsin 1/2 were embryonic lethal, and the lethality was due to impaired vascular development (Chen et al. 2009). These phenotypes were quite similar to those caused by defective Notch signaling. Notch is a single-pass transmembrane protein, and its signal transduction is initiated when the transmembrane ligand (Delta and Serrate in flies and LAG-2 in nematodes) on the neighboring cell binds to the Notch receptor. Upon ligand stimulation, Notch is sequentially endoproteolyzed at its extracellular domain and then at the transmembrane domain, freed from the membrane, and translocates to the nucleus where it functions as a transcriptional regulator (Kovall et al. 2017). Detailed analyses demonstrated that *Lqf* mutants have impaired endocytosis of Delta (Overstreet et al. 2003, 2004; Wang and Struhl 2004; Tian et al. 2004). Moreover, several mutants defective in endocytosis also showed the Notch phenotype. These results clearly demonstrated the inevitable role of the endocytosis of Delta in Notch signaling.

Interestingly, *Lqf* mutants did not show any accumulation of Delta on ligand-presenting cells despite their essential role in endocytosis, suggesting that Lqf regulates the internalization of a minor population of Delta proteins that participates in the signaling event (Wang and Struhl 2004). In addition, through biophysical experiments using optic tweezers and atomic force microscopy, it is observed that a pulling force applied to Notch receptor induces a conformational change of its extracellular domain in a way that causes the cleavage of the domain and downstream signaling (Meloty-Kapella et al. 2012; Stephenson and Avis 2012). This has raised an attractive hypothesis that Epsin generates a pulling force to enable Notch to undergo a conformational change for its activation through its activity of membrane-bending and/or regulation of endocytosis of the ligand/receptor complex (Meloty-Kapella et al. 2012). Moreover, *Neuralized* (fly) and *Mind bomb* (zebrafish), null mutants which provided other examples of defective Notch signaling, were discovered to code ubiquitin ligases for Delta, and the ubiquitination activity was essential for the endocytosis of Delta (Kovall et al. 2017). These results clearly indicate that the sorting of ubiquitinated cargos is the primary function of Epsin.

Epsin R is distantly related to the other Epsin proteins. Epsin R has an ENTH domain and clathrin-binding motifs, but it lacks AP2- and EPS15-binding motifs. Instead, Epsin R has a DFxD[W/F] motif, which is recognized by the adaptor protein 1 complex. Similar to AP2, adaptor protein 1 has a crucial role in the clathrin-mediated budding from the *trans*-Golgi network (TGN) to endosomes. In addition, the ENTH domain of Epsin R has different lipid specificity: it recognizes phosphatidylinositol 4-phosphate, which exists densely in the TGN membrane (Wasiak et al. 2002; Mills et al. 2003; Hirst et al. 2003). These features coordinately favor the functioning of Epsin R at the TGN. In fact, Epsin R is a major component of TGN-derived CCVs and is present at a 1:1 molar ratio with the clathrin triskelion (Borner et al. 2012). Moreover, the ENTH domain of Epsin R has a specific binding interface for Q-SNARE VTI1A and B,

which functions in the heterotypic fusion between late endosomes and lysosomes (Hirst et al. 2004; Miller et al. 2007), suggesting that the primary function of Epsin R is trafficking these SNAREs to their proper destinations. Therefore, Epsin R can be viewed as a Golgi counterpart of CALM, rather than as a member of classical Epsins.

4 ANTH/ENTH Domain: Structure and Biochemical Functions

4.1 Structural Basis of Phosphoinositide Recognition

The structure of the N-terminal half of the ANTH domain is closely related to that of the ENTH domain, which is composed of a super-helix of six α-helices (Fig. 3). Despite this structural similarity, ANTH and ENTH recognize phosphoinositides in different ways: for the ANTH domain of CALM, the lipid-binding moiety $(Kx_9Kx[K/R][H/Y])$ is exposed on the protein surface where a small number of basic residues (Lys38, Lys40, and His41) participate in the recognition of the 4'- and 5'-phosphates of the inositol ring (Fig. 3d). Of note, all the binding residues are located in helix 2 (Ford et al. 2001). In contrast, the ENTH domain of Epsin 1 binds to $PtdIns(4,5)P_2$ in a groove, which makes multiple contacts with the phosphate groups of the 4'- and 5'-positions, as well as the phosphodiester bond connecting the inositol ring to the glycerol backbone (Ford et al. 2002) (Fig. 3e). Interestingly, the binding residues that are completely different from those of the ANTH domain, namely, helices 1, 3, 4, and their adjacent loops, all contribute to the binding, whereas helix 2 does not. In addition, the N-terminal helix 0, which functions in bending the membrane, also participates in the recognition of $PtdIns(4,5)P_2$.

4.2 Specificity and Affinity

The ANTH/ENTH domains show a relatively high specificity to various phosphoinositide species. For example, the ANTH domains of AP180

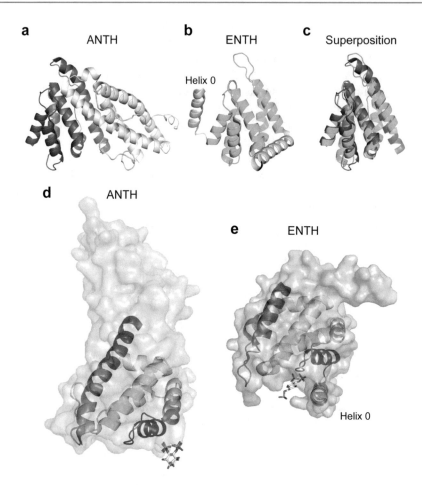

Fig. 3 Crystal structures of the ANTH and ENTH domains. Ribbon schematics of rat CALM ANTH domain (**a**, PDB accession 1HG2) and rat Epsin ENTH domain (**b**, PDB accession 1H0A). The N-terminal six helices of the ANTH domain (highlighted in magenta) constitute a solenoid that structurally resembles the ENTH domain (green). Their superposition is shown in (**c**). Co-crystal structures of the domains (molecular surfaces are rendered transparently) bound to substrate mimics (depicted as sticks) are shown for the ANTH domain with inositol 4,5-bisphosphate (**d**) and the ENTH domain with inositol 1,4,5-trisphosphate (**e**). To highlight their equivalent helices, helix 1 to 6 are superimposed and colored from blue to red in a similar manner for both the proteins. Note that the two proteins use completely different interfaces for the recognition of lipids, and that the Epsin ENTH domain has an additional α-helix (helix 0) at its N terminus (pink). These structures were illustrated using PyMOL

and CALM show the highest specificity to PtdIns$(4,5)P_2$, and although they can bind to phosphatidylinositol 4-phosphate and phosphatidylinositol (3,5)-bisphosphate to a lesser extent, they do not bind to phosphatidylinositol, PtdIns$(3,4)P_2$, or phosphatidylinositol (3,4,5)-trisphosphate. In addition, substitutions of lysine residues in the binding site abolishes these interactions (Ford et al. 2001; Kanatsu et al. 2016).

Whereas most ENTH/ANTH proteins bind most strongly to PtdIns$(4,5)P_2$, there are at least two exceptions: first, Epsin R specifically interacts with phosphatidylinositol 4-phosphate, but not with PtdIns$(4,5)P_2$. Some residues in its ENTH domain that are important for the recognition of phosphate groups are not conserved. In particular, Arg8 and Asn30 of Epsin 1 are substituted to negatively charged aspartic residues, which might favor the exclusion of the $5'$-phosphate through electrostatic repulsion (Mills et al. 2003). Second, HIP1/HIP1R bind to PtdIns$(3,4)$P$_2$ (Hyun et al. 2004). The consensus sequence of the ANTH domain, Kx$_9$Kx[K/R] [H/Y], is conserved in the cases of HIP1/HIP1R, suggesting that it is important for the recognition of two adjacent phosphate groups found in both PtdIns$(4,5)$P$_2$ and PtdIns$(3,4)$P$_2$. However, the mechanism of how they discriminate the two similar lipids remains unknown. Further studies are required to understand the structural basis determining such tight substrate specificity.

Regarding the affinity of the ANTH/ENTH domains to their substrate lipids, the dissociation constants are approximately in the range of 10^{-7}–10^{-5} M for ANTH, and 10^{-8}–10^{-6} M for ENTH, although the results considerably vary among reports (Ford et al. 2001, 2002; Itoh et al. 2001; Stahelin et al. 2003; Hom et al. 2007; Miller et al. 2015). This is partially due to the differences of the types of substrate, methods, and conditions used for the measurements. For example, one study estimated the K_d value of Epsin ENTH as 80 nM against a lipid monolayer containing 0.5% PtdIns$(4,5)$P$_2$, whereas another study reported it as 3.6 μM using inositol 1,4,5-trisphosphate as a headgroup mimic (Ford et al. 2002; Stahelin et al. 2003). This raises the possibility that the membrane-penetrating activity of the ENTH domain slows the protein's dissociation rate from the membrane. Furthermore, the binding affinity also depends on the coexistence of negatively charged lipids, such as phosphatidylserine, and buffer pH (Hom et al. 2007). These results suggest that nonspecific effects, such as electrostatic attraction, also contribute to the binding affinity to the membrane.

4.3   Membrane Deformation Activity and Its Structural Basis

AP180, CALM and Epsin all interact with the membrane via binding PtdIns$(4,5)$P$_2$. However, the effects of the binding on morphology or curvature of the membrane are different. Epsin has a unique activity to deform the lipid membrane into

tubules and vesicles (Ford et al. 2002). For this activity, an N-terminal sequence of the ENTH domain plays an indispensable role. This region, which comprises the N-terminal 16 residues, is unstructured in the absence of lipids. In contrast, the interaction with PtdIns$(4,5)$P$_2$ causes the formation of an amphipathic helix, in which hydrophobic residues are positioned on one side. This helix, termed helix 0, can penetrate like a wedge into the cytoplasmic leaflet of the lipid bilayer, thereby inducing a positive curvature in the membrane (Ford et al. 2002). From a biophysical viewpoint, Epsin renders tubule formation energetically more favorable, because it reduces the bending rigidity of the membrane (Gleisner et al. 2016).

In contrast to Epsin, AP180 and CALM were originally documented as exhibiting no membrane-bending activity (Ford et al. 2001). In the original crystal structure of CALM, this protein did not show any structure corresponding to the helix 0, implying the importance of this helix in the membrane deformation. However, according to another report from Owen lab, CALM exhibited a structure similar to helix 0, and also showed a membrane tabulation activity through a mechanism dependent on the existence of the helix (Miller et al. 2015). Although it has been not clear what causes this obvious discrepancy, one possible explanation is that earlier studies might have missed the existence of this helix, possibly due to its susceptibility to proteolysis (Wood and Royle 2015). This report also found that CALM could interact with the membrane through the hydrophobic residues in the helix, but unlike Epsin, the helix 0 of CALM adopted its helical shape even in the absence of PtdIns$(4,5)$P$_2$ (Miller et al. 2015). Currently, the mechanism as to how this helix interacts with the membrane remains unknown, and further studies are required to understand the precise roles of the helix in the function of CALM.

4.4   Interaction with Other Proteins

Besides phosphoinositides, the ANTH and ENTH domains interact with proteins on their own.

Interestingly, the binding interfaces that they utilize for their interaction are quite different from each other, although they are evolutionarily interrelated. Simply speaking, however, their interaction modes can be classified into the following three types.

Firstly, the ANTH domain of CALM can bind to SNAREs, such as VAMP2, 3, and 8. The binding interface resides in the ANTH domain, specifically at its C-terminal 100–150 amino-acid region, which is structurally not related to the ENTH domain (Fig. 4a). Crystallographic analyses demonstrated that VAMP8 binds to the hydrophobic "trough" in this region, which is formed by helix 9 and 10 of the ANTH domain (Miller et al. 2011). In detail, the SNARE motif of VAMP8 binds to these helices in a manner that closely mimics the SNARE complex. Furthermore, the binding of VAMP8 to the ANTH domain and to the SNARE complex are mutually exclusive, suggesting that CALM can bind to only the monomeric form of SNARE. It is possible that this mechanism ensures that CALM efficiently recruits free SNARE proteins for endocytosis and facilitates their recycling to intracellular compartments.

Secondly, the core structure shared between ANTH and ENTH also participates in protein interactions. Epsin R recognizes its cargo in this manner, where the ENTH domain directly acts in the physical association with the H_{abc} domain of VTI1B (Miller et al. 2007) (Fig. 4b). In this interaction, numerous hydrogen bonds are observed between Epsin R and VTI1B. Therefore, it is obvious that the underlying mechanism is completely different to that involved in the interaction between the ANTH domain and VAMP proteins.

Lastly, there is an example that the ANTH and ENTH domains bind each other: SLA2p and ENT1p can coassemble in the presence of PtdIns(4,5)P$_2$ to form an ordered lattice on the membrane (Skruzny et al. 2015). Interestingly, this coassembly can reshape flat membranes into tubules. Therefore, this may explain the mechanism by which Epsin can act in both the recruitment of HIP1R and the deformation of the membrane.

Taken together, the ways by which proteins use their ANTH and ENTH domains for binding are diverse, not only for their binding to lipids, but also to proteins. It is possible that this diversity of binding modes may stem from their evolutionary origin: the ANTH/ENTH domains show structural similarity to the Vps27/Hrs/STAM domain, armadillo repeats, and HEAT repeats, all of which are involved in protein-protein interactions (Hyman et al. 2000; De Craene et al. 2012). The structural features of the ANTH/ENTH domains may have provided a favorable template scaffold for protein-protein interactions, and it may have acted advantageously for the ANTH/ENTH proteins to acquire lipid-binding ability.

5 Involvement in Human Diseases

5.1 Involvement of PICALM in Leukemia and Alzheimer Disease

In this section, we discuss the involvement of ANTH/ENTH proteins in diseases, to understand their physiological importance in humans. Among the ANTH/ENTH proteins, only CALM has been confirmed to have strong genetic links with human diseases, except for Epsin R being suggested to be associated with the risk of schizophrenia (Pimm et al. 2005). PICALM, the CALM-encoding gene, was originally identified as a component of the PICALM-MLLT10 fusion gene, which is the product of chromosomal translocation t(10;11)(p13;q14) found in a patient with diffuse histiocytic lymphoma (Dreyling et al. 1996). Similar PICALM-MLLT10 fusion genes have been found frequently in patients with T-cell acute lymphoblastic leukemia, and in most of these cases, fusion genes encode almost full-length proteins of CALM and MLLT10 as their N- and C-terminal parts, respectively. Previous studies have shown that MLLT10 interacts with histone methyltransferase DOT1L, and PICALM-MLLT10 fusions have been shown to upregulate a number of leukemia-associated

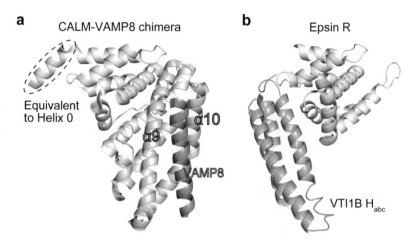

a CALM-VAMP8 chimera

Equivalent to Helix 0

α9 α10

VAMP8

b Epsin R

VTI1B H$_{abc}$

Fig. 4 ANTH and ENTH domains bound to their protein binding partners. (**a**) Structure of the CALM-VAMP8 chimera is shown, where VAMP8 (orange) is bound to the "trough" that is formed by helix 9 and 10 of CALM (PDB accession number 3ZYM). Note that, unlike Fig. 4a and d, CALM clearly shows an extra helix at its N terminus, reminiscent of the helix 0 of Epsin. (**b**) Co-crystal structure of Epsin R (grey) and VTI1B (green) is depicted and viewed from the same direction as that of **a**, to highlight the difference of the regions that they use for interacting with the cargo proteins (PDB accession number 2V8S)

gene expression in a DOT1L activity-dependent manner (Okada et al. 2005, 2006). This suggests that misregulation of DOT1L by MLLT10 fusion proteins is crucial for leukemogenesis. Importantly, a nuclear export signal within CALM was found to be necessary and sufficient for the leukemogenesis by *PICALM-MLLT10* (Conway et al. 2013; Suzuki et al. 2014), although it remains unclear how the nuclear export is related to the malfunction of DOT1L. This also suggests that the endocytic function of CALM is not important in the pathomechanisms of *PICALM-MLLT10*.

In more recent years, it has been revealed that polymorphisms of *PICALM* is associated with susceptibility of Alzheimer disease (AD). Although the precise role of CALM in the pathogenesis of AD is currently unclear, it is likely that abnormalities in physiologically important functions of CALM are caused by its genetic alterations and are related to the onset of the disease. Therefore, understanding the pathomechanisms of CALM will provide an insight into the physiological importance of this protein in human beings. Hence, we will focus on

AD and outline our current knowledge about the involvement of CALM, particularly focusing on the functional changes of CALM that might be related to disease mechanisms.

5.2 Amyloid Hypothesis of Alzheimer Disease

AD is the most common neurodegenerative disease, and is characterized by progressive cognitive decline. In the AD patient's brain, marked neuronal loss is observed in broad brain areas, including the neocortex and hippocampus. Pathologically, AD is characterized by (i) an exacerbated deposition of senile plaques in the brain, which are comprised of aggregated amyloid β (Aβ) peptides, and (ii) the appearance of dystrophic neurites, caused by neuronal inclusions composed of aggregated and hyperphosphorylated tau. Most cases of AD are of late onset, and do not show a Mendelian inheritance. On the other hand, rare but genetic forms of the disease (familial AD; FAD) also exist, and genetic analyses have demonstrated that abnormalities of Aβ production is crucial for the pathogenesis of FAD.

Aβ is generated from a single-pass transmembrane protein, named Aβ precursor protein (APP) in the following two-step proteolysis: first, β-secretase cleaves APP in its extracellular domain at a site close to the membrane. The membrane-embedded fragment termed C99 is subjected to the second cleavage by γ-secretase. γ-Secretase is a transmembrane protein complex, which is composed of Presenilin, Nicastrin, APH1, and PEN2 (Takasugi et al. 2003). γ-Secretase possesses a unique enzymatic activity that cleaves type-I transmembrane proteins, including C99, within their membrane-spanning region. γ-Secretase cleaves C99 at several positions within its transmembrane domain, yielding Aβ species of variable length. Whereas the 40 amino-acid peptide (Aβ40) accounts for the majority of the species, a less abundant Aβ peptide of 42 amino-acid residues (Aβ42) is also formed, and this species is highly prone to form neurotoxic aggregates. Interestingly, all the FAD mutations have been identified in APP, Presenilin-1, and Presenilin-2, and they are found to cause either an increase in the ratio of Aβ42 or the promotion of its aggregation, suggesting that aggregated Aβ plays a causative role in the pathogenesis of FAD.

Various species of Aβ aggregates have been reported, from oligomers to fibrillar Aβ. Although many lines of evidence support that aggregated Aβ species are neurotoxic, there is no consensus about the specific species of aggregated Aβ that is associated with the neurodegeneration of AD, and how it exerts its effect. Aβ has been thought to act upstream of tau aggregation and hyperphosphorylation, and therefore it is possible that Aβ exerts its neurotoxic effects through a tau-associated mechanism.

In the case of sporadic AD, which accounts for the majority of cases of this disease, although the deposition of Aβ is indistinguishable from those seen in FAD cases, it remains unclear whether Aβ is a prerequisite for pathogenesis. The unsuccessful results of the clinical trials of γ-secretase inhibitors have led many researchers to question the reliability of the Aβ hypothesis. However, it is possible that, because γ-secretase inhibitors have many adverse effects, it might compromise the efficacy of this treatment. In addition, Aβ pathology is found to precede the clinical manifestations of AD by several tens of years, and therefore, therapeutics based on Aβ formation should be started at an earlier stage of AD to successfully suppress disease progression. In fact, there are several lines of evidence that support this notion. The most important is the finding that a rare *APP* gene variant protects against AD (Jonsson et al. 2012). Namely, the amino acid substitution Ala673Thr was found to reduce the processing of APP by β-secretase. Furthermore, anti-Aβ immunotherapy, performed on AD patients at an early stage of the disease, reduced Aβ deposits in the brain and slowed the cognitive decline (Sevigny et al. 2016). Taken together, the Aβ hypothesis is still important to understand the pathology and to develop future therapeutics for AD.

5.3 Identification of *PICALM* Variants as Risk-Susceptible Loci for Alzheimer Disease

What type of changes in Aβ triggers the pathogenesis in sporadic patients, who do not have any mutations in APP or Presenilin genes? Generally speaking, the amount of Aβ in the brain reflects the relative ratio of its production to its degradation or clearance. Interestingly, there is a report that the clearance of Aβ from the brain is impaired in AD patients. Defects in Aβ metabolism can cause its accumulation in the brain and may contribute to disease onset. In addition, considering that Aβ also accumulates in the brains of cognitively normal elderly people, individual differences in the sensitivity to Aβ toxicity may be associated with disease susceptibility. To identify genetic factors underlying such individual differences, genome-wide association studies have been performed, and many risk susceptible loci have been identified.

Several single nucleotide polymorphisms (SNPs) in *PICALM* were identified as such genetic susceptibility loci for late-onset AD

(Harold et al. 2009; Lambert et al. 2009). Harold and colleagues reported that carriers of the A allele of the SNP rs3851179 showed a slightly decreased risk for AD (odds ratio: 0.85) (Harold et al. 2009). Rs3851179 is located 88.5 kilobase pairs upstream of *PICALM* on chromosome 11. Other SNPs in the *PICALM* gene were also found as susceptible loci, including rs541458, which is in linkage disequilibrium with rs3851179 (Harold et al. 2009). The association of *PICALM* has been replicated in various independent studies (Seshadri et al. 2010; Carrasquillo et al. 2010; Corneveaux et al. 2010; Lambert et al. 2011; Naj et al. 2011).

Because these SNPs are located in the 5′ region of *PICALM*, it is possible that they affect the expression levels of gene. Indeed, the protective allele rs3851179(A) is associated with a modest increase in *PICALM* expression level (Parikh et al. 2014). Zhao and colleagues observed a similar result by experimentally introducing the protective allele into induced pluripotent stem cells (Zhao et al. 2015). Furthermore, a rare nonsynonymous variation of *PICALM* (rs117411388; His465Arg according to the numbering of *PICALM* isoform 1) was found in late-onset AD cohorts (Vardarajan et al. 2015). It is currently not clear whether this amino acid substitution affects the function of CALM, but further biochemical studies are expected to contribute towards understanding its role in AD pathogenesis.

5.4   PICALM Variants and Alzheimer Disease Pathology

Besides being implicated in disease risk, genetic variations of *PICALM* modulate AD pathology. For example, AD patients carrying the risk allele rs541458(T) showed cognitive decline at a younger age than non-carriers (Sweet et al. 2012). Similarly, a lower age-at-onset was reported in patients carrying the risk allele rs3851179 (G) than non-carriers, although the effect was small (0.33–0.55 years earlier than non-carriers) (Thambisetty et al. 2013; Naj et al. 2014). In addition, in patients with Down syndrome,

several variants of *PICALM* were found to be associated with an earlier age-at-onset of dementia (Jones et al. 2013).

There is also a possible association between *PICALM* variants and Aβ metabolism: the risk allele rs541458 is associated with a decreased level of Aβ42 in cerebrospinal fluid (CSF) (Schjeide et al. 2011), although the association was not replicated in another study (Kauwe et al. 2011). Decreased Aβ42 in CSF may be attributed to an impairment of its efflux from the brain parenchyma to CSF, or it may reflect that risk carriers develop a higher amount of Aβ aggregates in the brain than non-carriers, considering the fact that Aβ aggregates can accelerate the aggregation of soluble Aβ, by acting as a "seed", and thereby impeding its efflux into the CSF.

Neuroimaging studies have provided several lines of evidence that the protective alleles of *PICALM* are associated with less atrophy in the hippocampus and entorhinal cortex, which are the most susceptible brain regions in AD (Biffi et al. 2010; Furney et al. 2011; Melville et al. 2012). On the other hand, Xu and colleagues conducted longitudinal analyses and found that slower atrophy rates, particularly in the posterior cingulate region, were associated with multiple *PICALM* variants (Xu et al. 2016). Interestingly, one such allele, rs642949(C), was significantly associated with a larger baseline thickness of the posterior cingulate in healthy people, suggesting that CALM has a protective effect against AD by affecting brain reserve capacity. Furthermore, electroencephalographic analyses in cognitively normal volunteers showed that homozygous carriers of the non-protective allele rs3851179 (G) have increased beta power, which has previously been linked to cortical hyperexcitability (Ponomareva et al. 2017).

5.5   Functional Roles of CALM in Alzheimer Disease

The endocytosis pathway has long been implicated in Aβ generation. Therefore, the involvement of CALM in Aβ production has been analyzed by several groups. The first report found that CALM positively regulates the endocytosis and cleavage by β-secretase of APP, and that *Picalm* knockout mice showed the reduced accumulation of Aβ (Xiao et al. 2012). In addition, another report found that β-secretase activity is also affected in *Picalm*-depleted cells (Thomas et al. 2016). Furthermore, our group reported that CALM regulates the endocytosis of γ-secretase (Kanatsu et al. 2014). γ-Secretase is localized in acidic compartments, such as late endosomes and lysosomes (Hayashi et al. 2012). However, in *Picalm*-depleted cells, γ-secretase is not endocytosed and accumulated in the PM. Interestingly, *Picalm*-depleted cells produced less Aβ42 compared with control cells, suggesting that γ-secretase in the acidic compartments generates a larger amount of Aβ42 than that residing at the PM. In agreement with this notion, APP transgenic mice crossed with *Picalm* knockout mice showed reduced Aβ42 deposition (Kanatsu et al. 2016) (Fig. 5a).

The involvement of CALM in the clearance of Aβ has also been analyzed. It was reported that in the brain, the expression of CALM is robust in endothelial cells of microvessels but is modest in other cell types, including neurons and glial cells (Baig et al. 2010; Parikh et al. 2014; Zhao et al. 2015). Zhao and colleagues therefore analyzed the function of CALM in brain capillary endothelial cells (Zhao et al. 2015). In *Picalm* heterozygous knockout mice, Aβ clearance was reduced, and Aβ was accumulated in the brain parenchyma. CALM was found to regulate the transcytosis of LRP1, suggesting its involvement in the efflux of the Aβ-LRP1 complex through endothelial cells (Fig. 5b). Importantly, induced pluripotent stem cell-derived endothelial cells introduced with the protective allele rs3851179 (A) showed increased mRNA levels of PICALM and increased Aβ transcytosis. Furthermore, CALM protein levels were reduced in AD brains, suggesting that the reduced activity of Aβ transcytosis was involved in the pathogenesis of AD.

Abnormally aggregated proteins have been shown to accumulate in the neuronal cytoplasm of patients with various neurodegenerative diseases, including tau in AD. The mechanisms of the metabolic processes of these proteins have attracted much attention. Among such mechanisms, it was shown that the autophagy pathway is involved in the degradation of tau (Fig. 5c). Interestingly, the activity of autophagy was reduced in CALM-depleted cells, in which cytosolic aggregation, such as tau and Huntingtin, was accumulated (Moreau et al. 2014). VAMP3, a cargo protein of CALM, reportedly mediates the heterotypic fusion of ATG16L1-containing and ATG9-containing vesicles (Puri et al. 2013). In addition, another cargo, VAMP8, is involved in autophagosome-lysosome fusion (Furuta et al. 2010; Itakura et al. 2012). In fact, a CALM mutant lacking SNARE-binding ability was unable to rescue the deficiency of autophagy in CALM-depleted cells, suggesting that CALM is important for autophagy through maintaining the subcellular localization of VAMPs. Interestingly, the level of CALM protein was reduced in several tau-associated diseases (Ando et al. 2016). The expression level of CALM was inversely correlated with phosphorylated tau deposition in the brain; furthermore, it was significantly correlated with increased LC3II and decreased Beclin-1, both of which are indicative of impaired autophagy flux.

In summary, whereas CALM increases the production of Aβ, particularly that of Aβ42 in neurons, it facilitates the efflux of Aβ in vascular endothelial cells. In addition, CALM may have a protective function against the neuronal accumulation of tau. The involvement of CALM in AD pathology is very complicated, and future studies are necessary to compare in detail the extent to which each of these mechanisms contributes to the onset and disease modification of AD.

6   Concluding Remarks

The ANTH/ENTH domains have striking similarities, not only in terms of structure but also in terms of many other factors, such as binding to lipids and cargo proteins, functions in the CME, etc. However, when looking at the details, these proteins also show surprising differences,

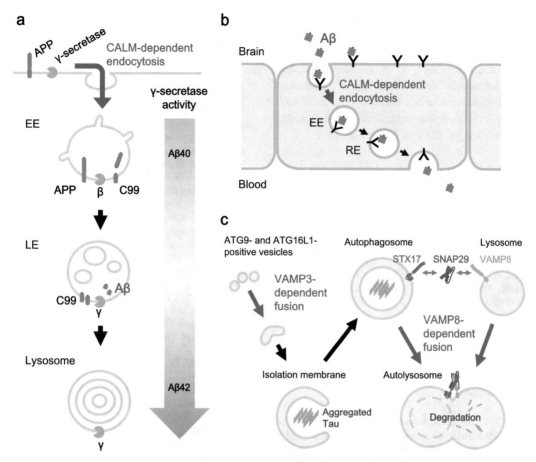

Fig. 5 Possible roles of CALM in AD. (**a**) CALM regulates the endocytosis of APP and γ-secretase. The Aβ peptide is generated through sequential cleavages of APP by β- and γ-secretases. Because β-secretase is more active in endosomes than in the PM, CALM facilitates the Aβ generation by relocating APP to endosomes. In addition, CALM increases the ratio of Aβ42 to Aβ40, through internalizing γ-secretase into endosomes, biophysical properties of which possibly affect the substrate specificity of this enzyme. (**b**) In vascular endothelial cells, Aβ is captured by a cell surface receptor from the brain parenchyma and subjected to the CME. Internalized Aβ is transported via early and recycling endosomes (abbreviated as EE and RE, respectively), and is excreted in the blood by exocytosis. (**c**) CALM regulates autophagy through maintaining the subcellular localization of VAMPs. The intracellular aggregate of tau is wrapped by the isolation membrane, which is generated by VAMP3-dependent fusion of vesicles containing ATG9 and ATG16L1, and is eventually encapsulated in a double-membrane vesicle, called autophagosome. Autophagosome is then fused with lysosome in a VAMP8-dependent manner, and yields autolysosome, in which the protein aggregate is degraded

such as in their binding modes for similar ligands. As structural data accumulate in the future, more and more diverse molecular functions of these two domains may be found, contrary to their apparent similarities.

Whereas many of the biochemical functions of ANTH/ENTH proteins have been elucidated, there are still many unclear points regarding their *in vivo* roles. Specifically, the phenotypes of knockout mice of the ANTH/ENTH family gene differ greatly from the findings obtained in cell experiments. We therefore require a deeper understanding about how ANTH/ENTH proteins work in individual cell types. For this purpose, it will be very useful to analyze various cell type-specific knockout mice. Regarding AD, the roles of glial cells, in addition to nerve cells, have been regarded as being important, and the involvement of CALM in these cell types is also interesting. The elucidation of cell type-specific functions will become increasingly important also in these studies. Furthermore, it is possible that the functions of ANTH/ENTH proteins are controlled by cell-type specific factors. Identification of such factors will be highly promising for establishing therapeutics for AD with fewer side effects.

Acknowledgements This work was supported in part by Grant-in-Aid for Young Scientists (B) from the Japan Society for the Promotion of Science (JSPS) [17 K15446 to S.T.], Grants-in-Aid for Scientific Research (A) from the Japan Society for the Promotion of Science [15H02492 to T.T.], by the Brain Mapping by Integrated Neurotechnologies for Disease Studies (Brain/MINDS) from the Japan Agency for Medical Research and Development (AMED) [17dm0207014h0004 to T.T.] and Sunbor Grant from the Suntory Foundation for Life Sciences [to S.T.].

Conflicts of Interest None declared.

References

Ando K, Tomimura K, Sazdovitch V et al (2016) Level of PICALM, a key component of clathrin-mediated endocytosis, is correlated with levels of phosphotau and autophagy-related proteins and is associated with tau inclusions in AD, PSP and pick disease. Neurobiol Dis 94:32–43. https://doi.org/10.1016/j.nbd.2016.05.017

Baig S, Joseph SA, Tayler H et al (2010) Distribution and expression of picalm in Alzheimer disease. J Neuropathol Exp Neurol 69:1071–1077. https://doi.org/10.1097/NEN.0b013e3181f52e01

Barriere H, Nemes C, Lechardeur D et al (2006) Molecular basis of oligoubiquitin-dependent internalization of membrane proteins in mammalian cells. Traffic 7:282–297. https://doi.org/10.1111/j.1600-0854.2006.00384.x

Biffi A, Anderson CD, Desikan RS et al (2010) Genetic variation and neuroimaging measures in Alzheimer disease. Arch Neurol 67:677. https://doi.org/10.1001/archneurol.2010.108

Borner GHH, Antrobus R, Hirst J et al (2012) Multivariate proteomic profiling identifies novel accessory proteins of coated vesicles. J Cell Biol 197:141–160. https://doi.org/10.1083/jcb.201111049

Boucrot E, Pick A, Çamdere G et al (2012) Membrane fission is promoted by insertion of amphipathic helices and is restricted by crescent BAR domains. Cell 149:124–136. https://doi.org/10.1016/j.cell.2012.01.047

Boucrot E, Ferreira APA, Almeida-Souza L et al (2015) Endophilin marks and controls a clathrin-independent endocytic pathway. Nature 517:460–465. https://doi.org/10.1038/nature14067

Boulant S, Kural C, Zeeh J-C et al (2011) Actin dynamics counteract membrane tension during clathrin-mediated endocytosis. Nat Cell Biol 13:1124–1131. https://doi.org/10.1038/ncb2307

Bradley SV, Hyun TS, Oravecz-Wilson KI et al (2007) Degenerative phenotypes caused by the combined deficiency of murine HIP1 and HIP1r are rescued by human HIP1. Hum Mol Genet 16:1279–1292. https://doi.org/10.1093/hmg/ddm076

Brett TJ, Legendre-Guillemin V, McPherson PS, Fremont DH (2006) Structural definition of the F-actin-binding THATCH domain from HIP1R. Nat Struct Mol Biol 13:121–130. https://doi.org/10.1038/nsmb1043

Burston HE, Maldonado-Báez L, Davey M et al (2009) Regulators of yeast endocytosis identified by systematic quantitative analysis. J Cell Biol 185:1097–1110. https://doi.org/10.1083/jcb.200811116

Buss F, Arden SD, Lindsay M et al (2001) Myosin VI isoform localized to clathrin-coated vesicles with a role in clathrin-mediated endocytosis. EMBO J 20:3676–3684. https://doi.org/10.1093/emboj/20.14.3676

Carrasquillo MM, Belbin O, Hunter TA et al (2010) Replication of CLU, CR1, and PICALM associations with Alzheimer disease. Arch Neurol 67:961–964. https://doi.org/10.1001/archneurol.2010.147

Chen H, Fre S, Slepnev VI et al (1998) Epsin is an EH-domain-binding protein implicated in clathrin-mediated endocytosis. Nature 394:793–797. https://doi.org/10.1038/29555

Chen H, Ko G, Zatti A et al (2009) Embryonic arrest at midgestation and disruption of Notch signaling produced by the absence of both epsin 1 and epsin 2 in mice. Proc Natl Acad Sci U S A 106:13838–13843. https://doi.org/10.1073/pnas.0907008106

Cocucci E, Aguet F, Boulant S, Kirchhausen T (2012) The first five seconds in the life of a clathrin-coated pit. Cell 150:495–507. https://doi.org/10.1016/j.cell.2012.05.047

Conway AE, Scotland PB, Lavau CP, Wechsler DS (2013) A CALM-derived nuclear export signal is essential for CALM-AF10-mediated leukemogenesis. Blood 121:4758–4768. https://doi.org/10.1182/blood-2012-06-435792

Corneveaux JJ, Myers AJ, Allen AN et al (2010) Association of CR1, CLU and PICALM with Alzheimer's disease in a cohort of clinically characterized and neuropathologically verified individuals. Hum Mol Genet 19:3295–3301. https://doi.org/10.1093/hmg/ddq221

De Craene J-O, Ripp R, Lecompte O et al (2012) Evolutionary analysis of the ENTH/ANTH/VHS protein superfamily reveals a coevolution between membrane trafficking and metabolism. BMC Genomics 13:297. https://doi.org/10.1186/1471-2164-13-297

Dittman JS, Kaplan JM (2006) Factors regulating the abundance and localization of synaptobrevin in the plasma membrane. Proc Natl Acad Sci 103:11399–11404. https://doi.org/10.1073/pnas.0600784103

Doherty GJ, McMahon HT (2009) Mechanisms of endocytosis. Annu Rev Biochem 78:857–902. https://doi.org/10.1146/annurev.biochem.78.081307.110540

Dreyling MH, Martinez-Climent JA, Zheng M et al (1996) The t(10;11)(p13;q14) in the U937 cell line results in the fusion of the AF10 gene and CALM, encoding a new member of the AP-3 clathrin assembly protein family. Proc Natl Acad Sci U S A 93:4804–4809

Engqvist-Goldstein AE, Kessels MM, Chopra VS et al (1999) An actin-binding protein of the Sla2/Huntingtin interacting protein 1 family is a novel component of clathrin-coated pits and vesicles. J Cell Biol 147:1503–1518

Engqvist-Goldstein AEY, Zhang CX, Carreno S et al (2004) RNAi-mediated Hip1R silencing results in stable association between the endocytic machinery and the actin assembly machinery. Mol Biol Cell 15:1666–1679. https://doi.org/10.1091/mbc.E03-09-0639

Ford MG, Pearse BM, Higgins MK et al (2001) Simultaneous binding of PtdIns(4,5)P2 and clathrin by AP180 in the nucleation of clathrin lattices on membranes. Science 291:1051–1055. https://doi.org/10.1126/science.291.5506.1051

Ford MGJ, Mills IG, Peter BJ et al (2002) Curvature of clathrin-coated pits driven by epsin. Nature 419:361–366. https://doi.org/10.1038/nature01020

Furney SJ, Simmons A, Breen G et al (2011) Genome-wide association with MRI atrophy measures as a quantitative trait locus for Alzheimer's disease. Mol Psychiatry 16:1130–1138. https://doi.org/10.1038/mp.2010.123

Furuta N, Fujita N, Noda T et al (2010) Combinational soluble N-Ethylmaleimide-sensitive factor attachment protein receptor proteins VAMP8 and Vti1b mediate fusion of antimicrobial and canonical Autophagosomes with lysosomes. Mol Biol Cell 21:1001–1010. https://doi.org/10.1091/mbc.E09-08-0693

Gleisner M, Kroppen B, Fricke C et al (2016) Epsin N-terminal homology domain (ENTH) activity as a function of membrane tension. J Biol Chem 291:19953–19961. https://doi.org/10.1074/jbc.M116.731612

Hao W, Tan Z, Prasad K et al (1997) Regulation of AP-3 function by inositides. Identification of phosphatidylinositol 3,4,5-trisphosphate as a potent ligand. J Biol Chem 272:6393–6398

Hao W, Luo Z, Zheng L et al (1999) AP180 and AP-2 interact directly in a complex that cooperatively assembles clathrin. J Biol Chem 274:22785–22794

Harold D, Abraham R, Hollingworth P et al (2009) Genome-wide association study identifies variants at CLU and PICALM associated with Alzheimer's disease. Nat Genet 41:1088–1093. https://doi.org/10.1038/ng.440

Hawryluk MJ, Keyel PA, Mishra SK et al (2006) Epsin 1 is a polyubiquitin-selective clathrin-associated sorting protein. Traffic 7:262–281. https://doi.org/10.1111/j.1600-0854.2006.00383.x

Hayashi I, Takatori S, Urano Y et al (2012) Neutralization of the γ-secretase activity by monoclonal antibody against extracellular domain of nicastrin. Oncogene 31:787–798. https://doi.org/10.1038/onc.2011.265

Henne WM, Boucrot E, Meinecke M et al (2010) FCHo proteins are nucleators of clathrin-mediated endocytosis. Science 328:1281–1284. https://doi.org/10.1126/science.1188462

Hirst J, Motley A, Harasaki K et al (2003) EpsinR: an ENTH domain-containing protein that interacts with AP-1. Mol Biol Cell 14:625–641. https://doi.org/10.1091/mbc.E02-09-0552

Hirst J, Miller SE, Taylor MJ et al (2004) EpsinR is an adaptor for the SNARE protein Vti1b. Mol Biol Cell 15:5593–5602. https://doi.org/10.1091/mbc.E04-06-0468

Hom RA, Vora M, Regner M et al (2007) pH-dependent binding of the Epsin ENTH domain and the AP180 ANTH domain to PI(4,5)P2-containing bilayers. J Mol Biol 373:412–423. https://doi.org/10.1016/j.jmb.2007.08.016

Huang KM, D'Hondt K, Riezman H, Lemmon SK (1999) Clathrin functions in the absence of heterotetrameric adaptors and AP180-related proteins in yeast. EMBO J 18:3897–3908. https://doi.org/10.1093/emboj/18.14.3897

Hyman J, Chen H, Di Fiore PP et al (2000) Epsin 1 undergoes nucleocytosolic shuttling and its eps15 interactor NH(2)-terminal homology (ENTH) domain, structurally similar to Armadillo and HEAT repeats, interacts with the transcription factor promyelocytic leukemia Zn(2)+ finger protein (PLZF). J Cell Biol 149:537–546

Hyun TS, Rao DS, Saint-Dic D et al (2004) HIP1 and HIP1r stabilize receptor tyrosine kinases and bind 3-Phosphoinositides via Epsin N-terminal homology domains. J Biol Chem 279:14294–14306. https://doi.org/10.1074/jbc.M312645200

Ishikawa Y, Maeda M, Pasham M et al (2015) Role of the clathrin adaptor PICALM in normal hematopoiesis and polycythemia vera pathophysiology. Haematologica 100:439–451. https://doi.org/10.3324/haematol.2014.119537

Itakura E, Kishi-Itakura C, Mizushima N (2012) The hairpin-type tail-anchored SNARE Syntaxin 17 targets to Autophagosomes for fusion with endosomes/lysosomes. Cell 151:1256–1269. https://doi.org/10.1016/j.cell.2012.11.001

Itoh T, De Camilli P (2006) BAR, F-BAR (EFC) and ENTH/ANTH domains in the regulation of membrane-cytosol interfaces and membrane curvature. Biochim Biophys Acta 1761:897–912. https://doi.org/10.1016/j.bbalip.2006.06.015

Itoh T, Koshiba S, Kigawa T et al (2001) Role of the ENTH domain in phosphatidylinositol-4,5-bisphosphate binding and endocytosis. Science 291:1047–1051. https://doi.org/10.1126/science.291.5506.1047

Jackson LP, Kelly BT, McCoy AJ et al (2010) A large-scale conformational change couples membrane recruitment to cargo binding in the AP2 clathrin adaptor complex. Cell 141:1220–1229. https://doi.org/10.1016/j.cell.2010.05.006

Jones EL, Mok K, Hanney M et al (2013) Evidence that PICALM affects age at onset of Alzheimer's dementia in Down syndrome. Neurobiol Aging 34:2441.e1–2441.e5. https://doi.org/10.1016/j.neurobiolaging.2013.03.018

Jonsson T, Atwal JK, Steinberg S et al (2012) A mutation in APP protects against Alzheimer's disease and age-related cognitive decline. Nature 488:96–99. https://doi.org/10.1038/nature11283

Kaksonen M, Roux A (2018) Mechanisms of clathrin-mediated endocytosis. Nat Rev Mol Cell Biol 19:313. https://doi.org/10.1038/nrm.2017.132

Kalthoff C, Alves J, Urbanke C et al (2002) Unusual structural organization of the endocytic proteins AP180 and epsin 1. J Biol Chem 277:8209–8216. https://doi.org/10.1074/jbc.M111587200

Kanatsu K, Morohashi Y, Suzuki M et al (2014) Decreased CALM expression reduces Aβ42 to total Aβ ratio through clathrin-mediated endocytosis of γ-secretase. Nat Commun 5:3386. https://doi.org/10.1038/ncomms4386

Kanatsu K, Hori Y, Takatori S et al (2016) Partial loss of CALM function reduces Aβ42 production and amyloid deposition in vivo. Hum Mol Genet 25:3988–3997. https://doi.org/10.1093/hmg/ddw239

Kauwe JSK, Cruchaga C, Karch CM et al (2011) Fine mapping of genetic variants in BIN1, CLU, CR1 and PICALM for association with cerebrospinal fluid biomarkers for Alzheimer's disease. PLoS One 6: e15918. https://doi.org/10.1371/journal.pone.0015918

Kay BK, Yamabhai M, Wendland B, Emr SD (1999) Identification of a novel domain shared by putative components of the endocytic and cytoskeletal machinery. Protein Sci 8:435–438. https://doi.org/10.1110/ps.8.2.435

Kovall RA, Gebelein B, Sprinzak D, Kopan R (2017) The canonical notch signaling pathway: structural and biochemical insights into shape, sugar, and force. Dev Cell 41:228–241. https://doi.org/10.1016/j.devcel.2017.04.001

Lambert J-C, Heath S, Even G et al (2009) Genome-wide association study identifies variants at CLU and CR1 associated with Alzheimer's disease. Nat Genet 41:1094–1099. https://doi.org/10.1038/ng.439

Lambert J-C, Zelenika D, Hiltunen M et al (2011) Evidence of the association of BIN1 and PICALM with the AD risk in contrasting European populations. Neurobiol Aging 32:756.e11–7756.15. https://doi.org/10.1016/j.neurobiolaging.2010.11.022

Loerke D, Mettlen M, Yarar D et al (2009) Cargo and dynamin regulate clathrin-coated pit maturation. PLoS Biol 7:e1000057. https://doi.org/10.1371/journal.pbio.1000057

Ma L, Umasankar PK, Wrobel AG et al (2016) Transient Fcho1/2·Eps15/R·AP-2 nanoclusters prime the AP-2 clathrin adaptor for cargo binding. Dev Cell 37 (5):428–443. https://doi.org/10.1016/j.devcel.2016.05.003

Meinecke M, Boucrot E, Camdere G et al (2013) Cooperative recruitment of dynamin and BIN/amphiphysin/Rvs (BAR) domain-containing proteins leads to GTP-dependent membrane scission. J Biol Chem 288:6651–6661. https://doi.org/10.1074/jbc.M112.444869

Meloty-Kapella L, Shergill B, Kuon J et al (2012) Notch ligand endocytosis generates mechanical pulling force dependent on dynamin, epsins, and actin. Dev Cell 22:1299–1312. https://doi.org/10.1016/j.devcel.2012.04.005

Melville SA, Buros J, Parrado AR et al (2012) Multiple loci influencing hippocampal degeneration identified by genome scan. Ann Neurol 72:65–75. https://doi.org/10.1002/ana.23644

Messa M, Fernández-Busnadiego R, Sun EW et al (2014) Epsin deficiency impairs endocytosis by stalling the actin-dependent invagination of endocytic clathrin-coated pits. elife 3:e03311. https://doi.org/10.7554/eLife.03311

Metzler M, Li B, Gan L et al (2003) Disruption of the endocytic protein HIP1 results in neurological deficits and decreased AMPA receptor trafficking. EMBO J 22:3254–3266. https://doi.org/10.1093/emboj/cdg334

Meyerholz A, Hinrichsen L, Esk P et al (2005) Effect of clathrin assembly lymphoid myeloid leukemia protein depletion on clathrin coat formation. Traffic 6 (12):1225–1234. https://doi.org/10.1111/j.1600-0854.2005.00355.x

Miller SE, Collins BM, McCoy AJ et al (2007) A SNARE-adaptor interaction is a new mode of cargo recognition in clathrin-coated vesicles. Nature 450:570–574. https://doi.org/10.1038/nature06353

Miller SE, Sahlender DA, Graham SC et al (2011) The molecular basis for the endocytosis of small R-SNAREs by the clathrin adaptor CALM. Cell 147:1118–1131. https://doi.org/10.1016/j.cell.2011.10.038

Miller SE, Mathiasen S, Bright NA et al (2015) CALM regulates clathrin-coated vesicle size and maturation by directly sensing and driving membrane curvature. Dev Cell 33:163–175. https://doi.org/10.1016/j.devcel.2015.03.002

Mills IG, Praefcke GJK, Vallis Y et al (2003) EpsinR: an AP1/clathrin interacting protein involved in vesicle trafficking. J Cell Biol 160:213–222. https://doi.org/10.1083/jcb.200208023

Moreau K, Fleming A, Imarisio S et al (2014) PICALM modulates autophagy activity and tau accumulation. Nat Commun 5:4998. https://doi.org/10.1038/ncomms5998

Morgan JR, Prasad K, Jin S et al (2003) Eps15 homology domain-NPF motif interactions regulate clathrin coat assembly during synaptic vesicle recycling. J Biol Chem 278:33583–33592. https://doi.org/10.1074/jbc.M304346200

Naj AC, Jun G, Beecham GW et al (2011) Common variants at MS4A4/MS4A6E, CD2AP, CD33 and EPHA1 are associated with late-onset Alzheimer's disease. Nat Genet 43:436–441. https://doi.org/10.1038/ng.801

Naj AC, Jun G, Reitz C et al (2014) Effects of multiple genetic loci on age at onset in late-onset Alzheimer disease. JAMA Neurol 71:1394. https://doi.org/10.1001/jamaneurol.2014.1491

Nonet ML, Holgado AM, Brewer F et al (1999) UNC-11, a Caenorhabditis elegans AP180 homologue, regulates the size and protein composition of synaptic vesicles. Mol Biol Cell 10:2343–2360

Norris FA, Ungewickell E, Majerus PW (1995) Inositol hexakisphosphate binds to clathrin assembly protein 3 (AP-3/AP180) and inhibits clathrin cage assembly in vitro. J Biol Chem 270:214–217

Okada Y, Feng Q, Lin Y et al (2005) hDOT1L links histone methylation to leukemogenesis. Cell 121:167–178. https://doi.org/10.1016/j.cell.2005.02.020

Okada Y, Jiang Q, Lemieux M et al (2006) Leukaemic transformation by CALM–AF10 involves upregulation of Hoxa5 by hDOT1L. Nat Cell Biol 8:1017–1024. https://doi.org/10.1038/ncb1464

Oravecz-Wilson KI, Kiel MJ, Li L et al (2004) Huntingtin interacting protein 1 mutations lead to abnormal hematopoiesis, spinal defects and cataracts. Hum Mol Genet 13:851–867. https://doi.org/10.1093/hmg/ddh102

Overstreet E, Chen X, Wendland B, Fischer JA (2003) Either part of a Drosophila epsin protein, divided after the ENTH domain, functions in endocytosis of delta in the developing eye. Curr Biol 13:854–860

Overstreet E, Fitch E, Fischer JA (2004) Fat facets and liquid facets promote delta endocytosis and delta signaling in the signaling cells. Development 131:5355–5366. https://doi.org/10.1242/dev.01434

Parikh I, Fardo DW, Estus S (2014) Genetics of PICALM expression and Alzheimer's disease. PLoS One 9: e91242. https://doi.org/10.1371/journal.pone.0091242

Pimm J, McQuillin A, Thirumalai S et al (2005) The epsin 4 gene on chromosome 5q, which encodes the clathrin-associated protein Enthoprotin, is involved in the genetic susceptibility to schizophrenia. Am J Hum Genet 76:902–907. https://doi.org/10.1086/430095

Ponomareva NV, Andreeva TV, Protasova MS et al (2017) Quantitative EEG during normal aging: association with the Alzheimer's disease genetic risk variant in PICALM gene. Neurobiol Aging 51:177.e1–177.e8. https://doi.org/10.1016/j.neurobiolaging.2016.12.010

Posor Y, Eichhorn-Gruenig M, Puchkov D et al (2013) Spatiotemporal control of endocytosis by phosphatidylinositol-3,4-bisphosphate. Nature 499:233–237. https://doi.org/10.1038/nature12360

Prasad K, Lippoldt RE (1988) Molecular characterization of the AP180 coated vesicle assembly protein. Biochemistry 27:6098–6104

Puri C, Renna M, Bento CF et al (2013) Diverse autophagosome membrane sources coalesce in recycling endosomes. Cell 154:1285–1299. https://doi.org/10.1016/j.cell.2013.08.044

Rao DS, Chang JC, Kumar PD et al (2001) Huntingtin interacting protein 1 is a clathrin coat binding protein required for differentiation of late spermatogenic progenitors. Mol Cell Biol 21:7796–7806. https://doi.org/10.1128/MCB.21.22.7796-7806.2001

Robinson MS (2015) Forty years of clathrin-coated vesicles. Traffic 16:1210–1238. https://doi.org/10.1111/tra.12335

Rosenthal JA, Chen H, Slepnev VI et al (1999) The epsins define a family of proteins that interact with components of the clathrin coat and contain a new protein module. J Biol Chem 274:33959–33965. https://doi.org/10.1074/jbc.274.48.33959

Sato Y, Yoshikawa A, Mimura H et al (2009) Structural basis for speci c recognition of Lys 63-linked polyubiquitin chains by tandem UIMs of RAP80. EMBO J 28:1–8. https://doi.org/10.1038/emboj.2009.160

Schjeide B-MM, Schnack C, Lambert J-C et al (2011) The role of clusterin, complement receptor 1, and phosphatidylinositol binding clathrin assembly protein in Alzheimer disease risk and cerebrospinal fluid biomarker levels. Arch Gen Psychiatry 68:207. https://doi.org/10.1001/archgenpsychiatry.2010.196

Seshadri S, Fitzpatrick AL, Ikram MA et al (2010) Genome-wide analysis of genetic loci associated with Alzheimer disease. JAMA 303:1832–1840. https://doi.org/10.1001/jama.2010.574

Sevigny J, Chiao P, Bussière T et al (2016) The antibody aducanumab reduces Aβ plaques in Alzheimer's disease. Nature 537:50–56. https://doi.org/10.1038/nature19323

Sims JJ, Cohen RE (2009) Linkage-specific avidity defines the lysine 63-linked polyubiquitin-binding preference of Rap80. Mol Cell 33:775–783. https://doi.org/10.1016/j.molcel.2009.02.011

Skruzny M, Desfosses A, Skruzny M et al (2015) An organized co-assembly of clathrin adaptors is essential for endocytosis article an organized co-assembly of clathrin adaptors is essential for endocytosis. Dev Cell 33:150–162. https://doi.org/10.1016/j.devcel.2015.02.023

Stachowiak JC, Schmid EM, Ryan CJ et al (2012) Membrane bending by protein–protein crowding. Nat Cell Biol 14:944–949. https://doi.org/10.1038/ncb2561

Stahelin RV, Long F, Peter BJ et al (2003) Contrasting membrane interaction mechanisms of AP180 N-terminal homology (ANTH) and epsin N-terminal homology (ENTH) domains. J Biol Chem 278:28993–28999. https://doi.org/10.1074/jbc.M302865200

Stephenson NL, Avis JM (2012) Direct observation of proteolytic cleavage at the S2 site upon forced unfolding of the notch negative regulatory region. Proc Natl Acad Sci U S A 109:E2757–E2765. https://doi.org/10.1073/pnas.1205788109

Suzuki M, Tanaka H, Tanimura A et al (2012) The clathrin assembly protein PICALM is required for erythroid maturation and transferrin internalization in mice. PLoS One 7:e31854. https://doi.org/10.1371/journal.pone.0031854

Suzuki M, Yamagata K, Shino M et al (2014) Nuclear export signal within CALM is necessary for CALM-AF10-induced leukemia. Cancer Sci 105:315–323. https://doi.org/10.1111/cas.12347

Sweet RA, Seltman H, Emanuel JE et al (2012) Effect of Alzheimer's disease risk genes on trajectories of cognitive function in the cardiovascular health study. Am J Psychiatry 169:954–962. https://doi.org/10.1176/appi.ajp.2012.11121815

Sweitzer SM, Hinshaw JE (1998) Dynamin undergoes a GTP-dependent conformational change causing vesiculation. Cell 93:1021–1029. https://doi.org/10.1016/S0092-8674(00)81207-6

Takasugi N, Tomita T, Hayashi I et al (2003) The role of presenilin cofactors in the γ-secretase complex. Nature 422:438–441. https://doi.org/10.1038/nature01506

Takei K, Slepnev VI, Haucke V, De Camilli P (1999) Functional partnership between amphiphysin and dynamin in clathrin-mediatedendocytosis. Nat Cell Biol 1:33–39. https://doi.org/10.1038/9004

Taylor MJ, Perrais D, Merrifield CJ (2011) A high precision survey of the molecular dynamics of mammalian clathrin-mediated endocytosis. PLoS Biol 9:e1000604. https://doi.org/10.1371/journal.pbio.1000604

Thambisetty M, An Y, Tanaka T (2013) Alzheimer's disease risk genes and the age-at-onset phenotype. Neurobiol Aging 34:2696.e1–2692696.e5. doi: https://doi.org/10.1016/j.neurobiolaging.2013.05.028

Thomas RS, Henson A, Gerrish A et al (2016) Decreasing the expression of PICALM reduces endocytosis and the activity of β-secretase: implications for Alzheimer's disease. BMC Neurosci 17:50. https://doi.org/10.1186/s12868-016-0288-1

Tian X, Hansen D, Schedl T, Skeath JB (2004) Epsin potentiates Notch pathway activity in Drosophila and C. elegans. Development 131:5807–5815. https://doi.org/10.1242/dev.01459

Traub LM (2009) Tickets to ride: selecting cargo for clathrin-regulated internalization. Nat Rev Mol Cell Biol 10:583–596. https://doi.org/10.1038/nrm2751

Vardarajan BN, Ghani M, Kahn A et al (2015) Rare coding mutations identified by sequencing of Alzheimer disease genome-wide association studies loci. Ann Neurol 78:487–498. https://doi.org/10.1002/ana.24466

Waelter S, Scherzinger E, Hasenbank R et al (2001) The huntingtin interacting protein HIP1 is a clathrin and alpha-adaptin-binding protein involved in receptor-mediated endocytosis. Hum Mol Genet 10:1807–1817

Wang W, Struhl G (2004) Drosophila epsin mediates a select endocytic pathway that DSL ligands must enter to activate Notch. Development 131:5367–5380. https://doi.org/10.1242/dev.01413

Wanker EE, Rovira C, Scherzinger E et al (1997) HIP-1: a huntingtin interacting protein isolated by the yeast two-hybrid system. Hum Mol Genet 6:487–495

Wasiak S, Legendre-Guillemin V, Puertollano R et al (2002) Enthoprotin: a novel clathrin-associated protein identified through subcellular proteomics. J Cell Biol 158:855–862. https://doi.org/10.1083/jcb.200205078

Wood LA, Royle SJ (2015) Zero tolerance: amphipathic helices in endocytosis. Dev Cell 33:119–120. https://doi.org/10.1016/j.devcel.2015.04.007

Xiao Q, Gil S, Yan P et al (2012) Role of phosphatidylinositol clathrin assembly lymphoid-myeloid leukemia (PICALM) in intracellular amyloid precursor protein (APP) processing and amyloid plaque pathogenesis. J Biol Chem 287:21279–21289. https://doi.org/10.1074/jbc.M111.338376

Xu W, Wang H-F, Tan L et al (2016) The impact of PICALM genetic variations on reserve capacity of posterior cingulate in AD continuum. Sci Rep 6:24480. https://doi.org/10.1038/srep24480

Ye W, Ali N, Bembenek ME et al (1995) Inhibition of clathrin assembly by high affinity binding of specific inositol polyphosphates to the synapse-specific clathrin assembly protein AP-3. J Biol Chem 270:1564–1568

Zhao Z, Sagare AP, Ma Q et al (2015) Central role for PICALM in amyloid-β blood-brain barrier transcytosis and clearance. Nat Neurosci 18:978–987. https://doi.org/10.1038/nn.4025

Adv Exp Med Biol - Protein Reviews (2019) 20: 77–137
https://doi.org/10.1007/5584_2018_288
Published online: 28 November 2018

Polyphosphoinositide-Binding Domains: Insights from Peripheral Membrane and Lipid-Transfer Proteins

Joshua G. Pemberton and Tamas Balla

Abstract

Within eukaryotic cells, biochemical reactions need to be organized on the surface of membrane compartments that use distinct lipid constituents to dynamically modulate the functions of integral proteins or influence the selective recruitment of peripheral membrane effectors. As a result of these complex interactions, a variety of human pathologies can be traced back to improper communication between proteins and membrane surfaces; either due to mutations that directly alter protein structure or as a result of changes in membrane lipid composition. Among the known structural lipids found in cellular membranes, phosphatidylinositol (PtdIns) is unique in that it also serves as the membrane-anchored precursor of low-abundance regulatory lipids, the polyphosphoinositides (PPIn), which have restricted distributions within specific subcellular compartments. The ability of PPIn lipids to function as signaling platforms relies on both non-specific electrostatic interactions and the selective stereospecific recognition of PPIn headgroups by specialized protein folds. In this chapter, we will attempt to summarize the structural diversity of modular PPIn-interacting domains that facilitate the reversible recruitment and conformational regulation of peripheral membrane proteins. Outside of protein folds capable of capturing PPIn headgroups at the membrane interface, recent studies detailing the selective binding and bilayer extraction of PPIn species by unique functional domains within specific families of lipid-transfer proteins will also be highlighted. Overall, this overview will help to outline the fundamental physiochemical mechanisms that facilitate localized interactions between PPIn lipids and the wide-variety of PPIn-binding proteins that are essential for the coordinate regulation of cellular metabolism and membrane dynamics.

The original version of this chapter was revised: Copyright holder was corrected. The correction to this chapter is available at https://doi.org/10.1007/5584_2019_390

J. G. Pemberton and T. Balla (✉)
Section on Molecular Signal Transduction, Eunice Kennedy Shriver National Institute of Child Health and Human Development, National Institutes of Health, Bethesda, MD, USA
e-mail: ballat@mail.nih.gov

Keywords

ANTH · BAR · C2 · Cellular trafficking · ENTH · FERM · FYVE · GLUE · GRAM · Lipid-binding domains · Membrane biology · Oxysterol-binding protein-related protein · PDZ · Phosphatidylinositol · Phosphatidylinositol-transfer protein · Phosphoinositides · Phox homology · Pleckstrin homology · PROPPINs · PTB · Signal transduction · Tubby

1   Introduction

The structural integrity and dynamic remodeling of biological membranes relies on reciprocal interactions between membrane proteins and smaller amphipathic lipids. In general, membrane components exist as part of an asymmetric bilayer consisting of functionally distinct inner and outer leaflets; although, some subcellular compartments may function as lipid monolayers (Holthuis and Menon 2014; Drin 2014). Even small modifications to the relative abundance or identity of either the protein or lipid constituents present can significantly alter the intrinsic physiochemical properties of cellular membranes with important consequences for the activities of integral as well as peripheral membrane proteins (van Meer et al. 2008; Drin 2014). Consequently, to perform specialized functions, eukaryotic cells have developed distinct membrane compartments with unique local properties that are characterized by specific protein and lipid compositions (Holthuis and Menon 2014). Despite the need for functional heterogeneity, throughout subcellular membranes, phospholipids are the most abundant structural components and are defined by the presence of a polar headgroup and two hydrophobic acyl tails, which can differ greatly in both their chain length as well as degree of hydrocarbon saturation across different membrane environments (Bigay and Antonny 2012; Barelli and Antonny 2016). Modifications to this general phospholipid structure, especially alterations to the surface-exposed headgroup, endow certain species with unique biophysical characteristics that have been shown to directly influence general membrane features such as fluidity, thickness, lipid-packing density, and surface charge (van Meer et al. 2008; Holthuis and Menon 2014; Jackson et al. 2016). Overall, cellular membrane dynamics relies on the functional diversity of membrane lipids and the phospholipid composition, in particular, plays an important role in the coordinate regulation of signaling and trafficking functions throughout subcellular membrane compartments.

Of the known membrane lipid species, phosphatidylinositol (PtdIns) is unique in that is not only functions as an essential structural phospholipid, but it also serves as the precursor for important low-abundance regulatory lipids that are collectively referred to as polyphosphoinositides (PPIn; Balla 2013). An essential mechanism involved in the regulation of diverse cellular functions depends on the reversible recruitment of peripheral cytosolic proteins or macromolecular complexes to the surface of specific subcellular membranes with high temporal resolution. Dynamic recruitment of peripheral protein effectors is orchestrated, in large part, by the local production of PPIn lipids through the addition of phosphate moieties to PtdIns at one or more of the hydroxyl-groups present at the 3-, 4-, or 5-position of the inositol ring. PtdIns-specific phosphorylation events are tightly controlled by highly conserved substrate-selective, as well as position-specific, lipid kinases and phosphatases that function to generate seven distinct membrane-embedded PPIn species; including mono-, bis-, or tris-phosphorylated derivatives (Balla 2013). Although variability exists, unique PPIn lipids appear to localize to overlapping, as well as distinct, membrane surfaces and can recruit different intracellular effectors that not only contribute to the initiation of signaling responses, but can also function to define membrane identity or control local membrane dynamics (Hammond and Balla 2015; Schink et al. 2016). In addition, many of the regulatory functions attributed to PPIn species can occur indirectly as a result of actions on cytoskeletal remodeling (Saarikangas et al. 2010; Bezanilla et al. 2015; Senju et al. 2017) or through the allosteric regulation of transmembrane-spanning receptors, ion channels, or transporters (Hilgemann et al. 2001; Gamper and Shapiro 2007; Barrera et al. 2013; Hille et al. 2015; Hedger and Sansom 2016).

Due to the expansive cellular roles performed by PPIn lipids, both as structural components and sites for protein-membrane interactions, it is not surprising that many human pathologies are the result of perturbations in PPIn production or clearance, including direct contributions of altered PPIn dynamics to: cancer, diabetes, degenerative myopathies and neuropathies (Pendaries et al. 2003; Wymann and Schneiter 2008; McCrea and De Camilli 2009; Bunney and Katan 2010; Balla 2013; Thapa et al. 2016), as well as being part of the invasion or evasion strategies employed by infectious agents (Altan-Bonnet and Balla 2012; Payrastre et al. 2012; Pizarro-Cerda et al. 2015; Altan-Bonnet 2017). To better understand how PPIn production regulates cellular functions, recent studies have tried to define the biosynthetic and inter-conversion pathways that modulate PPIn turnover as well as characterize the PPIn-binding domains responsible for recognizing distinct PPIn species within membrane compartments. Overall, this chapter will attempt to summarize the general mechanisms controlling PPIn recognition by peripheral membrane proteins and introduce the structural diversity of PPIn-interacting protein domains found in both prokaryotic and eukaryotic organisms. Collectively, by comparing the PPIn recognition systems used by peripheral proteins from bacterial and animal models, we hope to provide a foundation for understanding the general molecular processes that allow for inositol-containing lipids to function as membrane recognition sites that contribute to the dynamic regulation of diverse biological processes throughout evolution.

2 Synthesis and Subcellular Distribution of PPIn Lipids

The numerous PPIn kinases and phosphatases, as well as the reversible recruitment of numerous PPIn-binding effectors, all contribute to the steady-state availability of membrane PPIn species. The complex processes governing PPIn metabolism begins with the synthesis of PtdIns by a single enzyme, PtdIns synthase (PIS); which is present as an integral membrane protein within the endoplasmic reticulum (ER) and catalyzes the conjugation of the *myo-* stereoisomer of inositol to a cytidine diphosphate (CDP)-activated diacylglycerol (DAG) backbone (Agranoff et al. 1958; Agranoff et al. 1969; Agranoff 2009).

Despite localizing to membranes of the ER, work from our group has shown that catalytically-active PIS is concentrated within a mobile ER-derived sub-compartment that may function to actively distribute PtdIns to subcellular membranes (Kim et al. 2011). Within cellular membranes, PtdIns represents roughly 10–20 mol % of total phospholipids; whereas, despite their important cellular roles, PPIn species only represent an estimated 2–5% of the available PtdIns and therefore only contributes to 0.2–1 mol% of membrane phospholipids (Lemmon 2008; Balla 2013; Vance 2015). However, it is important to mention that the relative amounts of PPIn lipids found within cells shows significant variations across species and even between cell types. Downstream of PtdIns production, PPIn lipids are continuously being turned over, but can also be concentrated within discrete subcellular compartments. Consequently, the rapid and localized production of PPIn lipids from the abundant membrane precursor PtdIns, can drastically increase the ratio of the target PPIn relative to the amount of a membrane-binding effector; making it possible to recruit a large amount of peripheral proteins without saturating the available PPIn headgroups. Sequential interconversion of PPIn species using substrate-selective enzymatic modifications may also confer a degree of biochemical processivity to the control of dynamic membrane signaling events (Cullen et al. 2001; Balla 2005; Botelho 2009). The coordinate production and targeted recognition of PPIn species is thought to be enhanced by recruiting macromolecular complexes containing PPIn kinases or phosphatases in close proximity to PPIn substrates or downstream effectors. An excellent example of this regulatory paradigm comes from a recent description of metabolic channeling of PPIn-mediated signaling by the multi-domain scaffold protein IQGAP1 (IQ motif-containing GTPase-activating protein 1; Choi et al. 2016). Specifically, IQGAP1 regulates the assembly and substrate presentation for three distinct PPIn kinases at the PM, which facilitates the sequential phosphorylation of PtdIns, to produce the important second messenger PtdIns 3,4,5-trisphosphate (PtdIns(3,4,5)P_3), and controls the activation of additional PtdIns(3,4,5) P$_3$-sensitive effectors that are also associated with IQGAP1 (Choi et al. 2016). Though of central importance for understanding PPIn biology, the regulation and cellular functions of the wide-variety of PPIn-modifying enzymes responsible for the production and inter-conversion of subcellular PPIn species will not be discussed at length in this chapter, but have been reviewed in depth elsewhere (Sasaki et al. 2009; Dyson et al. 2012; Balla 2013; Hsu and Mao 2015). However, it is clear that understanding the complexities associated with the control of PPIn metabolism will require additional investigations into the roles played by molecular scaffolds and regulatory protein-protein interactions on membrane surfaces.

Studies examining the subcellular localization of PPIn-modifying enzymes have provided some details on the potential landscape of PPIn species within subcellular compartments; however, to truly understand how dynamic changes in membrane PPIn composition occur, membrane-embedded PPIn species need to be visualized with high spatial and temporal resolution. In recent years, work from many laboratories, including our own, have contributed greatly to imaging breakthroughs that have been pivotal for the study of specific PPIn lipids in living cells. In particular, foundational studies using fluorescently-tagged constructs consisting of isolated PPIn-binding domains were able to selectively follow subcellular PtdIns 4,5-bisphosphate (PtdIns(4,5)P$_2$; Varnai and Balla 1998; Stauffer et al. 1998) or PtdIns(3,4,5)P$_3$ (Kontos et al. 1998; Varnai et al. 1999; Watton and Downward 1999; Servant et al. 2000) dynamics in real-time following receptor-dependent hydrolysis or class I PI3K activation, respectively. In addition to PtdIns(4,5)P$_2$ and PtdIns(3,4,5)P$_3$, selective lipid-binding probes have been established that can reliably visualize the PPIn species PtdIns 4-phosphate (PtdIns4P), PtdIns 3-phosphate (PtdIns3P), and PtdIns 3,4-bisphosphate (PtdIns (3,4)P$_2$); as well as for other important structural or signaling lipids such as DAG, phosphatidic acid (PtdOH), and phosphatidylserine (PtdSer; Hammond and Balla 2015; Varnai et al. 2017).

Taken together, these studies not only revealed important details regarding cellular PPIn metabolism and turnover, but also provided proof of concept for the use of membrane-binding domains as specific biosensors to map subcellular phospholipid compartments. The utility of PPIn-binding domains as biosensors for visualizing and quantifying membrane PPIn lipids has been discussed at length by our group previously (Hammond and Balla 2015; Varnai et al. 2017) and will not be the central focus of this chapter. Overall, work using selective lipid-binding probes, as well as more traditional biochemical approaches, reveal that PPIn lipids show a restricted subcellular distribution and that the enrichment of specific PPIn species occurs within distinct membrane compartments. Defining the localization of distinct PPIn species is of obvious importance for understanding the specialized functions of these lipids, and therefore we will briefly outline the PPIn-specific territories that have been mapped to discrete membrane compartments or larger organelles. Please be aware that although roles for nuclear PPIn lipids are emerging (Irvine 2003; Barlow et al. 2010; Martelli et al. 2011; Shah et al. 2013; Crowder et al. 2017), and certainly represent an exciting new area of PPIn biology for investigation, we will restrict our discussion to the distribution of PPIn lipids in cytosolic membranes as these PPIn pools are more clearly defined and also appear to function independently from the unique system of PPIn metabolism that functions within the nucleus (Hammond and Balla 2015).

Within mammalian cells, PtdIns4P and PtdIns (4,5)P$_2$ are the most abundant PPIn species, constituting approximately 2–5% of the total cellular pool of PtdIns-containing lipids (Balla et al. 1988; Xu et al. 2003). The majority of PtdIns(4,5) P$_2$ is found within the PM, although evidence for PtdIns(4,5)P$_2$-mediated regulation of effectors at the Golgi (Watt et al. 2002; De Matteis et al. 2002) and within the endolysosomal system has also been presented (Choi et al. 2015; Tan et al. 2015). PtdIns4P pools appear to be generated in various membrane compartments using distinct PI4K enzymes (Balla and Balla 2006; Boura and Nencka 2015; Dornan et al. 2016). Specifically, the PM pool of PtdIns4P that serves as the precursor for PtdIns(4,5)P$_2$ synthesis is generated primarily by the PI4KIIIα isoform (Balla et al. 2008; Nakatsu et al. 2012; Bojjireddy et al. 2014). Alternatively, Golgi pools of PtdIns4P are produced by PI4KIIIβ (Godi et al. 1999), with additional contributions from both type II enzymes, PI4KIIα and PI4KIIβ (Weixel et al. 2005; Wang et al. 2003), which are also responsible for producing PtdIns4P in the late endosomes (Hammond et al. 2014). Outside of PtdIns(4,5)P$_2$ and PtdIns4P, the remaining PPIn species only contribute a small amount to the total cellular fraction of inositol-containing lipids. Minor amounts of PtdIns(3,4,5)P$_3$ are found within the inner leaflet of the PM and only increases upon receptor-mediated activation of class I PI3Ks (Vanhaesebroeck et al. 2010; Burke and Williams 2015); but, even at maximal levels, PtdIns(3,4,5)P$_3$ only represent 2–5% of the PM PtdIns(4,5)P$_2$ (Hawkins et al. 1992; Toker and Cantley 1997). Dephosphorylation of PtdIns(3,4,5)P$_3$ by 5-phosphatases is thought to generate PtdIns(3,4)P$_2$ within the PM (Erneux et al. 2011; Ooms et al. 2015; Posor et al. 2013) and recent studies also suggest that PtdIns(3,4)P$_2$ may persist within PM-derived vesicles internalized during endocytosis (Posor et al. 2013; Ketel et al. 2016; Marat et al. 2017; Malek et al. 2017). Interestingly, results *in vitro* as well as *in vivo* indicate that class II PI3Ks preferentially phosphorylate PtdIns and, to a lesser extent, PtdIns4P to generate PtdIns3P and PtdIns(3,4)P$_2$, respectively (Arcaro 1998; Misawa et al. 1998; Falasca et al. 2007; Franco et al. 2014; Braccini et al. 2015). Despite much debate about the relative importance of the kinase- and phosphatase-dependent metabolic pathways, cellular studies strongly suggest that local production of PtdIns(3,4)P$_2$ in the endosomal system contributes to coordinate control of cellular signaling and membrane dynamics (Li and Marshall 2015; Hawkins and Stephens 2016; Marat and Haucke 2016); including important roles during clathrin- and endophilin-mediated endocytosis (Posor et al. 2013; Boucrot et al. 2015; Renard et al. 2015). Mono-phosphorylated PtdIns3P represents 20–30% of the cellular PtdIns4P and is found

primarily within the early endosomes (Gillooly et al. 2000) and in autophagosomes (Funderburk et al. 2010); with an additional report describing the presence of PtdIns3P in membranes of the Golgi and ER (Sarkes and Rameh 2010). The more enigmatic mono-phosphorylated PPIn species PtdIns 5-phosphate (PtdIns5P) is estimated at only 1% of the PtdIns4P levels in mammalian cells and subcellular fractionation suggests that the highest amount is localized to the PM (Sarkes and Rameh 2010). Additional enrichments of PtdIns5P may also be found in ER and Golgi membrane fractions (Sarkes and Rameh 2010) as well as in the early endosomes (Ramel et al. 2011). Lastly, PtdIns 3,5-bisphosphate (PtdIns(3,5)P$_2$) has the lowest abundance of the PPIn species found within mammalian cells, making up 1% of cellular PtdIns(4,5)P$_2$ (Zolov et al. 2012; Sbrissa et al. 2012). Attempts to localize PtdIns(3,5)P$_2$ has been hampered by a lack of specificity in the available biosensors (Hammond et al. 2015), but functional studies suggest that this minor PPIn species is important for the proper sorting of cargoes within the late endosome (Gary et al. 1998; Bonangelino et al. 2002). Taken together, the unique spatial distribution and selective enrichment of subcellular PPIn species highlights the utility of these regulatory lipids as sites for coordinating the dynamic recruitment of peripheral proteins to discrete membrane compartments. The rapid inter-conversion of PPIn species by PPIn-modifying enzymes may also enhance the spatiotemporal specificity of cellular responses that are controlled by localized PPIn production.

3 General Features of Membrane Binding by Peripheral Proteins

Biological membranes contain a variety of lipids, including the seven distinct PPIn species, which function to coordinate reciprocal interactions with diverse families of intracellular proteins and any associated small molecules or ions. The interfacial regions surrounding membrane bilayers consist of a complex mixture of water, lipid headgroups, backbone phosphates, and any polar portions of the fatty acyl chains (Lee 2003; Cho and Stahelin 2005; Pasenkiewicz-Gierula et al. 2016). The kinetics and energetics of membrane interactions are locally governed by the physiochemical properties of both the membrane and protein surfaces (Marsh 2008). In general, initial membrane association of proteins are driven by diffusional as well as electrostatic forces to establish transient collisional intermediates that can be reinforced to form tightly-bound intermolecular complexes by additional hydrogen-bonding or electrostatic interactions (Cho and Stahelin 2005; Whited and Johs 2015). While non-specific interactions with membrane surfaces based on charge complementarity are unlikely to be sufficient to anchor proteins with high affinity, the initial membrane adsorption during these relatively weak associations facilitate specific membrane-binding events by orienting the geometry of peripheral proteins relative to the interface and reducing the dimensionality of the interaction space to the simple two-dimensional membrane surface; effectively increasing the local protein concentration (Cho and Stahelin 2005; Mulgrew-Nesbitt et al. 2006; Fernandes et al. 2015). In some instances, initial membrane attachment can also facilitate interfacial penetration of hydrophobic or aromatic residues that surround the lipid-binding pocket into the hydrocarbon core of the bilayer (Yau et al. 1998; Killian and von Heijne 2000; Lomize et al. 2007). Without the added affinity provided by interactions with lipid headgroups, peripheral protein domains are unable to penetrate the interfacial or hydrocarbon regions of membrane leaflets due to the high energetic penalty of desolvation (Pogozheva et al. 2013; Stahelin et al. 2014). Although a combination of these membrane-targeting mechanisms are required, specific lipid coordination and any associated electrostatic or hydrophobic interactions are essential components that drive the membrane recruitment and activation of peripheral membrane proteins; especially those that are coordinated by the anionic and structurally-distinct PPIn lipids. However, it should also be mentioned that, in addition to selective interactions with membrane-embedded lipids, bulk compositions

or structural features, such as the charge or degree of membrane curvature, also contribute to the recognition of specific membrane surfaces by peripheral proteins (Lee 2003; McMahon and Gallop 2005; Zimmerberg and Kozlov 2006; Marsh 2008; Baumgart et al. 2011). As we will discuss below, PPIn lipids have emerged as an essential platform for the specific interaction between a wide array of lipid-binding protein domains in almost all membrane compartments.

4   Principles of PPIn-Protein Interactions

Given the variety of regulatory mechanisms that contribute to the local control of PPIn metabolism and turnover, it is not difficult to imagine that the PPIn-binding surfaces utilized by protein effectors for the reversible recruitment to specific PPIn isomers are similarly diverse. The unique structures that have been described for peripheral membrane protein domains have revealed many diverse modes for membrane binding that result in different PPIn specificities and membrane-binding orientations. Despite these complexities, it is possible to uncover specific themes that control the dynamic regulation of cytosolic effectors by membrane-embedded PPIn lipids. As the title of this chapter would suggest, the predominant structural features of PPIn-regulated proteins are specialized membrane-binding modules that allow for the selective recognition of individual PPIn species. However, before discussing the molecular diversity of PPIn-interacting protein domains in more detail, we will first highlight some of the general principles that guide protein interactions with membrane PPIn lipids. Fundamental features contributing to communication between PPIn lipids and peripheral proteins were recently detailed by our group (Hammond and Balla 2015), as well as others (Kutateladze 2010; Moravcevic et al. 2012; Stahelin et al. 2014; Choy et al. 2017), and will be summarized below.

To reliably regulate protein functions in time and space, the interactions between membrane PPIn species and proteins should be governed by high-affinity and stoichiometric PPIn-binding; most characteristically through a dedicated PPIn-recognition domain. However, this principle is not universal and not all PPIn-binding domains have the requisite affinity to sufficiently dictate protein localization in isolation. Consequently, although some PPIn interactions that have been identified are not solely responsible for dictating membrane localization, PPIn-binding may add the necessary avidity to a secondary or co-incident interaction(s) that can act together to facilitate peripheral membrane protein recruitment. The idea that PPIn- or other lipid-binding modules can function to complement other protein-protein or protein-lipid interactions at membrane surfaces has been characterized in a variety of signaling contexts (Balla 2005; Carlton and Cullen 2005; Moravcevic et al. 2012). In particular, PPIn-assisted membrane binding, which capitalizes on combinatorial interactions or scaffolding functions, is best exemplified by the regulation of the Arf (ADP-ribosylation factor) and Arl (Arf-like) superfamily of small guanine nucleotide-binding proteins (Godi et al. 2004; DiNitto et al. 2007; Liu et al. 2014; Jian et al. 2015). In diverse membrane compartments, the integration of coincident signals from PPIn- and protein-interactions can be used to effectively tune the regulatory functions of peripheral proteins or macromolecular complexes bound at the membrane interface. However, in addition to simple roles as membrane scaffolds, PPIn-coordination can also contribute to the complex control of protein conformational dynamics. Allosteric regulation of protein effectors by PPIn lipids has been characterized in detailed mechanistic studies of protein kinase B (Akt; Calleja et al. 2007, 2009a, b, 2012; Ebner et al. 2017), PTEN (phosphatase and tensin homolog deleted on chromosome ten; Campbell et al. 2003; Iijima et al. 2004; Walker et al. 2004; Redfern et al. 2008; Wei et al. 2015), BTK (Bruton's tyrosine kinase; Joseph et al. 2017), Arf GTPase-activating proteins (GAPs; Kam et al. 2000; Campa et al. 2009), and Arf guanine nucleotide exchange factors (GEFs; Malaby et al. 2013); as well as several examples describing regulatory interactions between PPIn lipids and

transmembrane-spanning ion channels or receptors (Hilgemann and Ball 1996; Huang et al. 1998; Rohacs et al. 2003; Whorton and MacKinnon 2011; Barrera et al. 2013; Laganowsky et al. 2014). Conformational gating by membrane PPIn species is an emerging field that might be most important for controlling the functions of lipid transfer proteins that use PPIn lipids for membrane recognition and as transport cargoes. However, the regulatory role for PPIn recognition in non-vesicular lipid transport is not yet fully understood, but the communication between membrane PPIn species and the binding domains found within the cellular lipid transfer machinery will be addressed further in Sect. 7 of this chapter.

Independent of stereospecific lipid coordination, anionic membrane lipids, including PPIn species and PtdSer, can contribute to the membrane targeting of proteins possessing functionalized regions enriched with basic amino acid residues through non-specific electrostatic interactions (Heo et al. 2006; Hammond et al. 2012). Classical examples of proteins that interact with PPIn lipids using unstructured polybasic stretches, which are not organized within a characteristic motif, include the MARCKS (myristoylated alanine-rich C-kinase substrate) proteins (Wang et al. 2002; Wang et al. 2004; Gambhir et al. 2004), c-Src (cellular-sarcoma non-receptor protein tyrosine kinase; Yeung et al. 2006), K-Ras (Kirsten-rat sarcoma; Heo et al. 2006; Gulyas et al. 2017), and GAP43 (growth-associated protein 43; McLaughlin and Murray 2005). Interestingly, more recently, a unique structured membrane-binding module found at the C-terminus of the eukaryotic MARK (MAP/microtubule affinity-regulating kinases) family of kinases, called the KA1 (kinase-associated 1) domain, has also been shown to effectively sense membrane charge through cooperation between distinct basic regions on the membrane-binding surface (Moravcevic et al. 2010; Emptage et al. 2017a, b). Similar to unstructured polybasic segments, KA1 domains do not appear to distinguish between different anionic phospholipids *in vitro* or *in vivo* (Moravcevic et al. 2010; Hammond et al. 2012). Consequently, although important for controlling protein localization during diverse cellular processes, particularly within the unique electrostatic environment of the PM, these simple charge-based interactions capitalize on the general character of PPIn headgroups and will not be discussed at length in this chapter. Overall, using a combination of electrostatic interactions and PPIn-specific recognition, membrane-embedded PPIn species, though rare among phospholipids, function as central regulators of cellular physiology by orchestrating the dynamic recruitment and activation of diverse families of protein effectors at the membrane interface.

5 The Diversity of Eukaryotic PPIn-Binding Domains

There are a wide variety of well-folded modular domains that have evolved to selectively interact with PPIn-containing membranes through a combination of non-specific electrostatic interactions and the stereoselective coordination of PPIn headgroups. In this section, we will introduce the diversity of PPIn-recognizing protein folds, including descriptions of the PH (Pleckstrin Homology), PTB (phosphotyrosine-binding), PDZ (PSD-95, Discs Large, and ZO-1), GRAM (glucosyltransferases, Rab-like GTPase activators, and myotubularins), GLUE (GRAM-like ubiquitin-binding in EAP45), FERM (4.1, ezrin, radixin, and moesin), PX (phox homology), FYVE (Fab1p, YOTB, Vac1p, and EEA1), C2 (protein kinase C (PKC) conserved 2), Tubby, PROPPINs (β-propellers that bind phosphoinositides), ENTH (Epsin N-terminal homology), ANTH (AP180 N-terminal homology), and BAR (Bin, Amphiphysin, and Rvs) domain families. Importantly, detailed discussions of the structural and biophysical characteristics of the PH, PX, ENTH, ANTH, and BAR domains will be provided within other chapters of this volume; therefore, our goal with this brief overview is to demonstrate the diversity of PPIn-interacting modules found in peripheral membrane proteins and highlight some of the molecular properties that contribute to the specificity exhibited by

these domains during interactions with PPIn lipids.

5.1   PH Domains

PH domains typically consist of 100–120 amino acids and were the first protein fold shown to selectively recognize and coordinate membrane-embedded PPIn lipids (Harlan et al. 1994; Lemmon et al. 1995). Since their initial discovery based on sequence homology with two regions found within the major PKC substrate pleckstrin (Haslam et al. 1993; Mayer et al. 1993; Musacchio et al. 1993), PH domains have been identified in approximately 280 different human proteins; making them among the most commonly-occurring defined sequence motif within the eukaryotic proteome (Lemmon 2008). Subsequent studies have shown PH domains to be versatile structures that are not only involved in PPIn recognition, but are also used for mediating protein-protein interactions (Maffucci and Falasca 2001; Lemmon 2004, 2007; DiNitto and Lambright 2006). In fact, most PH domains weakly bind to PPIn lipids with limited specificity, and only a small fraction, estimated at between 10–15%, exhibit high affinity and selective binding to PPIn headgroups (Rameh et al. 1997; Takeuchi et al. 1997; Isakoff et al. 1998; Kavran et al. 1998; Yu et al. 2004). Of the seven PPIn species found within cells, to date, PH domains that specifically recognize PtdIns4P, PtdIns(4,5)P_2, PtdIns(3,4)P_2, and PtdIns(3,4,5)P_3 have been described; including detailed characterizations of the structural features that determine the PPIn-binding specificities (Cozier et al. 2004; Balla 2005; DiNitto and Lambright 2006; Kutateladze 2010). In addition to the recognition of anionic PPIn species, many PH domains have been shown to cooperatively target membranes through additional interactions with other lipid species (Knight and Falke 2009; Vonkova et al. 2015); as well as a growing number of examples defining independent binding sites for non-PPIn lipids, including PtdSer (Uchida et al. 2011; Jian et al. 2015).

In general, although PH domains show relatively low sequence homology (~30%; Lemmon et al. 2002), they adopt a characteristic fold consisting of two nearly orthogonal β-sheets formed by seven β-strands, splayed into a group of three (β5-β7) and four (β1-β4), that are capped by a C-terminal α-helix (Fig. 1a; Ferguson et al. 1994; Ferguson et al. 1995). Within the PH domain fold, there are six loops connecting the β-strands and, overall, the β-sheets are tightly packed, especially at the closed corners of the β-sandwich. Three extended loops connecting the β1-β2, β3-β4, and β6-β7 strands project into the membrane-binding interface, at the open end of the β-sandwich, and show considerable sequence variation across PH domains; which likely contributes to the differences observed in PPIn-binding specificities (Lemmon and Ferguson 2001; DiNitto and Lambright 2006). The canonical binding pocket for the PPIn headgroup is formed by the β1-β2 and β3-β4 strands, as well as the variable loops that connect them. In particular, a basic sequence motif in the β1-β2 loop, defined as $Kx_n(K/R)xR$, has been proposed to serve as a general interaction platform for PPIn headgroups by recognizing vicinal phosphate pairs present in stereospecific positions on the inositol ring (Lemmon 2007; Moravcevic et al. 2012). PH domains that contain the $Kx_n(K/R)xR$ motif all bind to PPIn lipids and this motif is retained in more complex sequence features that have previously been shown to determine the selective coordination of bis- and tris-phosphorylated PPIn species; specifically those with paired phosphate groups at adjacent 4- and 5- or 3- and 4-positions (Lietzke et al. 2000; Yu et al. 2004; Park et al. 2008). Interestingly, non-canonical binding modes of PPIn lipids have been demonstrated for PH domains lacking the $Kx_n(K/R)xR$ motif, including those from β-spectrin (Fig. 1b; Macias et al. 1994; Hyvonen et al. 1995), the p62 subunit of TFIIH (general transcription factor IIH; Di Lello et al. 2005), ArhGAP9 (Rho GTPase-activating protein 9; Ceccarelli et al. 2007), Tiam1 (T-lymphoma and metastasis 1; Ceccarelli et al. 2007), and the yeast protein Slm1 (synthetic lethal with MSS4 protein 1; Anand et al. 2012). Binding of PPIn

lipids to each of these non-canonical PH domains occurs on the opposite face of the β1-β2 loop, with the bound headgroup positioned on the side of the core β-barrel and between the loops that connect the β1-β2 and β5-β6 strands (Balla 2005; DiNitto and Lambright 2006). Outside of variations to the location of the PPIn-binding site, unique sequence features have also been shown to influence the selectivity of certain PH domains for PPIn species. In particular, a subclass of PH domains recognize one or both of PtdIns$(3,4)P_2$ and PtdIns$(3,4,5)P_3$ with a remarkable degree of specificity and affinity, including those found in BTK, PKB, and the cytohesin family Arf GEF Grp1 (general receptor for phosphoinositides 1; Lemmon 2008). The structural features contributing to this binding selectivity primarily involve sequence-specific elaborations of the variable loops. Specifically, the solved structure of Grp1 reveals a long twenty-residue insertion within the β6-β7 loop, which adopts a twisted β-hairpin structure and essentially extends the β-barrel from 7 to 9 strands (Fig. 1c; Lietzke et al. 2000; Ferguson et al. 2000). The headgroup of PtdIns$(3,4,5)P_3$ is contacted by conserved residues within the canonical PH domain pocket that is formed at the top of the β1-β2 and β3-β4 strands and lined by the β1-β2 loop (Lietzke et al. 2000; Ferguson et al. 2000). However, residues from the β-hairpin insertion replace the β3-β4 loop to form the second wall of the PPIn-binding pocket and provide two additional hydrogen bonds with the 5-phosphate that account for the high specificity of the Grp1 PH domains towards PtdIns$(3,4,5)P_3$ (Lietzke et al. 2000; Ferguson et al. 2000). A clear pocket for the 5-phosphate group is also seen in the PH domain of BTK, but, unlike the unique insertion found within the Grp1 PH domain, this pocket forms from an extension of the β1-β2 loop that is able to envelop the 5-phosphate (Baraldi et al. 1999; Ferguson et al. 2000). In addition to structural studies, recent efforts using molecular dynamics simulations of diverse PH domains have revealed important new insights into the binding orientation (Psachoulia and Sansom 2008; Lumb and Sansom 2012; Lenoir et al. 2015; Naughton et al. 2016; Yamamoto et al. 2016) and the dynamics

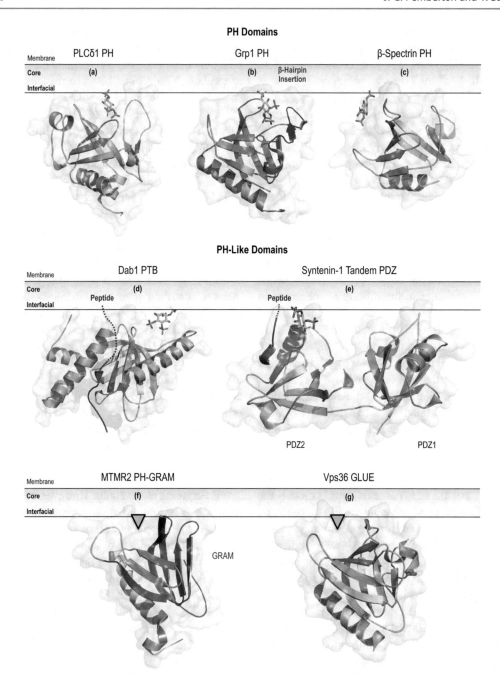

Fig. 1 Structural variations on the PH superfold. Crystal structures of representative PH domains are shown, including examples of canonical (**a**; PLCδ1 PH domain in complex with Ins(1,4,5)P₃; PDB entry 1MAI), elaborated (**b**; Grp1 PH domain in complex with Ins (1,3,4,5)P₄; PDB entry 1FHX), and atypical (**c**; β-Spectrin PH domain in complex with Ins(1,4,5)P₃; PDB entry 1BTN) PPIn-recognition modes. Coincident peptide (highlighted in red) and PPIn coordination is depicted for the PH-like PTB (**d**; Dab1 PTB domain

ternary complex with ApoER2 peptide and PtdIns(4,5) P₂; PDB entry 1NU2) and PDZ domains (**e**; Syntenin PDZ1 and PDZ2 tandem domains in a ternary complex with the Frizzled 7 C-terminal fragment and PtdIns(4,5)P₂; PDB entry 4Z33). The structurally-related GRAM (**f**; isolated from within the structure of MTMR2; PDB entry 1LW3) and GLUE domains (**g**; Vps36 N-terminal domain; PDB entry 2CAY) are shown with their putative membrane-binding pose and PPIn-coordinating pockets highlighted by the green arrowhead. Notice that both

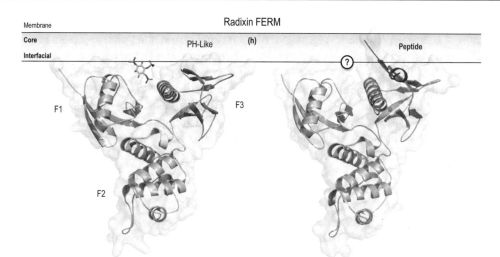

Fig. 1 (continued) may adopt an atypical PPIn-binding mode that is similar to that shown above in (**c**). Lastly, the unique inter-domain PPIn-binding surface of the Radixin FERM domain is depicted in association with either the PPIn lipid (left side; complex with Ins(1,4,5)P₃; PDB entry 1GC6) or cognate peptide (shown in red) ligand (right side; complexed with ICAM-2 cytoplasmic peptide; PDB entry 1 J19). For further details, please refer to Sects. 5.1 (PH domains) and 5.2 (PH-like superfolds) of the text. Prepared using the PyMOL Molecular Graphics System, Version 2.0 Schrödinger, LLC

associated with the diffusivity of the domain (Yamamoto et al. 2017) during PPIn-dependent membrane recognition events. Taken together, it is apparent that sequence differences in the PPIn-binding interface are responsible for the observed heterogeneity in the PPIn selectivity of distinct PH domains and may also influence membrane residency or conformational dynamics during PPIn interactions.

Regardless of the sequence-specific variations that have been documented, overall, PH domains that exhibit PPIn-binding generally show pronounced electrostatic polarization; with strongly positive amino acid residues located at the membrane-binding surface (Macias et al. 1994; Blomberg and Nilges 1997; Moravcevic et al. 2012). Following initial electrostatic interactions at the membrane interface, specific PPIn-binding is likely the primary tool for membrane-selective targeting and increased membrane residence. Interestingly, despite a general lack of hydrophobic or aromatic residues around the PPIn-coordinating pocket (Cho and Stahelin 2005), evidence for penetration of surface-oriented hydrophobic residues into the interfacial region has been presented for the PLCδ1 PH domain using solid state NMR (Tuzi et al. 2003) and

surface plasmon resonance (Flesch et al. 2005). Molecular dynamics simulations of other PH domains also suggest varying degrees of insertion of the PH domain fold into PPIn-containing monolayers (Manna et al. 2007; He et al. 2008; Lumb et al. 2011). However, overall, significant interfacial penetration does not seem to be a general feature of PH domains and is not a major driving force for membrane binding by PH domains; which are clearly more reliant on electrostatic attraction. Additionally, as mentioned above, despite examples of monovalent membrane recruitment, it is important to recognize that membrane binding of PH domains commonly require coincident binding to protein effectors. Coincidence detection by PH domains is not always a simple membrane localization signal, as growing evidence suggests that PH domains can also function at membrane interfaces as highly specific modules for communicating allosteric regulatory signals; including PPIn-mediated conformational switches (DiNitto and Lambright 2006; Nawrotek et al. 2016; Roy et al. 2016). Both of these unique regulatory features exhibited by PH domains have been recently proposed to participate in the dynamic regulation of the Dbl superfamily Rho-GEF

PREX1   (PtdIns(3,4,5)P$_3$-dependent   Rac
exchanger 1). Briefly, allosteric regulation of
PREX1-dependent GTP exchange activity by
PtdIns(3,4,5)P$_3$ binding is thought to occur in a
stepwise fashion. Initial recruitment of PREX1 to
the membrane involves transient electrostatic
interactions with basic residues in the β3-β4
loop of the PH domain, which are stabilized by
the coincident association of the fold to
PM-anchored Gβγ heterodimers that are released
from activated GPCRs (Cash et al. 2016). Gβ-
γ-binding unmasks the PtdIns(3,4,5)P$_3$-binding
site within the PH domain fold, while subsequent
conformational changes induced by PtdIns(3,4,5)
P$_3$ binding are transmitted through a hinge region
at the junction between the PH and catalytic Dbl
homology (DH) domain that results in the full
activation of PREX1 (Cash et al. 2016). These
mechanistic insights are consistent with previous
data demonstrating synergistic regulation of
PREX1 by PtdIns(3,4,5)P$_3$- and Gβγ-mediated
signals (Welch et al. 2002; Hill et al. 2005).
Understanding how protein dynamics and inter-
domain communication can be regulated at the
membrane interface are extremely important
areas for ongoing research. Descriptions of the
molecular interactions between PPIn species and
PH domains have provided some of the best
demonstrations of stereoselective PPIn recognition as well as reveal how membrane-binding
domains can function as integration centers that
relay coincident protein- and lipid-derived
signals.

5.2   PH Domain-Like Folds: PTB, PDZ, GRAM, GLUE, and FERM Domains

The slightly splayed β-sandwich superfold originally described for the PH domain has since been
found in a series of structural homologs that are
collectively referred to here as PH-like domains
(Blomberg et al. 1999; Balla 2005; Scheffzek and
Welti 2012). Despite limited sequence
similarities, the conserved structural core of
PH-like domains is commonly found in modular
proteins implicated in the regulation of signal
transduction and similarly possess binding sites

for PPIn lipids, as well as surfaces that are
involved in mediating protein-protein interactions
(Lemmon 2007; Scheffzek and Welti 2012). A
growing number of PH-like domains appear
within proteins with activities in diverse subcellular compartments show variable lipid- as well as
protein-binding partners. The identification of the
structural conservation of the PH domain
superfold included the early descriptions of the
EVH1 (Enabled/VASP homology 1; Prehoda
et al. 1999) and Ran-binding domain families
(Vetter et al. 1999); however, neither of these
folds are reported to exhibit PPIn-dependent regulation. Consequently, for the sake of this chapter, we will focus our discussion on representative
PH-like domains with established PPIn-binding
sites.

Originally characterized as protein modules
that   interact   with   tyrosine-phosphorylated
peptides, specifically those containing the consensus sequence NPxY or other variants of this
motif, a subset of PTB domains can also independently   or   simultaneously   recognize   PPIn
headgroups with a broad range of affinities
(DiNitto and Lambright 2006; Kaneko et al.
2012). Compared to canonical PH domains, the
PTB domain fold possesses a variable helical
loop inserted between the β1 and β2 strands
(Zhou et al. 1995; Zhou and Fesik 1995). Solution
structures of the Dab1 (disabled-1) PTB domain
show that stereospecific PtdIns(4,5)P$_2$ binding
occurs at the outer surface of the helical insertion,
distinct from the binding pockets described for
both the canonical and atypical PH domains;
whereas phosphorylated peptides associate within
an elongated hydrophobic cleft contoured by the
C-terminal α3 helix and the parallel β5 strand
(Fig. 1d; Yun et al. 2003; Stolt et al. 2003).
Rationale   mutagenesis   and   biophysical
investigations of the Dab1 PTB domain demonstrate that PPIn-binding is requisite for the membrane localization and catalytic function of Dab1
in vitro as well as *in vivo* (Stolt et al. 2005; Huang
et al. 2005; Xu et al. 2005). Furthermore, PPIn
and peptide binding to the Dab1 PTB domain are
energetically independent, and therefore do not
exhibit any apparent cooperativity (Stolt et al.
2004). Outside of Dab1, the PTB domains of

IRS-1 (insulin receptor substrate-1; Takeuchi et al. 1998; Dhe-Paganon et al. 1999) and Shc (Src homology 2 domain-containing transforming protein; Rameh et al. 1997; Ravichandran et al. 1997) have also been shown to selectively bind PPIn lipids, with some apparent selectivity for PtdIns(4,5)P$_2$. In terms of the location for PPIn headgroup coordination, interactions with the Shc PTB domain are thought to involve a cluster of exposed basic residues that are located on the same side of the domain as the inserted helical loop; however, the PPIn-binding sites mapped in the Dab1 and Shc PTB domains do not appear to overlap and the residues implicated in PPIn recognition are not conserved across PTB domains (DiNitto and Lambright 2006). Where investigated in depth, the ability of some PTB domains to coordinate PPIn lipids clearly contributes to the spatial organization and membrane adsorption of PTB domain-containing, particularly at the PM.

Similar to the PTB domain, the PDZ domain is another PH-like fold that typically functions during protein-protein interactions involving adaptor proteins and short peptide sequences, generally the last four to five residues, at the C-terminus of transmembrane proteins (Saras and Heldin 1996; Nourry et al. 2003). However, more recent studies have shown that many PDZ domains can also bind internal peptide sequences as well as membrane phospholipids (Chang et al. 2011; Ivarsson 2012; Mu et al. 2014); including an important role for PtdIns(4,5)P$_2$ (Zimmermann 2006; Wawrzyniak et al. 2013). Early studies of PPIn-PDZ interactions demonstrated that PtdIns(4,5)P$_2$-binding to conserved tandem PDZ domains controlled the cellular localization of the molecular scaffolds syntenin-1 (Zimmermann et al. 2002) and syntenin-2 (Mortier et al. 2005). More recently, a series of large-scale screens of human PDZ domains, using a combination of *in silico* analyses and high-throughput binding assays, found that membrane association is a common property of PDZ domains, found in roughly 20–40%, and that PPIn lipids likely contribute to the cellular localization of a broad collection of PDZ domains; including roles for PPIn interactions with PDZ domain-containing effectors within the nucleus (Mortier et al. 2005; Wu et al. 2007; Chen et al. 2012; Ivarsson et al. 2013). Additionally, a small subset of PDZ domains were shown to bind PPIn lipids with relatively high-affinity; although *in vitro* binding studies, as well as prior characterizations of PDZ function, suggest that the stereospecificity for PPIn headgroups is limited (Zimmermann et al. 2002; Mortier et al. 2005; Ivarsson et al. 2011). Despite the generally low affinity interactions that have been described, in the few cases investigated in detail, PPIn-binding does appear to be important for controlling the cellular functions of PDZ domain-containing protein adaptors (Wawrzyniak et al. 2013). Additionally, where established, it is apparent that PDZ domains can interact with PPIn lipids through different and complex membrane-binding modes (Gallardo et al. 2010); including biophysical investigations identifying surface-exposed electrostatic or hydrophobic residues that facilitate competitive as well as cooperative binding of PDZ domains to PPIn and peptide ligands (Ivarsson 2012; Wawrzyniak et al. 2013; Ernst et al. 2014). Although generally thought to lack a well-defined PPIn-binding pocket, a recent crystal structure of the tandem PDZ domains of syntenin in complex with both PtdIns(4,5)P$_2$ and a cognate C-terminal peptide fragment, shows that the backbone of the bound peptide actually provides direct contacts that help to form the PtdIns(4,5)P$_2$-binding interface and function to stabilize the inositol headgroup (Fig. 1e; Egea-Jimenez et al. 2016). These new structural studies support evidence for synergistic binding of PPIn lipids and peptides to the tandem PDZ domain and suggest that peptide binding likely reinforces the interaction with membrane-embedded lipids (Egea-Jimenez et al. 2016). Additional structures of intact PDZ domain complexes will be extremely informative for understanding the extent to which the seemingly variable PPIn-binding sites can communicate with the well-mapped peptide-binding groove. Overall, as highlighted previously for the classical PH domains, coincident recognition of lipid headgroups and membrane-localized binding partners represents an important regulatory principle that is utilized by diverse families of adaptor proteins to mediate a wide range of cellular processes. The presence of unique variations on the PH superfold, and the PTB and PDZ scaffolds in particular, facilitate diverse PPIn- and peptide-binding activities during membrane-initiated signaling events.

In addition to membrane-binding domains that simply incorporate the PH superfold, other examples of PH-like domains include the reorganization or assembly of the canonical PH module from unique sequence-specific variats. For instance, the GRAM domain was originally identified as small motif predicted based on sequence homology to occur in approximately 180 eukaryotic proteins, including several important membrane-associated proteins such as the myotubularin (MTM) family of PPIn phosphatases (Doerks et al. 2000). The solved structure of MTMR2 (myotubularin-related protein 2) subsequently revealed that the GRAM domain motif consists of five β-strands, but is part of a larger protein fold that incorporates adjacent sequence features to form a fold with a topology that was extremely similar to the canonical PH domain from pleckstrin (Fig. 1f; Begley et al. 2003). Sequence alignments of other representative members of the GRAM domain family showed high conservation of the residues involved in forming the hydrophobic core of the extended GRAM-PH motif, which is also referred to by some as the PH/G domain (Begley et al. 2003). Binding studies have shown that the PH/G domains of MTM1 (Tsujita et al. 2004) and MTMR3 (Lorenzo et al. 2005) preferentially bind to PtdIns$(3,5)$P$_2$ and PtdIns5P, respectively; whereas the PH/G domain of MTMR2 interacts with both PtdIns$(3,5)$P$_2$ and PtdIns5P (Berger et al. 2003). Crystallographic and deuterium exchange studies of MTMR2 show that the PH/G domain is strongly electropositive along the surface-exposed β5-β6 and β7-α1 loops (Begley et al. 2006). Although this electrostatically polarized surface represents the most likely interface for the recognition of PPIn-containing membranes, structural studies have yet to detect PPIn lipids associated with the PH/G domain. Alternatively, independent of PPIn-binding, the PH/G domain clearly plays important roles for

mediating protein-protein interactions as well as during the allosteric control of MTM catalytic activity (Begley and Dixon 2005; Clague and Lorenzo 2005; Hnia et al. 2012). Interestingly, following the description of the PH/G domain, a novel GRAM-like motif dubbed the GLUE domain, was identified as a conserved sequence feature within the N-terminal region of the metazoan Vps36 (vacuolar protein-sorting-associated protein 36) family of ubiquitin-binding proteins (Slagsvold et al. 2005). Notably, Vps36 and its mammalian ortholog, Eap45 (ELL-associated protein of 45 kDa), are components of the ESCRT-II (endosomal sorting complex required for transport II) complex, which plays an essential role during diverse membrane trafficking events, including the biogenesis of multi-vesicular bodies (MVBs; Saksena et al. 2007; Williams and Urbe 2007; Hurley 2008). Functional studies revealed that the GLUE domain binds to both ubiquitin and various 3-phosphorylated PPIn species; suggesting a possible role for PPIn lipids during the coordination of membrane-targeting and cargo recognition within the endosomal system (Slagsvold et al. 2005). The solved structure of the GLUE domain of Vps36 shows that it has a split PH domain architecture, with a yeast-specific insertion of two NZF (Npl4-like zinc finger) domains that are oriented away from the membrane-binding surface (Teo et al. 2006). The walls of the PPIn-binding pocket, which shows high selectivity for mono-phosphorylated PtdIns3P, are built by the β1-β2, β5-β6, and β7-α1 loops and therefore forms outside of the canonical PH domain binding pocket (Fig. 1g; Teo et al. 2006), on the opposite face of the β1-β2 loop, in a manner analogous to the atypical PPIn-binding site originally characterized for the β-spectrin PH domain (Macias et al. 1994; Hyvonen et al. 1995). The location of the PPIn-binding pocketed has since been confirmed by subsequent structures of human Eap45; however, the missing insertion of the NZF domains in the mammalian GLUE domain, reveals an alternative site for ubiquitin binding that lies along one edge of the core β-sandwich and distinct from the PPIn-binding pocket (Alam et al. 2006; Hirano et al. 2006). Overall, these collected structures of the GLUE domain help to demonstrate how recognition of ubiquitinated cargoes and endosomal membranes can be coupled during protein sorting in MVBs; once again pointing to the PH-like superfold as a common substrate for coincidence detection throughout biological systems. In fact, additional protein-protein contacts have been mapped between the GLUE domain and adjacent components of the ESCRT-II machinery, which highlight the critical role for multivalent membrane binding initiated by the PPIn-binding GLUE domain during ESCRT-II actions on protein and lipid sorting (Im and Hurley 2008).

Lastly, unlike the other PH-like domains discussed, the FERM domain incorporates the PH superfold into a much larger multi-domain structure with clear deviations in site utilized for PPIn-binding. Briefly, FERM domains are present is a variety of mammalian proteins (Chishti et al. 1998) that function as important macromolecular scaffolds that link the PM with the cytoskeleton through complex binding interactions with both proteins and lipids (Frame et al. 2010; Moleirinho et al. 2013; Baines et al. 2014). FERM domains are organized by intimate interdomain contacts and consist of three globular lobes (F1, F2, and F3), including a PH-like domain fold that forms the F3 subdomain and participates in the selective recognition of PtdIns(4,5)P$_2$ (Hamada et al. 2000). Binding to PtdIns(4,5)P$_2$ is thought to release an auto-inhibitory intermolecular interaction between the FERM domain and the C-terminal tail of ERM (ezrin, radixin, and moesin) proteins (Pearson et al. 2000; Edwards and Keep 2001; Jayasundar et al. 2012). Interestingly, the PtdIns(4,5)P$_2$ headgroup is coordinated within a shallow basic cleft located between the F1 and F3 subdomains, distinct from any of the binding surfaces mapped on other PH-like domains, by a relatively small number of hydrogen bonds that primarily target the 4-position phosphate group (Fig. 1h; Hamada et al. 2000; Smith et al. 2003). The relative lack of stereospecificity observed within this binding cleft, coupled to reports of FERM domains with an altered F1-F3 cleft (Ceccarelli et al. 2006) or possessing multiple non-specific PtdIns(4,5)P$_2$-binding motifs (Bompard et al. 2003; Zhao et al.

2010), raised the possibility that FERM domains may actually sense the density of PPIn lipids within membranes. Subsequent studies using fluorescence anisotropy measurements and molecular dynamics simulations suggest that the FERM domain from moesin can associate simultaneously with multiple PPIn headgroups and provide no evidence for the presence of discrete PPIn-binding pockets or for interactions of the FERM domain with the acyl chains of the lipid bilayer (Senju et al. 2017). These data support the idea that membrane binding of FERM domains is dependent on multivalent electrostatic interactions, potentiated by anionic PtdIns(4,5)P$_2$, and is also in agreement with the general absence of hydrophobic or aromatic residues at the binding interface (Cho and Stahelin 2005). Consequently, although interactions with PtdIns(4,5)P$_2$ are requisite for PM anchoring, the membrane-targeting of the FERM domain module does not appear to require a defined stereospecific PPIn-binding pocket.

5.3 PX Domains

PX domains were originally identified in the p40^{phox} and p47^{phox} subunits of the phagocyte NADPH oxidase complex (Ponting 1996; Sato et al. 2001; Wishart et al. 2001) and have subsequently been described in a variety of proteins involved in membrane trafficking and cellular signaling; including numerous sorting nexins (SNXs) as well as the class II PI3Ks (Seet and Hong 2006). Functional studies quickly characterized PX domains as short membrane-binding modules, consisting of 100–140 residues, which show specificity for 3-phosphorylated PPIn lipids (Cheever et al. 2001; Ellson et al. 2001; Hiroaki et al. 2001; Kanai et al. 2001; Song et al. 2001; Yu and Lemmon 2001) and a general preference for PtdIns3P (Seet and Hong 2006). Explicit examples of PX domain recognition of substrates outside of PtdIns3P include: PtdIns(3,4)P$_2$ binding to p47^{phox} (Karathanassis et al. 2002), PtdIns(3,4,5)P$_3$ selectivity for CISK (cytokine-independent survival kinase; Xu et al. 2001; Xing et al. 2004), as well as preferential recognition of PtdIns4P by the yeast protein Bem1 (bud emergence Protein 1; Ago et al. 2001; Stahelin et al. 2007) and PtdIns(4,5)P$_2$-specific coordination by Class II PI3K-C2α (Song et al. 2001; Stahelin et al. 2006; Parkinson et al. 2008). Despite some heterogeneity in the PPIn species that are recognized, crystal structures of the p40^{phox} (Bravo et al. 2001) and p47^{phox} (Karathanassis et al. 2002) subunits both revealed a characteristic PX domain fold, which consists of an N-terminal three-stranded β-meander that is followed by a C-terminal α-helical subdomain consisting of three or four α helices; two of which are linked by an elongated poly-proline loop (Fig. 2a). The PPIn isomer binds within a relatively narrow and positively charged groove that is formed by a β-bulge in the β1 strand that twists the β-sheet to form one wall of the binding pocket, as well as through specific contacts with the elongated loop joining the α1 and α2 helices (Cheever et al. 2006; Moravcevic et al. 2012). Specific recognition of the PPIn headgroup is facilitated by acidic membrane environments and is accompanied by the insertion of hydrophobic and aromatic residues within the flexible α1-α2 loop, also referred to as the membrane insertion loop (MIL), into the bilayer (Seet and Hong 2006). Although the alignment of PX domains shows considerable variability in the sequence of the MIL, the presence of a clear hydrophobic motif and membrane penetration of this region appears to be highly conserved across PX domains (Seet and Hong 2006; Kutateladze 2010). Conserved basic residues surrounding the deep PPIn-binding groove and variable loop are involved in electrostatic interactions that facilitate substrate recognition and also enhance affinity of the PX domain for PPIn substrates by inducing insertion of the MIL (Malkova et al. 2006; Stahelin et al. 2003a, 2004, 2006). Three core motifs essential for PPIn binding, including RRYx$_2$Fx$_2$Lx$_3$L of the β3-α1 loop, Px$_2$PxK within the MIL, and RR/Kx$_2$L of α2 are present within most PX domain sequences (Kutateladze 2010). Interestingly, adjacent to the PPIn-binding pocket, an additional well-defined binding site for PtdSer or PtdOH has been described in the p47^{phox} PX domain (Karathanassis et al. 2002; Stahelin et al. 2003a).

Simultaneous occupation of both lipid-binding pockets is thought to modulate the local electrostatic potential and induce a conformational change within the MIL that promotes the insertion of hydrophobic residues into the membrane (Zhou et al. 2003; Cho and Stahelin 2005). This cooperative binding mechanism appears to be an exaggeration of the common non-specific electrostatic interactions that normally initiates the adsorption of PPIn-binding domains onto anionic membrane surfaces. Outside of interactions with additional lipid substrates, PX domains are also involved in protein-protein interactions; including structural descriptions of the intramolecular binding between a C-terminal SH3 domain and the conserved Px$_2$PxK motif within PX domain of p47^{phox} that functions to prevent PtdIns(3,4)P$_2$ binding (Hiroaki et al. 2001; Karathanassis et al. 2002). Similar intramolecular interactions have been demonstrated between the PX and SH3 domains of the fission yeast protein Scd2 (Endo et al. 2003), as well as intermolecular binding of p40^{phox} or p47^{phox} with the cytoskeletal scaffold moesin (Wientjes et al. 2001). An unbiased genome-wide two-hybrid screen using isolated PX domains from yeast was also able to identified several putative PX domain-binding proteins that included known membrane-interacting effectors with roles in vesicular trafficking (Vollert and Uetz 2004). However, overall, it remains unclear the extent to which protein-protein interactions influence or reinforce the PPIn-binding roles of PX domains.

5.4   FYVE Domains

FYVE domains are highly homologous cysteine-rich domains of 70–80 amino acids that are found in around 30 human proteins that have been shown to broadly participate in vacuolar sorting or endocytosis through direct binding to PtdIns3P (Stenmark et al. 1996; Gaullier et al. 1998; Simonsen et al. 1998; Burd and Emr 1998); although the FYVE domain from protrudin has been proposed to associate with PtdIns(4,5)P$_2$, PtdIns(3,4)P$_2$, and PtdIns(3,4,5)P$_3$ both *in vitro* as well as in cells (Gil et al. 2012). The overall architecture of the FYVE domain is comprised of two anti-parallel β-hairpins and a small C-terminal α-helix, which is stabilized by two zinc-binding clusters containing four CxxC motifs in a cross-braced topology (Fig. 2b; Kutateladze et al. 1999; Misra and Hurley 1999). A conserved basic motif, defined as RR/KHHCR, in the first β-strand surrounding the third zinc-coordinating cysteine forms a shallow positively-charged PtdIns3P-binding pocket (Dumas et al. 2001; Kutateladze 2006). Additional WxxD and RVC signature motifs not only help to distinguish FYVE domains within the larger family of zinc-coordinating RING fingers, but also, along with the RR/KHHCR motif, are centrally involved in the coordination of the PtdIns3P headgroup (Kutateladze 2010). Due to the relatively shallow PtdIns3P-binding pocket and coordination of only a single phosphate, FYVE domains bind the monomeric PtdIns3P headgroup rather weakly and FYVE domain-containing effectors tend to require multivalent mechanisms for membrane localization (Dumas et al. 2001; Kutateladze et al. 2004). Importantly, a variable-length turret loop next to the PtdIns3P-binding pocket contains hydrophobic residues that insert into the lipid bilayer and stabilize membrane-bound complexes (Misra and Hurley 1999; Kutateladze and Overduin 2001; Stahelin et al. 2002; Diraviyam et al. 2003; Kutateladze et al. 2004; Brunecky et al. 2005). Basic and polar residues flank the turret loop and play an important role in non-specific electrostatic interactions with acidic lipids, including PtdSer and PtdOH, at the membrane interface and, similar to the PX domain, can be used to drive initial membrane docking (Kutateladze 2010). Based on homology with the C1B domain of PKCδ, FYVE domains were originally proposed to bind with the long axis of the domain perpendicular to the membrane surface, which would facilitate the simultaneous recognition of PtdIns3P within the binding pocket and membrane insertion of the tip of the turret loop (Misra and Hurley 1999). Biophysical and computational studies provide support for this mechanism and show that electrostatic association and PtdIns3P binding drive the membrane insertion of the hydrophobic and aromatic residues within the turret loop (Stahelin et al.

2002; Diraviyam et al. 2003). Interestingly, membrane association of the FYVE domain exhibits pH sensitivity, which is regulated by the adjacent histidine residues found within the core RR/KHHCR motif responsible for coordinating the 3-position phosphate group of PtdIns3P (Lee et al. 2005; He et al. 2009). Protonation of the motif occurs at acidic pH and reinforces the interactions between the PtdIns3P headgroup and the positively charged histidine pair, whereas deprotonation promotes the release of the PtdIns3P ligand and causes rapid membrane dissociation (Lee et al. 2005; He et al. 2009). Membrane avidity of FYVE domain interactions can also be enhanced by dimerization (Callaghan et al. 1999; Lawe et al. 2000; Dumas et al. 2001; Mao et al. 2000; Hayakawa et al. 2004); including a structural characterization of concurrent binding to two PtdIns3P headgroups through a parallel coiled-coil homodimer that juxtaposes the two FYVE domains of EEA1 (early endosomal antigen 1; Dumas et al. 2001). However, sequence analysis of FYVE domains show heterogeneity in their relative hydrophobicity at the putative dimer interface, as well as in residues within the membrane-penetrating turret loop, suggesting that individual FYVE domains are likely to show substantial variance with regards to their propensity for dimerization and orientation at the membrane interface (Stahelin et al. 2014). Regardless, the exquisite selectivity of the FYVE domain helps to demonstrate the importance of coordinate electrostatic and hydrophobic interactions with membrane surfaces during the specific recognition of PPIn lipids. More generally, coincident or multivalent membrane recognition modes are likely requisite for establishing high-affinity binding interactions with mono-phosphorylated PPIns.

5.5   C2 Domains

Originally identified as one of two regulatory domains in PKC (Ono et al. 1989; Osada et al. 1990), C2 domains have since been characterized as versatile membrane-interacting modules that are found in close to 150 different human proteins

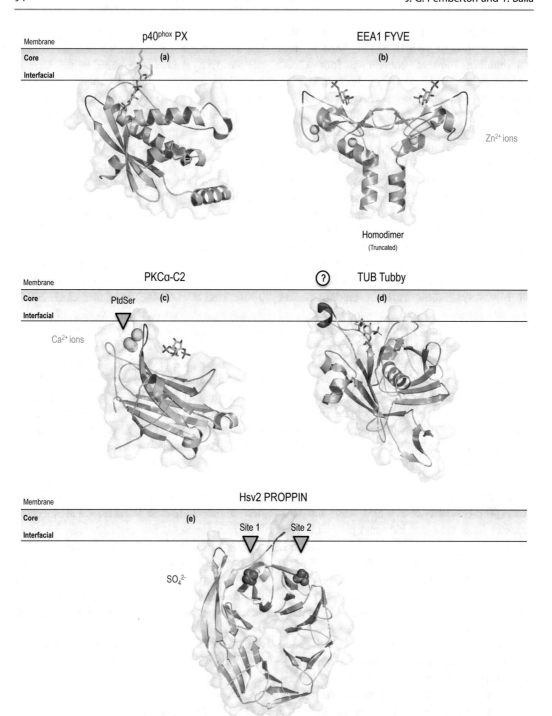

Fig. 2 Diverse domains within peripheral membrane protein exhibiting stereospecific PPIn-binding. Selected examples of unique PPIn-binding domains from the PX (**a**; p40^{phox} PX domain in complex with PtdIns3P; PDB entry 1H6H), FYVE (**b**; EEA1-FYVE domain homodimer bound to Ins(1,3)P$_2$; PDB entry 1JOC), C2 (**c**; PKCα-C2 domain in complex with Ca^{2+} and PtdIns(4,5)P$_2$; PDB entry 3GPE), Tubby (**d**; C-terminal domain of Tubby bound to PtdIns(4,5)P$_2$; PDB entry 1I7E), and PROPPINs (**e**; yeast PROPPIN Hsv2; PDB entry 4EXV) domain

(Cho and Stahelin 2006; Corbalan-Garcia and Gomez-Fernandez 2014a, b). Canonical C2 domains show a common fold that consists of an eight-stranded antiparallel β-sandwich that is connected by variable surface loops (Shao et al. 1996). The majority of C2 domains show Ca^{2+}-dependent binding to common anionic or zwitterionic membrane lipids, including PtdSer (Verdaguer et al. 1999; Stahelin et al. 2003c) or PtdCho (Perisic et al. 1998; Nalefski et al. 1998), through a lipid-binding site that is acidic in character rather than basic; as in PH, PX, and FYVE domains (Moravcevic et al. 2012). In general, Ca^{2+} ions influence C2 domain binding by enhancing the positive electrostatic potential around the Ca^{2+}-binding loops to accelerate association with anionic membranes (Rizo and Sudhof 1998; Murray and Honig 2002) and Ca^{2+} can also induce local structural rearrangements that facilitate membrane binding (Sutton et al. 1995; Grobler et al. 1996; Davletov et al. 1998; Bittova et al. 1999; Kulkarni et al. 2002; Lai et al. 2010; Alwarawrah and Wereszczynski 2017). Additionally, in some C2 domains, Ca^{2+} may slow membrane dissociation by directly coordinating lipid headgroups through Ca^{2+}-mediated bridging (Verdaguer et al. 1999) or induce partial membrane penetration of aromatic residues surrounding the binding interface (Frazier et al. 2002; Kulkarni et al. 2002; Kohout et al. 2003; Stahelin et al. 2003c; Morales et al. 2016). Irrespective of the Ca^{2+} involvement, membrane recognition by C2 domains appears to occur without a great degree of specificity, which is not surprising since C2 domains lack a well-defined lipid-binding pocket (Stahelin et al. 2014; Corbalan-Garcia and Gomez-Fernandez 2014a, b). However, outside of the canonical lipid- and Ca^{2+}-coordinating surface, a small patch of positively-charged residues on the concave side of the β-sandwich, termed the polybasic cluster or cationic β-groove, has been shown to play an important role in the specific recognition of membrane lipids, including PPIn species (Cho and Stahelin 2006; Guerrero-Valero et al. 2009; Corbalan-Garcia and Gomez-Fernandez 2014a, b). The length and net electrostatics of the cationic β-groove, as well as the surface loops, are highly variable across C2 domains and the relative contribution of the two lipid-coordinating sites can be altered as a function of the intracellular Ca^{2+} concentration (Cho and Stahelin 2006; Stahelin et al. 2014). The stereospecific recognition of PPIn species has been best characterized by studies of the PKCα-C2 domain, which binds predominantly to $PtdIns(4,5)P_2$, but also other PPIn lipids, with nanomolar affinity (Fig. 2c; Sanchez-Bautista et al. 2006; Manna et al. 2008). Binding of $PtdIns(4,5)P_2$ to the PKCα-C2 occurs through interactions of the inositol headgroup with three lysine (K197, K209, and K211), one asparagine (N253), and two aromatic (Y197 and W245) residues within the cationic β-groove (Guerrero-Valero et al. 2009). Structure-based alignments of C2 domains suggests that these six residues likely form a consensus $PtdIns(4,5)P_2$-interaction motif; although variability in the polybasic cluster exists, especially with regards to the conservation of K211 (Corbalan-Garcia and Gomez-Fernandez 2014a, b). Interestingly, despite some controversies in the sequence of interactions, PPIn-association with the C2 domain appears to augment PtdSer binding by increasing the duration of membrane residency (Manna et al. 2008; Honigmann et al. 2013). Cooperative binding as a result of coincident Ca^{2+} and $PtdIns(4,5)P_2$ signals has also been demonstrated for the C2B domain of synaptotagmin 1 (van den Bogaart et al. 2012; Guillen et al. 2013), C2C domain of ESyt1 (extended-synaptotagmin 1; Giordano et al. 2013), as well as for the C2A and C2B domains of rabphilin3A (Chung et al. 1998; Coudevylle et al. 2008; Montaville et al. 2008; Guillen et al.

Fig. 2 (continued) families that exhibit stereospecific coordination of the target PPIn lipid. In each of these structures, sequences features identified as defined membrane insertion elements are highlighted in red. For further details about each of these domains, please refer to Sects. 5.3 (PX), 5.4 (FYVE), 5.5 (C2), 5.6 (Tubby), and 5.7 (PROPPINs) of the text. Prepared using the PyMOL Molecular Graphics System, Version 2.0 Schrödinger, LLC

2013). In addition, Ca^{2+}-independent binding PtdIns(4,5)P_2 has been described for the C2C domains of ESyt2 and Esyt3 (Giordano et al. 2013). Other examples of C2 domain recognition of PPIn species include promiscuous recognition of PPIn lipids by the Rasal C2B domain (Sot et al. 2013) and Ca^{2+}-dependent interactions between mono-phosphorylated PPIn species and the KIBRA (KIdney/BRAin protein) C2 domain (Duning et al. 2013). Selective binding of PtdIns (3,4,5)P_3 has been demonstrated for the DHR-1 (dock homology region-1) domain of the Dock family of atypical Rho-GEFs; which uses an elaborated C2 domain scaffold (Premkumar et al. 2010). Interestingly, structural and functional studies suggest that coordination of the PtdIns(3,4,5)P_3 headgroup occurs within a basic pocket generated by extended surface loops, rather than through contacts with the cationic β-groove (Premkumar et al. 2010). Taken together, these studies clearly demonstrate the diversity of PPIn interactions with C2 domains and also highlight the how the complexity of multivalent binding events can be integrated at the membrane interface. Future studies should look to establish functional correlates between the diverse membrane-bound states of C2 domains, which appear to be highly sensitive to distinct combinations of anionic membrane lipids and Ca^{2+} ions, and their abilities to coordinate conformational dynamics within or between diverse macromolecular protein complexes.

5.6   Tubby Domains

The tubby domain consists of a roughly 260 amino acid module that is found within the C-terminus of members from the tubby-like protein (TULP) family of transcription factors (Kleyn et al. 1996; Noben-Trauth et al. 1996; Carroll et al. 2004). The isolated tubby domain displays PtdIns(4,5)P_2-dependent membrane association *in vitro* as well as PtdIns(4,5)P_2-mediated targeting to the PM within intact cells (Santagata et al. 2001; Szentpetery et al. 2009). Structural descriptions of the tubby C-terminal domain reveal a unique fold comprised of a closed β-barrel, consisting of 12 antiparallel

β-strands that surround a central hydrophobic α-helix (Fig. 2d; Boggon et al. 1999; Santagata et al. 2001). Coordination of the PtdIns(4,5)P_2 headgroup occurs within a relatively shallow and positively charged cavity that results in a general lack of stereospecificity for PtdIns(3,5) P_2 or monophosphorylated PPIn lipids relative to PtdIns(4,5)P_2, PtdIns(3,4)P_2, or PtdIns(3,4,5) P_5 (Santagata et al. 2001). Recognition of the bound PtdIns(4,5)P_2 requires specific interactions between the 4-position phosphate with conserved basic residues K330 and R332, as well as coordination of the inositol ring at the 3-position hydroxyl group by R363 (Santagata et al. 2001; Mukhopadhyay and Jackson 2011). Of these PPIn-coordinating residues, K330 is positioned to interact with adjacent phosphate groups, which may help to explain the high selectivity of the tubby domain for bis- or tris-phosphorylated PPIn lipids with adjacent phosphate groups; including clear selectivity for PPIn species phosphorylated at both the 4- and 5-positions (Santagata et al. 2001). An adjacent loop that flanks the binding cavity, as well as polybasic patches on the tubby protein surface, may assist with high-affinity membrane interactions by associating with the interfacial region or through inserting into the membrane (Moravcevic et al. 2012). It is also important to realize that in addition to the selective recognition of PtdIns(4,5)P_2, the tubby domain functions as an important transcriptional regulator by directly binding to double-stranded DNA; a process that once again capitalizes on the positively-charged binding surface described above (Boggon et al. 1999). Consequently, targeting of TULP proteins to PtdIns(4,5)P_2 within the PM has been suggested to prevent nuclear localization and sequester TULP away from effectors within the nucleus (Santagata et al. 2001; Carroll et al. 2004). Although unlikely to be subject to coincident-binding within the PM, given the growing roles for nuclear PPIn lipids, understanding the relationship between selective PPIn coordination and the DNA-binding activity of TULPs within the nucleus could be an interesting area to investigate.

5.7 PROPPINs

The PROPPINs fold was originally described within a family of eukaryotic membrane-binding proteins that includes the important yeast macroautophagy effector Atg18 (autophagy-related protein 18; Michell et al. 2006). In general, PROPPINs consist of a seven-bladed β-propeller (Krick et al. 2012; Baskaran et al. 2012) and contain a conserved FRRG motif that is responsible for the specific recognition of PtdIns3P or PtdIns(3,5)P_2 (Dove et al. 2004; Stromhaug et al. 2004; Krick et al. 2006; Obara et al. 2008). Recent solution structures show that PROPPINs contain two PPIn-binding sites, which are both localized at the rim of the β-propeller, and the side chains of each arginine within the conserved FRRG motif participate in the coordination of PPIn lipids within both binding pockets (Fig. 2e; Krick et al. 2012; Baskaran et al. 2012). Interestingly, PROPPINs are thought to bind to membranes with an edge-on geometry that involves the insertion of aromatic residues into the membrane from within a flexible and exposed loop that protrudes from the β-propeller core and connects the two outer strands of blade six (6CD loop; Baskaran et al. 2012). Due to penetration into the membrane bilayer, PROPPINs such as Atg18 have been shown to bind more strongly to membrane-embedded PPIn lipids compared to short chain analogs or isolated headgroups (Lemmon 2008). Membrane recognition is also thought to be curvature dependent and the initial targeting of the PROPPINs fold likely requires non-specific electrostatic interactions that are reinforced by the selective coordination of PPIn species and insertion of the flexible 6CD loop (Busse et al. 2015). Overall, the presence of two PPIn-binding sites, as well as a defined loop for membrane penetration, confer PROPPINs with the ability to interact with PtdIns3P- or PtdIns(3,5)P_2-containing membranes with high avidity and affinity. Following membrane association, the exposure of the relatively large PROPPINs fold beyond the membrane interface could facilitate protein-protein interactions, including a recent report describing oligomerization of Atg18 upon binding to the membrane surface (Scacioc et al. 2017), which are likely to contribute to the membrane-targeting and function of PROPPINs *in vivo* (Michell et al. 2006; Busse et al. 2015).

5.8 ENTH, ANTH, and BAR Domains

In addition to selective interactions with discrete lipid headgroups, a subset of PPIn-binding domains are able to coordinately recognize or directly influence the local degree of membrane curvature; an important general feature of biological membranes (Antonny 2011; Baumgart et al. 2011; Jarsch et al. 2016). Examples of these specialized curvature-sensitive binding modules include the ENTH, ANTH, and BAR domain families; which all play important roles during complex biological processes that involve substantial membrane deformation events (Itoh and De Camilli 2006; Lemmon 2008). The structurally-related ENTH, and later ANTH, domains were originally identified based on homology to an N-terminal region of epsin (Chen et al. 1998; Kay et al. 1999) and have since been identified within a small family of clathrin adaptor proteins that function as important regulators of membrane endocytosis as well as participate in additional aspects of vesicular trafficking (Itoh et al. 2001; De Camilli et al. 2002; Legendre-Guillemin et al. 2004). The ENTH and ANTH domains both consist of a superhelical solenoid of α-helices that are connected by loops of varying lengths; with the ANTH domain C-terminally extended by one or more α-helices compared with the ENTH domain (Itoh and De Camilli 2006). Despite structural similarities and a shared preference for PtdIns(4,5)P_2, the PPIn-binding modes observed for the ENTH and ANTH domains are quite distinct. Briefly, the structure of the epsin ENTH domain shows that binding of the PtdIns(4,5)P_2 occurs within a well-defined pocket that makes extensive contacts with both the PPIn headgroup and glycerol backbone (Fig. 3a, b; Ford et al. 2002); whereas the coordination of PtdIns(4,5)P_2 by the ANTH domain relies on interactions between the

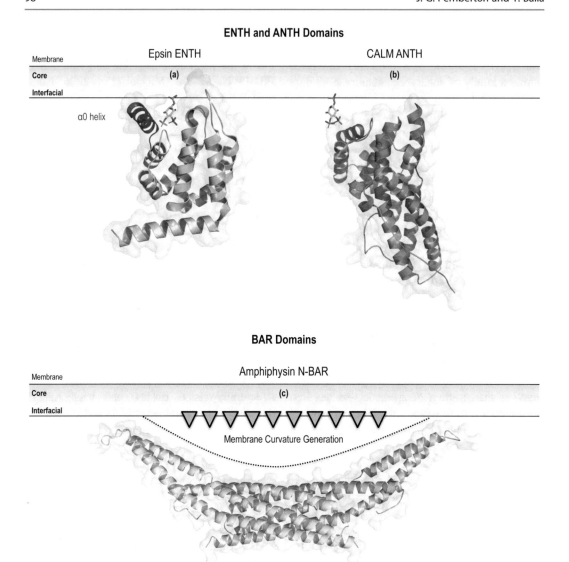

Fig. 3 PPIn-binding domains associated with the generation or recognition of membrane curvature. A comparison of the different PPIn-binding modes used by the membrane deforming ENTH (**a**; epsin ENTH bound to Ins (1,4,5)P$_3$; PDB entry 1H0A) and ANTH (**b**; N-terminal domain of CALM bound to PtdIns(4,5)P$_2$; PDB entry 1HFA) domains. The unstructured α0 helix that becomes structured upon interactions with targeted PPIn lipids, typically PtdIns(4,5)P$_2$, is highlighted by blue. The canonical N-BAR domain of amphyphysin (**c**; PDB entry 1URU) is shown as the activate homodimer. For further details about each of these domains, please refer to Sect. 5.8 of the text. Prepared using the PyMOL Molecular Graphics System, Version 2.0 Schrödinger, LLC

PPIn phosphate groups and a surface-exposed patch of basic residues, formed by helices α1 and α2, that appears to be part of a consensus Kx$_9$Kx(K/R)(H/Y) PPIn-binding motif (Ford et al. 2001; Mao et al. 2001; Itoh and De Camilli 2006). An amphipathic α-helix located at the N-terminus, referred to as helix α0, was unseen in previous crystal structures of the ANTH domain (Ford et al. 2001; Miller et al. 2011) and therefore, was thought to be a specific structural feature of the ENTH domain (Ford et al. 2002). However, recent studies have identified helix α0, which turns out to be an extension of helix α1, within a variety of ANTH domains (Silkov et al. 2011); including detailed functional characterizations of helix α0 in the ANTH domain of the ubiquitous mammalian clathrin adaptor CALM (clathrin-assembly lymphoid myeloid leukaemia protein; Miller et al. 2015). Interestingly, helix α0 is unstructured in both the ENTH and ANTH domains, but becomes ordered upon binding to PtdIns(4,5)P$_2$ (Ford et al. 2002; Miller et al. 2015). PtdIns(4,5)P$_2$-induced structural rearrangements allow the hydrophobic face of helix α0 to penetrate deeply into targeted membranes to promote positive curvature through localized deformation of the membrane leaflet (Ford et al. 2002; Stahelin et al. 2003b; Kweon et al. 2006; Yoon et al. 2010; Lai et al. 2012; Miller et al. 2015). Unlike the ANTH domain, the ordered helix α0 of the ENTH domain contacts the PPIn headgroup within the binding pocket; seemingly to confer additional stereospecificity and possibly slow membrane dissociation kinetics (Hyman et al. 2000; Ford et al. 2002; Stahelin et al. 2003b; Lemmon 2008). Consequently, relative to ENTH domains, the ANTH domain possesses relatively low affinity for PtdIns(4,5)P$_2$ and is also quite promiscuous in terms of PPIn selectivity (Lemmon 2003; Stahelin et al. 2003b). The apparent differences in the PPIn-binding modalities, as well as subtle discrepancies in the depth of the helix α0 insertion, support the idea that membrane-binding of proteins containing the ENTH and ANTH domains are likely to serve distinct functional roles during endocytosis. However, a recent cryo-electron microscopy study of clathrin adaptors from yeast suggests that ENTH and ANTH domains may co-assemble in a PtdIns(4,5)P$_2$-dependent manner and form an organized oligomeric lattice that links polymerized clathrin to the membrane during remodeling events essential for endocytosis (Skruzny et al. 2015). The regular patterning of the ENTH and ANTH domains appears to require ENTH-mediated contact with the membrane leaflet through both its amphipathic α0 helix and PtdIns(4,5)P$_2$-binding pocket, while the ANTH domain stabilizes the oligomer by contacting the ENTH domain but not the membrane (Skruzny et al. 2015). The extent to which this assembly occurs within other model systems has yet to be determined; although, in addition to a potential role in protein-protein interactions, membrane insertion of the ANTH domain from CALM appears to play an important role in promoting membrane curvature and defining the size of clathrin-coated vesicles in mammals (Miller et al. 2015). Overall, these studies identify an essential role for localized PPIn-binding by the ENTH and ANTH domains during the coordinate regulation of clathrin-mediated endocytosis. Future studies investigating the temporal relationship and interactions between ENTH and ANTH domain-containing effectors are still required to understand how the handling of PPIn lipids, and PtdIns(4,5)P$_2$ in particular, is controlled during the formation of a clathrin-coated vesicle.

In addition to the ENTH and ANTH domains, proteins possessing domains from the BAR superfamily are also thought to promote as well as sense membrane curvature (Frost et al. 2009; Qualmann et al. 2011; Mim and Unger 2012; Simunovic et al. 2015; Salzer et al. 2017). In general, several structurally-related groups exist within the BAR superfamily, with classifications based primarily on distinct elaborations of the classical BAR domain fold, and include the well-characterized N-BAR (N-terminal helix BAR), F-BAR (extended Fes/CIP4 homology BAR), and I-BAR (inverse-BAR) domain sub-types (Qualmann et al. 2011). Although these groupings share relatively little sequence similarities and lack signature motifs, all BAR superfamily domains possess a characteristic anti-parallel helical bundle of coiled-coils that interact to form a variety of curved dimeric modules (Fig. 3c; Salzer et al. 2017). Depending on the oligomerization properties of the domain and the shape of the binding surface, BAR superfamily domains can generate positive (N-BAR, and F-BAR; Peter et al. 2004; Shimada et al. 2007) or negative (I-BAR; Millard et al. 2005) membrane curvature, as well as, in relatively few cases, function to stabilize planar membrane sheets (I-BAR; Pykalainen et al. 2011). The membrane-binding interface of each BAR domain contains a series of positively-charged patches, each representing a relatively weak membrane-binding site, which only cooperate with one another if the geometry of the membrane conformers to the degree of curvature defined by the assembled BAR module (Moravcevic et al. 2012). This method of binding relies on delocalized electrostatic attraction, rather than specific coordination of lipid headgroups, and most BAR domains show a general preference for anionic lipids; including targeting to membranes enriched with PtdIns$(4,5)P_2$ and PtdSer (Moravcevic et al. 2012; Salzer et al. 2017). However, functionally-distinct binding modes and different lipid sensitivities are apparent across the BAR superfamily; including the description of a selective PPIn-binding site within the F-BAR domain from a conserved yeast RhoGAP (Moravcevic et al. 2015). Additionally, analogous to the ENTH and ANTH domains, the elaborated canonical BAR domain fold of N-BAR, as well as certain I-BAR domains (Saarikangas et al. 2009), possess amphipathic α-helices that can be inserted into membranes to potently induce curvature (Masuda et al. 2006; Gallop et al. 2006; Mim et al. 2012). Penetration of the amphipathic α-helix, and therefore the membrane deformation activity, is thought to be controlled by the local concentration of PtdIns $(4,5)P_2$ (Mattila et al. 2007; Yoon et al. 2012). Interestingly, the ability of BAR domains to function as diffusion barriers that can restrict membrane PPIn dynamics has also been proposed to be a general feature of the BAR superfamily (Zhao et al. 2013) and may serve to coordinate the scaffolding of additional effectors at sites of membrane deformation. Independent of the recruitment of additional effectors, the presence of flanking PPIn-binding PX (PX-BAR domains; Pylypenko et al. 2007) or PH (BAR-PH domains; Li et al. 2007; Zhu et al. 2007) domains can also direct a subset of BAR domains towards membranes enriched with anionic PPIn lipids (Frost et al. 2009). In fact, recent evidence for coincidence detection and intermolecular communication between BAR domains and the classical PPIn-coordinating modules in PX-BAR (Pylypenko et al. 2007; Daste et al. 2017; Lo

et al. 2017; Schoneberg et al. 2017) and BAR-PH domains (Pang et al. 2014; Chan et al. 2017), suggests that PPIn lipids are likely to function as complex regulators of BAR domain activities. Consequently, despite the relatively non-specific interactions of BAR superfamily domains with PPIn lipids, the ability of BAR domains to couple curvature-sensing and scaffolding roles with the recruitment of additional PPIn-binding elements allows for these domains to shape the complex inter-relationship between membrane PPIn levels and the local architecture of the membrane. Importantly, in addition to this short overview, the structural and regulatory features of the ENTH, ANTH, and BAR domains, as well as the relationship between these domain families and the regulation and sensing of membrane curvature, will be explored in further detail within other chapters of this volume.

6 PPIn-Interacting Domains from Prokaryotic Effector Proteins

The reversible recruitment of peripheral proteins using membrane-embedded PPIn lipids is not a unique feature of eukaryotes and many pathogens target host cell membranes using PPIn-mediated interactions (Ham et al. 2011; Altan-Bonnet and Balla 2012). The demonstrated ability of secreted bacterial effectors to target specific PPIn lipids within defined subcellular compartments highlights the need to better understand the structural features that control such selective and high affinity membrane interactions. Although new evidence for PPIn-dependent membrane targeting by prokaryotic peripheral proteins continues to emerge, the binding motifs identified in secreted bacterial proteins to date lack significant sequence or structural homology with the eukaryotic PPIn-binding domains described above. That said, it is interesting to note that a prokaryotic origin for the PH domain superfold has been suggested from sequence analysis and structural studies (Xu et al. 2010); however, the role of bacterial PH-like domains remains unclear and appears to primarily involve mediating protein-protein interactions rather than functioning as PPIn-targeting modules. Consequently, rather than exploring distant homology with known eukaryotic PPIn-coordinating domains, within this section we will describe the unique protein folds used by known PPIn-binding effectors from prokaryotes and focus specifically on those with descriptions of stereospecific PPIn coordination. In particular, virulence factors SidC and SidM of the intracellular parasite *Legionella pneumophila* have been shown to anchor to the replication-permissive *Legionella*-containing vacuole (LCV) through direct interactions with PtdIns4P (Ragaz et al. 2008; Brombacher et al. 2009; Schoebel et al. 2010). The exquisite specificity of the PtdIns4P-binding domains from SidC and SidM have both been exploited to generate unbiased probes that can be used to detect the major steady-state pools of PtdIns4P in living cells (Hammond et al. 2014; Luo et al. 2015). The recent descriptions of highly specific bacterial PPIn-binding domains, as well as their obvious therapeutic relevance, has reinforced the need to understand the unique features that define the selective recognition of host membranes by secreted bacterial effectors.

Structural analyses of SidC revealed a novel PtdIns4P-binding fold, called the P4C (P̲tdIns4̲P binding of SidC̲) domain, which was comprised of a four α-helical bundle with the PtdIns4P-coordinating pocket forming from a collection of cationic residues at one end of the bundled domain (Fig. 4a; Luo et al. 2015). Two conserved arginine residues, R652 and R638, significantly contribute to the overall charge of the P4C pocket and likely coordinate the PtdIns4P headgroup directly (Luo et al. 2015). In addition to the electrostatic potential, two hydrophobic patches present on the L1 (W642, W643, and F644) and L2 (W704 and F705) loops that surround the PtdIns4P-binding pocket may also facilitate membrane insertion of the P4C (Luo et al. 2015). Importantly, mutation of the electrostatic or hydrophobic features of the P4C significantly reduced membrane binding *in vitro* and could abolish localization to the LCV within cells (Luo et al. 2015). PtdIns4P-binding by the P4C domain was not only requisite for the localization of SidC to the LCV, but, interactions between P4C and PtdIns4P also stimulated the E3 ligase

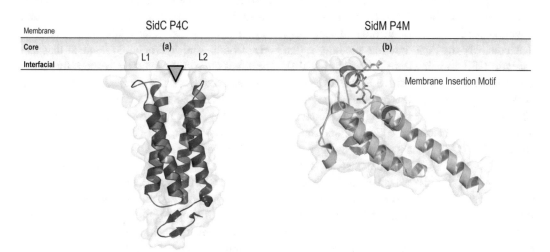

Fig. 4 Prokaryotic PPIn-binding domains. The prokaryotic PPIn-binding P4C (**a**; unbound structure isolated from within the full-length SidC structure; PDB entry 4ZUZ) and P4M (**b**; in complex with di-butyl-PtdIns4P; PDB entry 4MXP) modules are shown with their predicted membrane-bound orientations. The PPIn-binding site (green arrowhead) of the P4C has been mapped by functional and mutagenesis studies, whereas the structure of the P4M module has been solved in complex with the PtdIns4P headgroup. An elaborated membrane insertion motif that significantly penetrates the membrane, as well as contributes to the coordination of the PPIn headgroup within the binding pocket, is shown in red. For further details, please refer to Sect. 6 of the text. Prepared using the PyMOL Molecular Graphics System, Version 2.0 Schrödinger, LLC

activity of SidC; presumably through a conformational switch that functions to extend the P4C domain and uncover the ubiquitin ligase catalytic site of the SNL (SidC N-terminal E3 ligase) domain (Luo et al. 2015). As discussed previously, this type of allosteric regulation by interactions with membrane-embedded PPIn lipids is well characterized in eukaryotes; although, the need for secreted bacterial proteins to communicate directly with the host cell machinery appears to prioritize dynamic domain reorganizations and coincident-binding regulation during the membrane recruitment of many bacterial effectors.

Another secreted *Legionella* effector SidM, which functions as a GEF and adenylyltransferase that is specific for the host Rab1 GTPase (Machner and Isberg 2006; Ingmundson et al. 2007), also contains a novel PtdIns4P-binding module (P4M) that has high affinity and specificity for PtdIns4P (Schoebel et al. 2010; Zhu et al. 2010; Del Campo et al. 2014). Unlike the P4C domain, the structural basis for stereospecific recognition of PtdIns4P by the P4M has been determined explicitly (Fig. 4b; Del Campo et al. 2014). The structure of the P4M fold consists of six α-helices (α10- α15) and an ordered loop (L_C) that connects the lipid-binding module to the catalytic GEF domain (Del Campo et al. 2014). The base of the electropositive PtdIns4P-binding pocket is supported by three parallel helices (α11, α12, and α15), while, at the top of the domain, residues from helices α10, α13, α15, and the L_C contact the DAG backbone to envelope the PtdIns4P headgroup (Del Campo et al. 2014). Coordination of the 4-position phosphate includes contributions from basic and polar residues that define a deep and narrow cavity that shows significant complementarity for the PtdIns4P headgroup while also excluding optimal binding modalities for other PPIn species (Del Campo et al. 2014). PH and PX domains have also been identified with deep PPIn-binding pockets, while the stereospecific coordination of the 1- and 4-position phosphate groups by the P4M certainly resembles the basic and polar networks characterized for many PH domains (Moravcevic et al. 2012; Del Campo et al.

2014). Alongside the constricted PtdIns4P-binding pocket, the α14 helix also extends well above the binding pocket and contains several leucine residues (L610, L614, L615, and L617) that appear to function as an elaborated version of the putative membrane insertion elements found in the P4C domain or those present in other eukaryotic PPIn-coordinating domains (Del Campo et al. 2014); including similarity to the examples discussed above for the PX and FYVE domain families. Importantly, penetration of SidM into PtdIns4P-containing monolayers was dependent on the density of PtdIns4P present within the membrane and not the general electrostatic character of the membrane interface (Del Campo et al. 2014). This suggests that stereospecific recognition of the PtdIns4P headgroup by P4M likely determines the extent of interfacial insertion and subsequent hydrophobic anchoring of SidM within cellular membranes (Del Campo et al. 2014). Overall, the unique structure and exquisite selectivity of the helical P4M fold suggests that the added depth of the binding pocket, resulting from an exaggerated membrane insertion, may help to convey enhanced specificity during PPIn-binding interactions. Understanding the complex inter-relationships between electrostatic and hydrophobic features that support selective PPIn-binding, especially across diverse model systems, represents an important step for defining conserved sequence features that contribute to PPIn-selective membrane interactions during the dynamic regulation of peripheral protein functions.

Outside of the detailed investigations into SidC and SidM interactions with PtdIns4P, relatively few studies have characterized stereospecific coordination of PPIn species by bacterial protein domains. Though lacking structural descriptions, other *Legionella* effectors have also been shown to require PPIn-binding, including PtdIns3P-specific binding during the recruitment of SetA (subversion of eukaryotic traffic A) to early endosomes (Jank et al. 2012) as well as for dynamic associations of LpnE (*Legionella pneumophila* entry) with the LCV (Weber et al. 2009). The cytoskeletal effector ActA (actin assembly-inducing protein) from *Listeria monocytogenes*, which binds to monomeric actin (Skoble et al. 2000; Zalevsky et al. 2001) and the Arp2/3 actin nucleation complex (Welch et al. 1998) to drive actin-dependent motility of the bacteria inside of host cells, is also able to interact with PtdIns(4,5)P_2 or PtdIns(3,4,5)P_3 using a small sub-region within the N-terminal domain (Cicchetti et al. 1999; Steffen et al. 2000; Sidhu et al. 2005). More generally, a broad series of bioinformatic and functional analyses identified a putative family of bacterial PPIn-binding domains (BPDs) within functionally diverse effectors from the type III secretion systems of both animal and plant pathogens (Salomon et al. 2013). Secondary structure predictions and NMR analysis identified a common BPD fold consisting of two β-strands followed by two α-helices; which is somewhat similar to the topology of the eukaryotic PX domain structure of three short β-strands followed by three or four α-helices (Salomon et al. 2013). Predicted BPD domains from phylogenetically-distinct effectors were shown to bind diverse PPIn species, although the most thorough evidence for BPD-mediated interactions with PPIn lipids emerge from studies showing the selective recognition of PtdIns(4,5)P_2 by the *Vibrio parahaemolyticus* effectors VopR (*Vibrio* outer protein R) and VopS (*Vibrio* outer protein S). Interestingly, biophysical studies of VopR showed that the BPD domain is unfolded in solution and significantly increases in secondary structure in the presence of PtdIns(4,5)P_2, but not other PPIn species; outlining a possible role for PPIn lipids during the refolding of bacterial effectors after entering into host cells (Salomon et al. 2013). Another conserved bacterial domain capable of associating with PtdIns(4,5)P_2 was also described in the *Pseudomonas* cytotoxin ExoU (exoenzyme U; Tyson et al. 2015). Briefly, solution structures obtained in the absence of the PPIn headgroup show that the C-terminal membrane localization domain (MLD) of ExoU is organized as a four-helical bundle and possesses a conserved arginine residue (R616) protruding from the cap of the bundle that is required for PtdIns(4,5)P_2 binding *in vitro* as well as for PM

targeting of ExoU within cells (Tyson et al. 2015). Additional polar and charged residues that line surface-exposed pockets formed by the intervening loops of the helical bundle may also assist in PtdIns(4,5)P_2 coordination; however, it is interesting to note that any conserved hydrophobic residues are buried within the MLD and are not likely to assist with membrane binding (Tyson et al. 2015).

Taken together, the novel bacterial domain structures that have been described help to reinforce the importance of electrostatic and polar residues for coordination of the anionic PPIn headgroup, as well as highlight the potential for hydrophobic contacts and membrane insertion for enhanced PPIn-binding specificity. Future structural and biophysical studies of the putative membrane-binding regions described in other prokaryotic effectors should help to identify a wider array of PPIn-interacting folds and may also lead to the identification of conserved prokaryotic PPIn-binding domains. These efforts will require high-throughput screening to more efficiently identify PPIn-interacting bacterial effectors. Along these lines, a recent study has carried out a systematic characterization of the membrane-binding properties and subcellular localization of close to 200 different type III and IV secreted effectors from six bacterial pathogens using yeast as a cellular model (Weigele et al. 2017). This screen identified 57 membrane-associating effectors, including 23 proteins that exhibited changes in localization following isogenic knockout or conditional repression of endogenous yeast PPIn kinases (Weigele et al. 2017). Additional *in vitro* binding studies identified 10 effectors with high affinity for PPIn lipids, but most effectors associated with multiple PPIn species; suggesting non-specific electrostatic interactions rather than PPIn-selective recognition (Weigele et al. 2017). Indeed, of the identified PPIn-bindings proteins, detailed studies of the *Shigella flexneri* factor IpgB1, a known membrane effector and functional mimetic of Rho-family GTPases (Alto et al. 2006; Handa et al. 2007), showed that PPIn recognition occurred through an N-terminal amphipathic helix enriched with basic residues (Weigele et al. 2017). Consequently, although this screen failed to identify effectors with clearly defined PPIn-coordinating pockets, these studies continue to highlight the breadth and diversity of the PPIn-binding strategies employed by prokaryotic effectors, including a conserved strategy that capitalizes on polybasic targeting of anionic PPIn headgroups. More expansive screens that incorporate PPIn-specific immobilization strategies and mass spectrometry, as described for the unbiased characterization of the mammalian PPIn-interacting proteome (Jungmichel et al. 2014), may help to reveal additional bacterial PPIn-binding effectors with more selective membrane recognition strategies. Nonetheless, these collected examples help to demonstrate the convergence of membrane-targeting strategies employed by eukaryotic and prokaryotic peripheral proteins, which also reflects the absolute conservation of PPIn lipids as integral regulatory components within biological membranes.

7 Non-vesicular Lipid Transport and PPIn-Coordinating Lipid-Transfer Domains

The appearance and redistribution of unique PPIn species within distinct subcellular membranes strongly suggests that local production or dynamic trafficking of inositol-containing lipids must occur to maintain PPIn availability. How newly synthesized lipids, and PtdIns in particular, are transported between membrane compartments has become a central question with significant implications for understanding fundamental questions throughout cell biology. Communication between the ER, the primary site of phospholipid biosynthesis, and many distant membrane compartments, including the distal PM, was long thought to be regulated by bulk membrane transfer through the vesicular trafficking system; but, it has recently been demonstrated that specialized lipid-transfer proteins (LTPs) can act to transport lipid monomers across aqueous spaces and between membranes independent of budding

vesicles (Lev 2010; Prinz 2010; Drin 2014). Non-vesicular lipid transport occurs primarily at specialized membrane interfaces, enriched with high-specificity molecular tethers, which are referred to as membrane contact sites (MCSs; Eisenberg-Bord et al. 2016; Jain and Holthuis 2017; Muallem et al. 2017). The formation and dynamics of MCSs has become an increasingly important component of the regulatory mechanisms contributing to cellular membrane remodeling as they allow for the exchange of lipid isomers between organelles possessing different bulk properties. Directional movement of lipids at MCSs occurs by using intrinsic concentration gradients that are maintained through a combination of bulk vesicular trafficking and lipid transfer cycles, which ultimately function as part of the machinery used to define subcellular membrane identity. The lipid transport actions of LTPs are mediated by complex interactions with membrane surfaces and the associated transitions between distinct conformational states appear to confer the selectivity as well as directionality of the lipid transfer process (Chiapparino et al. 2016; Tong et al. 2016). In this section, we will describe the molecular features controlling the specific recognition of PPIn species by lipid transfer domains from eukaryotic families of LTPs. Despite the apparent structural diversity of LTPs, the functional lipid-binding cavities involved in lipid transfer exhibit similarities in their overall architecture and general mode of cargo recognition (Chiapparino et al. 2016). By examining the PPIn-binding modalities used across different families of LTPs, we will attempt to identify conserved strategies for PPIn recognition that emerge from comparing PPIn-selective lipid-transfer domains (LTDs) with the diversity of PPIn-binding folds already described.

7.1 PtdIns4P Transport by ORPs

PtdIns4P has emerged as an important cargo that can be used to drive the counter-exchange of other lipid species at MCSs, including foundational studies demonstrating lipid fluxes at ER-Golgi (Mesmin et al. 2013) and ER-PM contacts (Chung et al. 2015; Moser von Filseck et al. 2015a). Movement of PtdIns4P into the ER can be coupled with PtdIns recycling through the activity of the ER-resident PPIn phosphatase Sac1 (suppressor of actin mutations 1-like), which dephosphorylates PtdIns4P to produce PtdIns (Stefan et al. 2011). The extent to which Sac1 contributes to steady state PtdIns levels or metabolic-tunneling of PtdIns towards PPIn re-synthesis remains an active area of research. However, most importantly, the ability of LTPs to solubilize and transport PtdIns4P across membrane compartments that are defined by different PPIn-modifying enzymes provided a molecular mechanism for cells to generate PPIn lipid gradients (Kim et al. 2013b; Jackson et al. 2016; Wong et al. 2017). Throughout these studies, the molecular driving force responsible for the non-vesicular movement of PtdIns4P was shown to involve members of the oxysterol-binding protein (OSBP)-related protein (ORP) family of eukaryotic LTPs. As implied from their name, the ORP family of LTPs were originally thought to regulate the transport of sterols, but can also transfer other lipid cargoes; including a conserved role in the recognition and extraction of PtdIns4P (Kentala et al. 2016; Tong et al. 2016). Among the best-studies ORPs include the seven yeast ORP homologs (Osh1–Osh7), but more recent work has been focused on the repertoire of 12 ORP-encoding genes in humans; which give rise to 16 human ORP variants through alternative translation or splicing (Jaworski et al. 2001; Lehto et al. 2001). In general, ORPs are cytosolic proteins that possess a combination of sequence features that are collectively used to control protein-protein interactions or membrane targeting; including representative ORPs with N-terminal PH domains (Levine and Munro 1998) or centralized FFAT (two phenylalanines (FF) in an acidic tract) motifs (Loewen and Levine 2005), as well as some homologs possessing ankyrin repeats (Johansson et al. 2003; Tong et al. 2016). Not surprisingly, as discussed above in Sect. 5.1, the PH domains of ORPs are used for membrane targeting, including

specific binding to PPIn species (Tong et al. 2016); whereas the FFAT motif present in some ORPs directly binds to ER-resident proteins called VAPs (vesicle associated membrane protein (VAMP)-associated proteins; Loewen and Levine 2005; Weber-Boyvat et al. 2015; Murphy and Levine 2016) that function to attach ORPs to the surface of ER membranes. Interestingly, two closely-related human ORPs, ORP5 and ORP8, are unique in that they contain a C-terminal transmembrane domain that anchors them within the ER (Yan and Olkkonen 2008; Du et al. 2011); although the lipid-interacting PH and lipid-transfer domains are still capable of accessing lipids in adjacent membrane compartments when localized to MCSs (Chung et al. 2015).

Despite diversity in their domain organizations and recruitment to subcellular membranes, ORPs can all be identified at the sequence level by a C-terminal OSBP-related domain (ORD) containing a conserved N-terminal oriented signature motif, defined as EQVSHHPP, which is important for controlling cellular functions (Tong et al. 2016). The fold of the ORD from ORPs is unique among the known eukaryotic LTPs and although the structures determined to date are limited to Osh homologs from yeast, sequence conservation and functional studies of mammalian ORPs strongly support the ability of these proteins to facilitate lipid extraction and transport. The first solved structure reported from the yeast ORP homolog Osh4 shows the characteristic topology of the ORD, consisting of a hydrophobic tunnel that runs through a near complete central β-barrel that is built around an anti-parallel β-sheet of 15 β-strands (Fig. 5a; Im et al. 2005). The extreme N-terminus contains a small amphipathic α-helix (α1) that is connected to a flexible loop, which attaches to an elongated antiparallel bundle consisting of a two stranded β-sheet (β1-β2) and three α-helices (α2-α4), and functions to close the incomplete β-barrel by forming one wall of the central lipid-binding tunnel (Im et al. 2005). An elongated C-terminal subdomain follows the base of the central β-barrel and projects to exterior surface where the α5, α6, and α7 helices line one-side of the tunnel opening (Im et al. 2005). Interestingly, the region

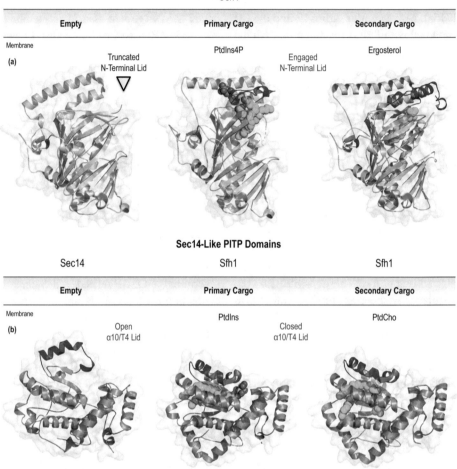

Fig. 5 Lipid-transfer domains with selectivity for PPIn species. For each of the known families of PPIn-coordinating LTPs, the structure of the open LTD fold is shown in relationship to structures with either the primary or secondary lipid cargo bound. Of note, each of the LTPs are shown with the opening of the LTD pocket located at the top of the molecule. Unlike the ORP or Sec14-like LTDs, the PPIn headgroup is buried within the domain, with the fatty acyl chains projecting upwards to the top of the binding pocket. (**a**) The well-studied family of ORPs are important regulators of non-vesicular lipid transport across eukaryotes; including a conserved function for transporting PtdIns4P (PDB entry 3SPW) and a variety of secondary lipid cargoes, including sterols (PDB entry 1ZHZ). Please note that the open fold (PDB entry 1ZI7) could only be crystallized following truncation of the N-terminal lid (shown in blue) that gates the hydrophobic lipid-binding pocket. (**b**) The large family of Sec14-like

PITPs might play more diverse roles outside of lipid transport to control intracellular signaling responses. The recognition of the conserved PtdIns (PDB entry 3B7N) and PtdCho (PDB entry 3B7Q) cargoes clearly involves unique binding surfaces within the PITP, as well as the dynamic reorganization of the α10/T4 lid (highlighted in blue) relative to the unbound structure (PDB entry 1AUA). (**c**) The prototypical StARkin-related PITP domain is also shown bound to PtdIns (PDB entry 1UW5) and PtdCho (PDB entry 1 T27). Compared to the open fold (PDB entry 1KCM), closure of the lipid exchange loop (shown in blue) also stabilizes the elongated C-terminus to pin the αG helix (cyan) in the closed conformation. For further details, please refer to Sects. 7.1 (ORPs) and 7.2 (7.2.1, Sec14-like; 7.2.2, StARkin-related) of the text. Prepared using the PyMOL Molecular Graphics System, Version 2.0 Schrödinger, LLC

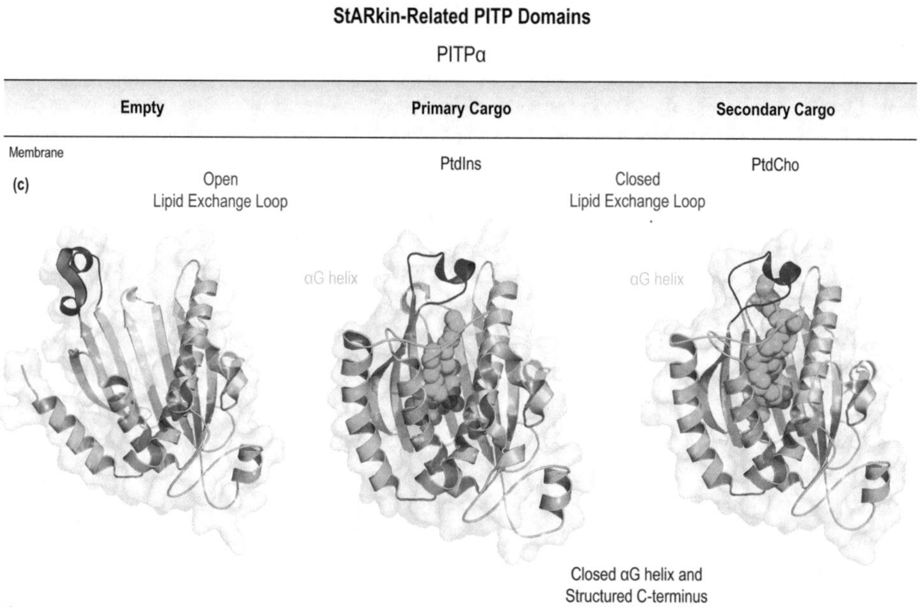

StARkin-Related PITP Domains

PITPα

Fig. 5 (continued)

comprising the N-terminal lid is thought to remain highly flexible in the unbound state, but is stabilized in a closed conformation through direct interactions with lipid cargoes (Im et al. 2005). Compared with the sterol-bound structure, the loss of interactions between the lipid cargo and residues lining the lid causes a conformational change that not only opens the tunnel but also reorganizes the α7 helix and β1 strand, which significantly shifts a conserved cluster of basic residues that is present at the tunnel entrance (Im et al. 2005). The conformational dynamics linking ligand-binding and the movements of the N-terminal lid help to define early models for membrane-binding cycles that would facilitate the acquisition and deposition of lipid cargoes by ORPs. However, these initial studies defining the ORD fold from Osh4 also showed that the sterol binding site was poorly conserved throughout the ORPs, but that the N-terminal lid and basic cluster were present in all ORP homologs and might function during the transport of non-sterol ligands (Im et al. 2005).

Subsequent experiments using *in vitro* liposome transfer assays demonstrated an interesting relationship between the rate of Osh4-mediated sterol transport and the presence of PPIn lipids

(Raychaudhuri et al. 2006; Schulz et al. 2009). PtdIns4P-dependent inhibition of sterol transfer activity led to the eventual discovery of competitive binding between PtdIns4P and sterols for the Osh4 internal binding site (de Saint-Jean et al. 2011). As predicted from the *in vitro* lipid transfer data, the solved structure of Osh4 bound to PtdIns4P showed that the PtdIns4P-coordinating pocket overlaps with the defined sterol-binding site (Fig. 5a; de Saint-Jean et al. 2011). PtdIns4P-recognition occurs with the backbone acyl chains occupying the central hydrophobic tunnel and the phosphorylated inositol headgroup being recognized by a shallow pocket at the tunnel entrance that is covered by the N-terminal lid, which also wraps the glycerol moiety (de Saint-Jean et al. 2011). The 4-position phosphate group makes direct hydrogen bonds with the side chains from a conserved arginine (R344) in the α7 helix as well as with H143 and H144 within the β4-β5 sheets; which are now known to define the conserved OSBP-specific EQVSHHPP signature motif (de Saint-Jean et al. 2011). Alternatively, the 1-position phosphate group that bridges the inositol ring and glycerol backbone forms hydrogen bonds to conserved lysine residues in the β4 sheet (K109) and α7 helix (K336; de Saint-Jean et al. 2011). This general model for the specific recognition of PtdIns4P is also observed in the subsequent PtdIns4P-bound ORD structures obtained from Osh3 (Tong et al. 2013) and Osh6 (Moser von Filseck et al. 2015a). The strict sequence conservation of the residues that contact the PtdIns4P headgroup in the ORP homologs strongly suggests that PtdIns4P is a common ligand for the ORP family of LTPs (de Saint-Jean et al. 2011; Tong et al. 2016). However, despite the demonstrated structural and functional conservation for PtdIns4P recognition, variations in the topology of the central hydrophobic binding pockets of Osh3, Osh4, and Osh6 together indicate that the identity of secondary ligands, such as sterols or other phospholipids, may be unique to individual ORPs (Tong et al. 2016). As outlined above, Osh4 shows clear structural features that facilitate the coordination of sterols along with PtdIns4P (Im et al. 2005). Alternatively, the Osh3 ORD shows a constricted pocket that excludes sterols (Tong et al. 2013), while Osh6 possesses a deeper hydrophobic tunnel that can accommodate the elongated acyl chains from an alternative secondary ligand, PtdSer (Moser von Filseck et al. 2015a). Regardless of the cargoes involved, the structural dynamics associated with the loading of lipids into the ORD remains unknown. Recent studies of Osh4 suggest that the directionality of lipid transfer is influenced by intermolecular interactions with the bound cargo, as well as the anionic character of the local membrane environment; which both regulate the movements of the N-terminal lid (Moser von Filseck et al. 2015b). Unfortunately, results from molecular dynamics simulations of Osh4 have been unable to investigate the role of dynamic gating of the ORD gating during membrane interactions, but these approaches do identify complex contacts with the membrane surface that involve multiple binding regions, including some that run along the length of the molecule and anti-parallel to the entrance of the ORD, which appear to result from a rotation of the fold once it is anchored at the membrane (Rogaski and Klauda 2012; Monje-Galvan and Klauda 2016). Clearly, additional biophysical descriptions of ORPs at the membrane interface will be required

to elucidate the orientation of the binding pocket as well as for defining the role of the N-terminal lid during cargo loading. The relative lack of information regarding the initial events contributing cargo recognition, in particular, make it difficult to understand how PtdIns4P binding can occur with the anionic headgroup oriented at the top of the binding pocket and fatty acyl chains buried within the hydrophobic core of the ORD.

As outlined above, while the majority of the early studies characterizing ORPs have emerged from yeast, PtdIns4P-specific binding and transport activity has been established for many mammalian ORP homologs, including: OSBP (Mesmin et al. 2013), ORP4 (Charman et al. 2014), ORP5 (Chung et al. 2015), ORP8 (Chung et al. 2015), and ORP9 (Liu and Ridgway 2014). However, the identity of the secondary ligands used by most ORPs remains unclear; especially considering the demonstrated roles for ligand-induced conformational dynamics in ORP functions as well as the importance of localized lipid gradients during the control of transfer activity. One of the best examples for the heterotypic exchange of lipid cargoes by mammalian ORPs comes from elegant studies demonstrating the bi-directional movement of PtdIns4P and PtdSer by ORP5 and ORP8 at MCSs between the ER and PM (Chung et al. 2015). This counter-transport mechanism was shown to be important for the homeostatic regulation of PM composition and revealed an intricate relationship between PtdIns4P gradients and PtdSer metabolism that was dependent on ORP5- and ORP8-mediated lipid exchange (Chung et al. 2015; Sohn et al. 2016). A more recent study of ORP5 and ORP8 functions has suggested that their ORD domains interact with multiple PPIn species and identify PtdIns(4,5)P_2, in particular, as the primary lipid-transfer substrate for ORP8 (Ghai et al. 2017). At this time, it is unclear how the structural descriptions of the ORD fold, including those from the closest yeast homolog Osh6, can be reconciled with the proposed PtdIns(4,5)P_2-transfer function of ORP8. It also remains unclear whether the transfer of PtdIns(4,5)P_2 is as universal as the ability of ORPs to bind and transport

PtdIns4P, as well as whether inter-domain communication within ORPs during the lipid-recognition process contributes to the regulation of the lipid exchange cycle.

7.2 PtdIns Interactions with LTPs

Studies examining PPIn metabolism and redistribution have been hampered by the fact that there is still a general lack of understanding with regards to how the synthesis as well as transport of PtdIns occurs. Interestingly, putative PtdIns-transfer proteins (PITPs) have been identified that are thought to rapidly exchange PtdIns, as well as other lipid cargoes, from donor to acceptor membranes and may also contribute to the allosteric control of other membrane-associated proteins or as modules for presenting bound cargo to PtdIns-modifying enzymes (Ile et al. 2006; Grabon et al. 2015; Wong et al. 2017). Outside of the molecular mechanisms involved, the proposed cellular roles for PITPs include the regulation of membrane trafficking and organelle biogenesis, as well as important roles within intracellular signal transduction networks (Kim et al. 2013b). In particular, two major families of eukaryotic PITPs have been identified based on either homology to the yeast protein Sec14 (Sec14-like; Grabon et al. 2015) or as part of a larger superfamily of related proteins that have a conserved StARkin (related (kin) to steroidogenic acute regulatory protein-related and Bet v 1) lipid-transfer domain (StARkin-related or Class I and II PITPs; Wong and Levine 2016; Wong et al. 2017). Within this section, we will review the unique structural characteristics of each of these families of PITPs and attempt to highlight the features responsible for the selective recognition of lipid cargoes.

7.2.1 Sec14-Like PITP Domains

Studies examining the structure and functions of PITPs have been centered around the Sec14-like homologs (Grabon et al. 2015). The yeast Sec14 protein was originally described as a major regulator of phospholipid composition in membranes of the *trans*-Golgi network, making them

permissive for vesicular transport (Bankaitis et al. 1989; Bankaitis et al. 1990; Salama et al. 1990; Ile et al. 2006). In addition to Sec14, five Sec14-like homologs are found in yeast (Sfh1-Sfh5) that all perform cellular functions related to lipid metabolism (Grabon et al. 2015). Many proteins containing Sec14-like domains are also present in higher eukaryotes and several phylogenetically-distinct proteins with sequence similarities to the conserved Sec14-like domain also show lipid transfer activities (Bankaitis et al. 2010; Kim et al. 2013b). Despite some controversy regarding their activities, the Sec14-like domain fold is thought to function as a dual lipid recognition module, similar in function to the ORPs, with selectivity and *in vitro* lipid transfer activity towards PtdIns and the amino-phospholipid phosphatidylcholine (PtdCho; Cleves et al. 1991; Bankaitis et al. 2010). Structural descriptions of the Sec14-like domain come predominantly from the yeast, including an open structure of the founding Sec14 protein, which lacks a lipid cargo, and lipid-bound structures from the Sec14 homolog Sfh1. The unbound Sec14 structure shows that the functional core of the fold is composed of two distinct subdomains consisting of a C-terminal hydrophobic pocket and helical N-terminal domain that are held together by hydrophobic stacking interactions (Fig. 5b; Sha et al. 1998). The large hydrophobic pocket at the C-terminus is formed by six β-strands (β1-β6) that constitute the floor of the binding pocket with the sides of the binding cavity being lined by helices α8 and α9 on one side, as well as helices α10, 3_{10} (T)4, and α11 on the other (Sha et al. 1998). The hydrophobic pocket is supported by an extended string motif that stretches around the floor of the central β-sheet (Sha et al. 1998). Within the N-terminal domain, a tripod-like motif comprised of helices α2, α3, and α4 is essential for membrane targeting, while the adjoining α5 helix may also help to support the hydrophobic core by surrounding central helices α8 and α9 (Sha et al. 1998). In this open conformation, core helices α9 and α10/T4 that line the lipid-binding pocket are bent away from one another to increase the volume of the cavity (Sha et al. 1998). Furthermore, a string of hydrophobic residues are found along the solvent-exposed surface of helices α10/T4, with their side chains oriented away towards the membrane interface (Sha et al. 1998). Movement of the α10/T4 helices away from the surface of Sec14 was proposed to facilitate insertion of the hydrophobic side chains into the inner leaflet of the membrane bilayer, which would open the hydrophobic pocket and allow for the deposition of phospholipid cargoes (Sha et al. 1998). Retraction of the α10/T4 helices from the membrane and toward the protein core would subsequently facilitate phospholipid extraction as a result of direct contacts between the inner face of the helices with the fatty acyl chains of the new lipid cargo (Sha et al. 1998).

Phospholipid-bound structures of the Sec14 homolog Sfh1 confirmed the dramatic repositioning of the α10/T4 helices (α9/T3 helices in Sfh1) to gate the hydrophobic cavity and explicitly define the closed conformation of the Sec14-like lipid-binding pocket (Fig. 5b; Schaaf et al. 2008). However, it is important to note that despite sequence similarities, Sfh1 lacks true Sec14-like biological functions in yeast (Schaaf et al. 2011) and possesses reduced lipid-transfer activities *in vitro* (Li et al. 2000). That said, sequence analyses suggest that Sfh1 and Sec14 likely share the structural motifs required for phospholipid recognition (Schaaf et al. 2008; Schaaf et al. 2011). Interestingly, relatively few changes are observed in the core backbone of the Sec14-like PITP domain with different lipid occupancies (Schaaf et al. 2008). Binding of two structurally similar amino-phospholipids, PtdCho and phosphatidylethanolamine (PtdEtn), show a conserved orientation within the binding pocket; supported by extensive contacts between the lipid cargo and residues along the entire face of the β-sheet as well as in the helical pillars on either side of the cavity (Schaaf et al. 2008). However, as expected, incorporation of PtdIns within the Sfh1 lipid-binding pocket occurred with an orientation that was distinct from the amino-phospholipids (Schaaf et al. 2008). Although the distal regions of the fatty acyl chains are configured similarly, rather than the headgroup of PtdIns binding deep within the pocket, the inositol headgroup is positioned near the protein surface; but is still shielded from the solvent by the closed helical gate (Schaaf et al. 2008). Extensive hydrogen bonding is distributed across the PtdIns backbone and coordination of the headgroup phosphate occurs through an electrostatic interaction with K241 (K239 in Sec14; Schaaf et al. 2008). The hydroxyl groups on the inositol ring also forms direct hydrogen bonds or H_2O-mediated contacts with side chains from the N-terminal tripod motif (R61 and K62, helix α2) and the helical walls of the cavity (D209, helix α8; D235, α9/T3 helical gate; Schaaf et al. 2008). Overall, the extensive hydrogen bonding network and Van der Waals contacts observed in the phospholipid-bound Sfh1 structures are in agreement with the experimental data demonstrating a much higher affinity of Sec14-like PITPs for PtdIns relative to PtdCho (Schaaf et al. 2008; Bankaitis et al. 2010).

Outside of the clear differences in the coordination of the lipid headgroups, a more general feature of the closed conformation of Sfh1, as well as the lipid-interacting surfaces mapped within Sec14 (Smirnova et al. 2006, 2007), is the apparent immobilization of the fatty acyl chains within the binding cavity. Interestingly, the intricate bonding network within hydrophobic pocket is much different than that observed for the ORD domain of ORPs, which show relatively loose nonspecific interactions between the acyl chains of lipid cargoes and the central hydrophobic tunnel (Tong et al. 2016). The top-down binding of the ORD, wherein the headgroup coordinating residues that line the top of the lipid-binding tunnel and lid confer selectivity towards PtdIns4P, but do not place as many restrictions on the identities of potential secondary ligands; including sequence-selective coordination of sterols and PtdSer that have already been described for different ORP homologs. Conversely, the much more restrictive hydrophobic pocket observed in Sec14 and Sfh1 clearly defines binding modalities for both the high-affinity primary ligand, PtdIns, and the secondary cargo; which is almost always PtdCho. These differences in binding are also reflected in the relatively small changes in backbone structure observed during the internal occupancy of different lipid cargoes in the Sec14-like PITP domains, compared to rather large changes in the conformation of the mobile lid observed in the PtdIns4P- and sterol-bound structures of the Osh4 ORD and other ORPs. That said, in terms of the functional communication between lipid cargoes, although the binding pocket of the Sec14-like PITP domain only accommodates a single bound phospholipid (Schaaf et al. 2008), studies using structure-guided design of headgroup-specific binding mutants showed that selective binding of PtdIns and PtdCho within the same molecule is required to maintain the cellular functions performed by intact Sec14 (Schaaf et al. 2008). These data suggest that the functions of Sec14 are defined by sequential lipid exchange cycles of PtdIns and PtdCho, rather than by the strict lipid-selectivity or conformational transitions associated with the lipid-transfer domain. However, the importance of conformational dynamics during Sec14-related lipid transfer activities have been highlighted by unique experiments that used rationale mutagenesis and directed evolution to alter the gating properties of the lipid-binding cavity to enhance phospholipid cycling and resurrect the Sec14-like biological functions of Sfh1; which, despite its utility for structural studies, is relatively inactive as a putative PITP (Schaaf et al. 2011). Molecular dynamics simulations have also been used to track conformational rearrangements during lipid binding (Ryan et al. 2007), but were unable to relate how the motions observed correlate with phospholipid exchange. Consequently, despite immense effort, further experiments are still required to uncover the relationship between the selective recognition of lipid cargoes and the explicit transport or presentation activities of Sec14-like PITPs bound to PtdIns.

7.2.2 StARkin-Related PITP Domains

Outside of the Sec14-like PITPs, the prototypical and first-described mammalian PITPs were the soluble Class I PITP isoforms PITPα and PITPβ from the StARkin-related superfamily (Cockcroft and Carvou 2007; Kim et al. 2013b; Wong et al. 2017). Details regarding the structure and potential PtdIns-transfer characteristics of the Class I PITPs have been uncovered, however the cellular function of these proteins remains obscure. In general, Class I PITPs are relatively small proteins that have high sequence identity, but show different cellular localizations; PITPα is mostly present within the cytosol and nucleus, whereas PITPβ is found associated with membranes of the Golgi (de Vries et al. 1995; De Vries et al. 1996; Larijani et al. 2003). The StARkin-like PITP domain of both PITPα and PITPβ are capable of transferring PtdIns, and to a lesser extent PtdCho, between natural membranes and artificial liposomes (Van Paridon et al. 1987; Wirtz 1991; Wirtz 1997; Segui et al. 2002); making them functionally similar to the Sec14-like PITP domains, despite almost no sequence homology (Hsuan and Cockcroft 2001; Kim et al. 2013b). Also analogous to the Sec14-like PITPs, independent of any lipid-transfer functions, PITPα and PITPβ have been proposed to present PtdIns to modifying enzymes such as PPIn kinases (Cockcroft 1999; Kular et al. 2002) and PPIn-specific phospholipases (Snoek et al. 1999). However, as their sequences would suggest, there are major differences in the overall structure of the Sec14-like and StARkin-related PITP domains; including clear deviations in the orientation of bound phospholipid cargoes.

Comprehensive descriptions of the StARkin-related PITP fold come from the solved structures of PITPα in the unbound conformation as well as in complex with PtdIns or PtdCho lipid cargoes; although, it should be mentioned, prior structures of other StAR-related lipid-transfer domains were also instructive (Tsujishita and Hurley 2000; Roderick et al. 2002; Romanowski et al. 2002). The overall architecture of the open PITP fold shows a single phospholipid-binding pocket that is formed by the surface of a concave β-sheet made by eight β-strands (β1-β8) and two long α-helices (αA and αF) that are tethered by a regulatory loop and flank the core of the fold (Fig. 5c; Schouten et al. 2002). Additional functional regions include a small lipid exchange loop, containing the short αB helix, that acts as a lid to gate access to the lipid-binding pocket as well as a C-terminal region containing the elongated αG helix (Schouten et al. 2002). The transport-competent conformations of PITPα, as well as PtdCho-bound PITPβ (Vordtriede et al. 2005), show the same overall orientation and accommodate a single phospholipid within the enclosed central cavity. Binding of PtdIns (Tilley et al. 2004) or PtdCho (Yoder et al. 2001; Vordtriede et al. 2005) to PITPα is characterized by similar structural rearrangements of the PITP fold; however, it is interesting to note that, opposite to the PPIn-binding LTDs of either the ORPs or Sec14-like PITPs, the phospholipid headgroups are buried within the hydrophobic core of the PITP fold, while the fatty acyl chains oriented outwards within two central channels (Fig. 5c; Yoder et al. 2001; Tilley et al. 2004). The similarities between the lipid-bound structures of PITPα have made it difficult to define the dynamics and intermediate structural steps that contribute to the lipid exchange mechanism. In general, based on the open and closed structures, lipid-binding results in a downward movement of the αB helix that is associated with flattening of the peripheral regions within the β2-β4 strands to close off the hydrophobic cavity at the membrane interface (Yoder et al. 2001). The movement of the lipid exchange loop is also accompanied by an inward swing of the αG helix towards the core β-sheet and stabilization of the C-terminal tail (Yoder et al. 2001; Tilley et al. 2004). Interestingly, within the unbound state, the widening of the core hydrophobic pocket between the central β-sheet and helices αA and αF, as well as the formation of a smaller entrance between the N-terminus of helix αG and the β2-β3 strands, results in the opening of both ends of the lipid-binding cavity; effectively creating a central channel that is supported by the relatively immobile half of the core PITP β-sheet (Yoder et al. 2001). The presence of this hydrophobic tunnel could be relevant during the exchange of lipid cargoes at the membrane interface, or perhaps for the presentation of bound PtdIns to PPIn-modifying enzymes.

Despite the similarities of the transport-competent conformations of PITPs bound to PtdIns and PtdCho, there are clear differences in the coordination of the phospholipid headgroups that appear to explain the differences observed in the binding affinities and transfer rates described for the class I PITPs. Coordination of the inositol ring occurs through specific hydrogen bonds, including selective recognition of the PtdIns headgroup by K61, N90, and Q22 (Tilley et al. 2004). Selective binding to PtdCho and PtdIns involves recognition of the shared phosphate moiety in both lipids by K195, which binds to one phosphate oxygen, and residues T114 and T97 that interact with the other (Tilley et al. 2004). Residues T59 and E86 also contact both PtdCho and PtdIns, but do so in distinctive ways; making only van der Waals contacts with PtdCho, but facilitating hydrogen-bonding to the PtdIns headgroup (Tilley et al. 2004). Overall, studies using site-directed mutagenesis are in agreement with the structural data and demonstrate inhibition of PtdIns and PtdCho binding within cells as well as in lipid transfer assays performed *in vitro*. Selective reductions of PtdIns binding or transfer, without detrimental effects on PtdCho, could be achieved by mutations to the important inositol-coordinating residues T59, K61, E86, and N90 (Tilley et al. 2004). However, mutations significantly inhibiting PtdCho coordination were always followed by similar reductions in PtdIns-related binding (Tilley et al. 2004). These data suggest that the higher affinity of the PITPs towards PtdIns likely results from the additional hydrogen bonding network that forms between the polar inositol headgroup and the residues that line the internal binding cavity. The mechanisms contributing to the specificity of PITPs towards PtdCho need to be investigated in more detail and it is possible that additional secondary lipid cargoes may be relevant *in vivo*. Nevertheless, the residues that line the PITP lipid-binding cavity are highly conserved, including a number aromatic residues that have the potential to contribute membrane interactions during the lipid-exchange process. In fact, two conserved aromatic residues (W203 and W204) located at the end of the αF helix and adjacent to helix αG, have been shown to modulate the membrane association of Class I PITPs and are also thought to play a role during lipid exchange (Tilley et al. 2004; Phillips et al. 2006; Shadan et al. 2008;

Yadav et al. 2015). Specifically, membrane insertion of this hydrophobic motif has also been postulated to initiate the opening of the binding cavity by disrupting the C-terminal tail and dislodging the αG helix; but no biophysical evidence for this mechanism has been provided (Shadan et al. 2008). A more recent study using molecular dynamics has attempted to further resolve features of the lipid exchange cycle utilized by the StARkin-related PITPs (Grabon et al. 2017). All-atom simulations using the open or membrane-docked conformation of PITPα identified overlapping as well as unique regions of the PITP domain interaction surface during binding to PtdCho or PtdIns (Grabon et al. 2017). Interestingly, after membrane association, bilayer insertion of the lipid exchange loop was shown to facilitate the partial loading of a single PtdCho into the hydrophobic pocket and coincident shielding of the fatty acyl chains from the bulk membrane (Grabon et al. 2017). These studies were unable to simulate the complete extraction of the phospholipid cargo into the PITP binding pocket, but do highlight the inherent complexities associated with the membrane recruitment and headgoup-specific coordination that are required for the selective transfer of lipids by PITPs.

Outside of the Class I PITPs, studies from *Drosophila* (Vihtelic et al. 1991, 1993) and, more recently, mammals have characterized three homologs of class II PITPs: PITPNC1 (Class IIB), Nir2 (PITPNM1, Class IIA), and Nir3 (PITPNM2, Class IIA; Ocaka et al. 2005; Wyckoff et al. 2010). The only member of the Class IIB PITPs, PITPNC1, is a soluble protein that possesses a Class II-like PITP domain, but more closely resemble the Class I PITPs in overall architecture (Cockcroft 2012; Kim et al. 2013b). Alternatively, the Class IIA PITPs are multi-domain proteins that possess an N-terminal PITP domain that is flanked by an acidic stretch containing an FFAT motif, which, similar to some ORPs, tethers the proteins to the ER membrane through interactions with ER-resident VAPs (Amarilio et al. 2005). The Class IIA PITPs also contain a DDHD domain that possesses some homology to sequence features found within a small group of intracellular phospholipase A1 (PLA1) proteins (PA-PLA1/DDHD1, p125/Sec23-Interacting Protein, and KIAA0725p/DDHD2; Inoue et al. 2012; Tani et al. 2012), as well as a C-terminal LNS2 (Lipin, Ned1, and Smp2) domain that was originally described in the well-characterized family of PtdOH phosphatases called lipins (Reue 2009). Unfortunately, unlike the Class I PITPs, there are currently no structural descriptions of the PITP domains from the Class II PITPs; however extensive functional analyses have identified important roles for these proteins in the control of cellular signal transduction. In particular, the Class IIA PITPs, Nir2 and Nir3, have been shown to maintain PPIn signaling competence in response to PM PtdIns(4,5)P_2 hydrolysis through the counter-exchange of PtdIns and PtdOH at ER-PM contact sites (Chang et al. 2013; Kim et al. 2013a, 2015; Chang and Liou 2015; Yadav et al. 2015). The soluble Class IIB PITP, PITPNC1, has also been suggested to bind and transfer PtdOH *in vitro* (Garner et al. 2012); although our own studies using intact cells were unable to detect an enhanced clearance of PtdOH from the PM following over-expression of PITPNC1 (Kim et al. 2015, 2016). The identification of PtdOH transfer activity by the Class IIA PITPs, in particular, provides important evidence that phospholipids other than PtdIns or PtdCho can be used as cargoes by StARkin-related PITP domains. There are also reports of Class I PITPs binding to and transporting sphingomyelin *in vitro* (Li et al. 2002; Vordtriede et al. 2005), however, cellular studies do not support a role for PITPβ in the regulation of sphingomyelin biosynthesis or trafficking (Segui et al. 2002). Despite the possibility of binding to alternate cargoes, it is important to note that the overall topology of the distinct sequence motifs present within the StARkin-related PITP domains (Wyckoff et al. 2010), as well as the important residues directly involved in coordinating the inositol headgroup (T59, K61, E86, and N90; PITPα numbering), are conserved across the eukaryotic Class I and II PITPs (Tilley et al. 2004; Grabon et al. 2017). Taken together, these sequence features and the accompanying functional analyses suggest that the primary cargo of the Class I and II PITPs is likely to be PtdIns; whereas, similar to the ORPs, heterogeneity in the identity of the secondary cargo during the lipid exchange cycle could result from the limited specific contacts formed between phospholipids lacking the inositol headgroup. Consequently, although PtdIns transfer activity by Class II PITPs has been demonstrated *in vitro* and inferred *in vivo*, the complexities of the lipid exchange cycle, including the molecular determinants for cargo selectivity during lipid loading and unloading, still need to be determined.

8   Summary and Perspectives

PPIn lipids function as universal regulators of metabolism and membrane biology in part by orchestrating the spatial organization or activity of proteins within defined subcellular compartments. The goal of this chapter is to provide an overview of the structurally-diverse PPIn-binding protein folds and highlight the unique binding modalities that can be used for the specific recognition of PPIn lipids by peripheral binding proteins. Specific domain families exhibit clear preferences in their general methods for membrane association, although all of the examples presented here require that stereospecific headgroup recognition is coupled to a combination of electrostatic attraction and interfacial penetration of membrane-anchoring structural elements. Additionally, binding specificity is influenced not only by interactions with PPIn headgroups, but is also sensitive to the physical properties of the targeted membrane; including clear roles for membrane charge and curvature. It is still not clear to what extent heterogeneity in acyl chain composition contributes to the regulation of binding interactions between peripheral membrane proteins and PPIn species. The requirement for membrane insertion of specific residues or subdomains in many PPIn-binding folds suggests that features of the lipid backbone might be sampled during binding events. In particular, differences in acyl chain length could influence the presentation of the PPIn headgroup

relative to other lipids within the bilayer (Choy et al. 2017), while the degree saturation within the local lipid environment may alter the relative ability of specific domain features to penetrate into the membrane leaflet. As outlined in the introduction, lipid saturation also has the potential to function as a general feature of organelle membrane identity (Bigay and Antonny 2012; Barelli and Antonny 2016) that likely functions in concert with PPIn-selective recognition modules for the selective targeting of proteins to specific subcellular compartments. Independent of general alterations to membrane composition, most of the PPIn-binding folds that have been identified show a high degree of conservation in their strategies for PPIn headgroup coordination. However, despite clear examples of high-affinity and univalent recognition of specific PPIn isomers, many of the domains discussed here interact with membrane-embedded PPIn species too weakly to drive membrane association of proteins in isolation. This reality highlights the need to understand how PPIn-binding domains communicate with other structural features to integrate multivalent interactions that not only increase membrane avidity of multi-domain proteins, but may also function to regulate the catalytic activity or coincident binding of additional protein effectors or lipids. New biophysical approaches, including advances in mass spectrometry (Konermann et al. 2014; Vadas and Burke 2015) and spectroscopy (Chergui 2016; Liang and Tamm 2016), as well as the adoption of more sophisticated computational approaches (Lindahl and Sansom 2008; Dror et al. 2012; Hospital et al. 2015; Hertig et al. 2016), will be required to map the conformational dynamics that relay PPIn-induced movements within individual proteins or membrane-binding macromolecular complexes. Investigations targeted at identifying molecular transitions will be vital for describing the structural intermediates associated with the coordinate regulation of membrane recognition and lipid exchange by the variety of PPIn-binding LTPs and should also provide novel functional insights into the metabolic coupling of PtdIns trafficking and PPIn production. Considering that many LTPs also possess additional membrane-targeting motifs or other structured folds, including some with PPIn-binding PH domains, LTPs offer the unique opportunity to study the potential role of PPIn-sensing during the conformation gating of the LTD for targeted lipid exchange. More generally, the presence of distinct PPIn-binding motifs within proteins with PPIn-modifying activity provides an interesting platform for investigating how patterns of PPIn recognition and metabolism function to regulate the spatiotemporal organization of PPIn species within subcellular compartments. Given the central importance of PPIn lipids for controlling membrane trafficking and signal transduction, and the clear links already identified between the dysregulation of PPIn metabolism and numerous human diseases, additional mechanistic insights into the interactions between PPIn species and peripheral protein effectors will be essential for defining the molecular pathways controlling cellular PtdIns metabolism. Just as importantly, these types of investigations, which uncover detailed molecular information regarding protein and membrane association, have the potential to inform novel therapeutic approaches that might be able to selectively target PPIn-dependent binding interactions or influence defined PPIn-mediated allosteric switches within specific membrane contexts.

Disclosures Statement The authors have nothing to disclose.

Sources of Research Support TB is supported by the National Institutes of Health (NIH) Intramural Research Program (IRP). JGP is supported by an NICHD Visiting Fellowship and a Natural Sciences and Engineering Research Council of Canada (NSERC) Banting Postdoctoral Fellowship.

References

Ago T, Takeya R, Hiroaki H, Kuribayashi F, Ito T, Kohda D, Sumimoto H (2001) The PX domain as a novel phosphoinositide- binding module. Biochem Biophys Res Commun 287(3):733–738. https://doi.org/10.1006/bbrc.2001.5629

Agranoff BW (2009) Turtles all the way: reflections on myo-Inositol. J Biol Chem 284(32):21121–21126. https://doi.org/10.1074/jbc.X109.004747

Agranoff BW, Bradley RM, Brady RO (1958) The enzymatic synthesis of inositol phosphatide. J Biol Chem 233(5):1077–1083

Agranoff BW, Benjamin JA, Hajra AK (1969) Biosynthesis of phosphatidylinositol. Ann N Y Acad Sci 165 (2):755–760

Alam SL, Langelier C, Whitby FG, Koirala S, Robinson H, Hill CP, Sundquist WI (2006) Structural basis for ubiquitin recognition by the human ESCRT-II EAP45 GLUE domain. Nat Struct Mol Biol 13 (11):1029–1030. https://doi.org/10.1038/nsmb1160

Altan-Bonnet N (2017) Lipid tales of viral replication and transmission. Trends Cell Biol 27(3):201–213. https://doi.org/10.1016/j.tcb.2016.09.011

Altan-Bonnet N, Balla T (2012) Phosphatidylinositol 4-kinases: hostages harnessed to build panviral replication platforms. Trends Biochem Sci 37(7):293–302. https://doi.org/10.1016/j.tibs.2012.03.004

Alto NM, Shao F, Lazar CS, Brost RL, Chua G, Mattoo S, McMahon SA, Ghosh P, Hughes TR, Boone C, Dixon JE (2006) Identification of a bacterial type III effector family with G protein mimicry functions. Cell 124 (1):133–145. https://doi.org/10.1016/j.cell.2005.10.031

Alwarawrah M, Wereszczynski J (2017) Investigation of the effect of bilayer composition on PKCalpha-C2 domain docking using molecular dynamics simulations. J Phys Chem B 121(1):78–88. https://doi.org/10.1021/acs.jpcb.6b10188

Amarilio R, Ramachandran S, Sabanay H, Lev S (2005) Differential regulation of endoplasmic reticulum structure through VAP-Nir protein interaction. J Biol Chem 280(7):5934–5944. https://doi.org/10.1074/jbc.M409566200

Anand K, Maeda K, Gavin AC (2012) Structural analyses of the Slm1-PH domain demonstrate ligand binding in the non-canonical site. PLoS One 7(5):e36526. https://doi.org/10.1371/journal.pone.0036526

Antonny B (2011) Mechanisms of membrane curvature sensing. Annu Rev Biochem 80:101–123. https://doi.org/10.1146/annurev-biochem-052809-155121

Arcaro A (1998) The small GTP-binding protein Rac promotes the dissociation of gelsolin from actin filaments in neutrophils. J Biol Chem 273(2):805–813

Baines AJ, Lu HC, Bennett PM (2014) The Protein 4.1 family: hub proteins in animals for organizing membrane proteins. Biochim Biophys Acta 1838 (2):605–619. https://doi.org/10.1016/j.bbamem.2013.05.030

Balla T (2005) Inositol-lipid binding motifs: signal integrators through protein-lipid and protein-protein interactions. J Cell Sci 118(Pt 10):2093–2104. https://doi.org/10.1242/jcs.02387

Balla T (2013) Phosphoinositides: tiny lipids with giant impact on cell regulation. Physiol Rev 93

(3):1019–1137. https://doi.org/10.1152/physrev.00028.2012

Balla A, Balla T (2006) Phosphatidylinositol 4-kinases: old enzymes with emerging functions. Trends Cell Biol 16(7):351–361. https://doi.org/10.1016/j.tcb.2006.05.003

Balla T, Baukal AJ, Guillemette G, Catt KJ (1988) Multiple pathways of inositol polyphosphate metabolism in angiotensin-stimulated adrenal glomerulosa cells. J Biol Chem 263(9):4083–4091

Balla A, Kim YJ, Varnai P, Szentpetery Z, Knight Z, Shokat KM, Balla T (2008) Maintenance of hormone-sensitive phosphoinositide pools in the plasma membrane requires phosphatidylinositol 4-kinase IIIalpha. Mol Biol Cell 19(2):711–721. https://doi.org/10.1091/mbc.E07-07-0713

Bankaitis VA, Malehorn DE, Emr SD, Greene R (1989) The Saccharomyces cerevisiae SEC14 gene encodes a cytosolic factor that is required for transport of secretory proteins from the yeast Golgi complex. J Cell Biol 108(4):1271–1281

Bankaitis VA, Aitken JR, Cleves AE, Dowhan W (1990) An essential role for a phospholipid transfer protein in yeast Golgi function. Nature 347(6293):561–562. https://doi.org/10.1083/347561a0

Bankaitis VA, Mousley CJ, Schaaf G (2010) The Sec14 superfamily and mechanisms for crosstalk between lipid metabolism and lipid signaling. Trends Biochem Sci 35(3):150–160. https://doi.org/10.1016/j.tibs.2009.10.008

Baraldi E, Djinovic Carugo K, Hyvonen M, Surdo PL, Riley AM, Potter BV, O'Brien R, Ladbury JE, Saraste M (1999) Structure of the PH domain from Bruton's tyrosine kinase in complex with inositol 1,3,4,5-tetrakisphosphate. Structure 7(4):449–460

Barelli H, Antonny B (2016) Lipid unsaturation and organelle dynamics. Curr Opin Cell Biol 41:25–32. https://doi.org/10.1016/j.ceb.2016.03.012

Barlow CA, Laishram RS, Anderson RA (2010) Nuclear phosphoinositides: a signaling enigma wrapped in a compartmental conundrum. Trends Cell Biol 20(1):25–35. https://doi.org/10.1016/j.tcb.2009.09.009

Barrera NP, Zhou M, Robinson CV (2013) The role of lipids in defining membrane protein interactions: insights from mass spectrometry. Trends Cell Biol 23(1):1–8. https://doi.org/10.1016/j.tcb.2012.08.007

Baskaran S, Ragusa MJ, Boura E, Hurley JH (2012) Two-site recognition of phosphatidylinositol 3-phosphate by PROPPINs in autophagy. Mol Cell 47(3):339–348. https://doi.org/10.1016/j.molcel.2012.05.027

Baumgart T, Capraro BR, Zhu C, Das SL (2011) Thermodynamics and mechanics of membrane curvature generation and sensing by proteins and lipids. Annu Rev Phys Chem 62:483–506. https://doi.org/10.1146/annurev.physchem.012809.103450

Begley MJ, Dixon JE (2005) The structure and regulation of myotubularin phosphatases. Curr Opin Struct Biol 15(6):614–620. https://doi.org/10.1016/j.sbi.2005.10.016

Begley MJ, Taylor GS, Kim SA, Veine DM, Dixon JE, Stuckey JA (2003) Crystal structure of a phosphoinositide phosphatase, MTMR2: insights into myotubular myopathy and Charcot-Marie-Tooth syndrome. Mol Cell 12(6):1391–1402

Begley MJ, Taylor GS, Brock MA, Ghosh P, Woods VL, Dixon JE (2006) Molecular basis for substrate recognition by MTMR2, a myotubularin family phosphoinositide phosphatase. Proc Natl Acad Sci U S A 103(4):927–932. https://doi.org/10.1073/pnas.0510006103

Berger P, Schaffitzel C, Berger I, Ban N, Suter U (2003) Membrane association of myotubularin-related protein 2 is mediated by a pleckstrin homology-GRAM domain and a coiled-coil dimerization module. Proc Natl Acad Sci U S A 100(21):12177–12182. https://doi.org/10.1073/pnas.2132732100

Bezanilla M, Gladfelter AS, Kovar DR, Lee WL (2015) Cytoskeletal dynamics: a view from the membrane. J Cell Biol 209(3):329–337. https://doi.org/10.1083/jcb.201502062

Bigay J, Antonny B (2012) Curvature, lipid packing, and electrostatics of membrane organelles: defining cellular territories in determining specificity. Dev Cell 23(5):886–895. https://doi.org/10.1016/j.devcel.2012.10.009

Bittova L, Sumandea M, Cho W (1999) A structure-function study of the C2 domain of cytosolic phospholipase A2. Identification of essential calcium ligands and hydrophobic membrane binding residues. J Biol Chem 274(14):9665–9672

Blomberg N, Nilges M (1997) Functional diversity of PH domains: an exhaustive modelling study. Fold Des 2(6):343–355. https://doi.org/10.1016/S1359-0278(97)00048-5

Blomberg N, Baraldi E, Nilges M, Saraste M (1999) The PH superfold: a structural scaffold for multiple functions. Trends Biochem Sci 24(11):441–445

Boggon TJ, Shan WS, Santagata S, Myers SC, Shapiro L (1999) Implication of tubby proteins as transcription factors by structure-based functional analysis. Science 286(5447):2119–2125

Bojjireddy N, Botyanszki J, Hammond G, Creech D, Peterson R, Kemp DC, Snead M, Brown R, Morrison A, Wilson S, Harrison S, Moore C, Balla T (2014) Pharmacological and genetic targeting of the PI4KA enzyme reveals its important role in maintaining plasma membrane phosphatidylinositol 4-phosphate and phosphatidylinositol 4,5-bisphosphate levels. J Biol Chem 289(9):6120–6132. https://doi.org/10.1074/jbc.M113.531426

Bompard G, Martin M, Roy C, Vignon F, Freiss G (2003) Membrane targeting of protein tyrosine phosphatase PTPL1 through its FERM domain via binding to phosphatidylinositol 4,5-biphosphate. J Cell Sci 116 (Pt 12):2519–2530. https://doi.org/10.1242/jcs.00448

Bonangelino CJ, Nau JJ, Duex JE, Brinkman M, Wurmser AE, Gary JD, Emr SD, Weisman LS (2002) Osmotic stress-induced increase of phosphatidylinositol 3,5-bisphosphate requires Vac14p, an activator of the lipid kinase Fab1p. J Cell Biol 156(6):1015–1028. https://doi.org/10.1083/jcb.200201002

Botelho RJ (2009) Changing phosphoinositides "on the fly": how trafficking vesicles avoid an identity crisis. BioEssays 31(10):1127–1136. https://doi.org/10.1002/bies.200900060

Boucrot E, Ferreira AP, Almeida-Souza L, Debard S, Vallis Y, Howard G, Bertot L, Sauvonnet N, McMahon HT (2015) Endophilin marks and controls a clathrin-independent endocytic pathway. Nature 517 (7535):460–465. https://doi.org/10.1038/nature14067

Boura E, Nencka R (2015) Phosphatidylinositol 4-kinases: function, structure, and inhibition. Exp Cell Res 337 (2):136–145. https://doi.org/10.1016/j.yexcr.2015.03.028

Braccini L, Ciraolo E, Campa CC, Perino A, Longo DL, Tibolla G, Pregnolato M, Cao Y, Tassone B, Damilano F, Laffargue M, Calautti E, Falasca M, Norata GD, Backer JM, Hirsch E (2015) PI3K-C2gamma is a Rab5 effector selectively controlling endosomal Akt2 activation downstream of insulin signalling. Nat Commun 6:7400. https://doi.org/10.1038/ncomms8400

Bravo J, Karathanassis D, Pacold CM, Pacold ME, Ellson CD, Anderson KE, Butler PJ, Lavenir I, Perisic O, Hawkins PT, Stephens L, Williams RL (2001) The crystal structure of the PX domain from p40(phox) bound to phosphatidylinositol 3-phosphate. Mol Cell 8(4):829–839

Brombacher E, Urwyler S, Ragaz C, Weber SS, Kami K, Overduin M, Hilbi H (2009) Rab1 guanine nucleotide exchange factor SidM is a major phosphatidylinositol 4-phosphate-binding effector protein of Legionella pneumophila. J Biol Chem 284(8):4846–4856. https://doi.org/10.1074/jbc.M807505200

Brunecky R, Lee S, Rzepecki PW, Overduin M, Prestwich GD, Kutateladze AG, Kutateladze TG (2005) Investigation of the binding geometry of a peripheral membrane protein. Biochemistry 44(49):16064–16071. https://doi.org/10.1021/bi051127+

Bunney TD, Katan M (2010) Phosphoinositide signalling in cancer: beyond PI3K and PTEN. Nat Rev Cancer 10 (5):342–352. https://doi.org/10.1038/nrc2842

Burd CG, Emr SD (1998) Phosphatidylinositol(3)-phosphate signaling mediated by specific binding to RING FYVE domains. Mol Cell 2(1):157–162

Burke JE, Williams RL (2015) Synergy in activating class I PI3Ks. Trends Biochem Sci 40(2):88–100. https://doi.org/10.1016/j.tibs.2014.12.003

Busse RA, Scacioc A, Krick R, Perez-Lara A, Thumm M, Kuhnel K (2015) Characterization of PROPPIN-Phosphoinositide binding and role of loop 6CD in PROPPIN-Membrane binding. Biophys J 108 (9):2223–2234. https://doi.org/10.1016/j.bpj.2015.03.045

Callaghan J, Simonsen A, Gaullier JM, Toh BH, Stenmark H (1999) The endosome fusion regulator early-endosomal autoantigen 1 (EEA1) is a dimer. Biochem J 338(Pt 2):539–543

Calleja V, Alcor D, Laguerre M, Park J, Vojnovic B, Hemmings BA, Downward J, Parker PJ, Larijani B (2007) Intramolecular and intermolecular interactions of protein kinase B define its activation in vivo. PLoS Biol 5(4):e95. https://doi.org/10.1371/journal.pbio.0050095

Calleja V, Laguerre M, Larijani B (2009a) 3-D structure and dynamics of protein kinase B-new mechanism for the allosteric regulation of an AGC kinase. J Chem Biol 2(1):11–25. https://doi.org/10.1007/s12154-009-0016-8

Calleja V, Laguerre M, Parker PJ, Larijani B (2009b) Role of a novel PH-kinase domain interface in PKB/Akt regulation: structural mechanism for allosteric inhibition. PLoS Biol 7(1):e17. https://doi.org/10.1371/journal.pbio.1000017

Calleja V, Laguerre M, Larijani B (2012) Role of the C-terminal regulatory domain in the allosteric inhibition of PKB/Akt. Adv Biol Regul 52(1):46–57. https://doi.org/10.1016/j.advenzreg.2011.09.009

Campa F, Yoon HY, Ha VL, Szentpetery Z, Balla T, Randazzo PA (2009) A PH domain in the Arf GTPase-activating protein (GAP) ARAP1 binds phosphatidylinositol 3,4,5-trisphosphate and regulates Arf GAP activity independently of recruitment to the plasma membranes. J Biol Chem 284 (41):28069–28083. https://doi.org/10.1074/jbc.M109.028266

Campbell RB, Liu F, Ross AH (2003) Allosteric activation of PTEN phosphatase by phosphatidylinositol 4,5-bisphosphate. J Biol Chem 278(36):33617–33620. https://doi.org/10.1074/jbc.C300296200

Carlton JG, Cullen PJ (2005) Coincidence detection in phosphoinositide signaling. Trends Cell Biol 15 (10):540–547. https://doi.org/10.1016/j.tcb.2005.08.005

Carroll K, Gomez C, Shapiro L (2004) Tubby proteins: the plot thickens. Nat Rev Mol Cell Biol 5(1):55–63. https://doi.org/10.1038/nrm1278

Cash JN, Davis EM, Tesmer JJG (2016) Structural and biochemical characterization of the catalytic core of the metastatic factor P-Rex1 and its regulation by Ptdins (3,4,5)P3. Structure 24(5):730–740. https://doi.org/10.1016/j.str.2016.02.022

Ceccarelli DF, Song HK, Poy F, Schaller MD, Eck MJ (2006) Crystal structure of the FERM domain of focal adhesion kinase. J Biol Chem 281(1):252–259. https://doi.org/10.1074/jbc.M509188200

Ceccarelli DF, Blasutig IM, Goudreault M, Li Z, Ruston J, Pawson T, Sicheri F (2007) Non-canonical interaction of phosphoinositides with pleckstrin homology domains of Tiam1 and ArhGAP9. J Biol Chem 282 (18):13864–13874. https://doi.org/10.1074/jbc.M700505200

Chan KC, Lu L, Sun F, Fan J (2017) Molecular details of the PH domain of ACAP1(BAR-PH) protein binding to PIP-containing membrane. J Phys Chem B 121 (15):3586–3596. https://doi.org/10.1021/acs.jpcb.6b09563

Chang CL, Liou J (2015) Phosphatidylinositol 4,5-bisphosphate homeostasis regulated by Nir2 and Nir3 proteins at endoplasmic reticulum-plasma membrane junctions. J Biol Chem 290(23):14289–14301. https://doi.org/10.1074/jbc.M114.621375

Chang BH, Gujral TS, Karp ES, BuKhalid R, Grantcharova VP, MacBeath G (2011) A systematic family-wide investigation reveals that ~30% of mammalian PDZ domains engage in PDZ-PDZ interactions. Chem Biol 18(9):1143–1152. https://doi.org/10.1016/j.chembiol.2011.06.013

Chang CL, Hsieh TS, Yang TT, Rothberg KG, Azizoglu DB, Volk E, Liao JC, Liou J (2013) Feedback regulation of receptor-induced Ca2+ signaling mediated by E-Syt1 and Nir2 at endoplasmic reticulum-plasma membrane junctions. Cell Rep 5(3):813–825. https://doi.org/10.1016/j.celrep.2013.09.038

Charman M, Colbourne TR, Pietrangelo A, Kreplak L, Ridgway ND (2014) Oxysterol-binding protein (OSBP)-related protein 4 (ORP4) is essential for cell proliferation and survival. J Biol Chem 289(22):15705–15717. https://doi.org/10.1074/jbc.M114.571216

Cheever ML, Sato TK, de Beer T, Kutateladze TG, Emr SD, Overduin M (2001) Phox domain interaction with PtdIns(3)P targets the Vam7 t-SNARE to vacuole membranes. Nat Cell Biol 3(7):613–618. https://doi.org/10.1038/35083000

Cheever ML, Kutateladze TG, Overduin M (2006) Increased mobility in the membrane targeting PX domain induced by phosphatidylinositol 3-phosphate. Protein Sci 15(8):1873–1882. https://doi.org/10.1110/ps.062194906

Chen H, Fre S, Slepnev VI, Capua MR, Takei K, Butler MH, Di Fiore PP, De Camilli P (1998) Epsin is an EH-domain-binding protein implicated in clathrin-mediated endocytosis. Nature 394(6695):793–797. https://doi.org/10.1038/29555

Chen Y, Sheng R, Kallberg M, Silkov A, Tun MP, Bhardwaj N, Kurilova S, Hall RA, Honig B, Lu H, Cho W (2012) Genome-wide functional annotation of dual-specificity protein- and lipid-binding modules that regulate protein interactions. Mol Cell 46(2):226–237. https://doi.org/10.1016/j.molcel.2012.02.012

Chergui M (2016) Time-resolved X-ray spectroscopies of chemical systems: new perspectives. Struct Dyn 3(3):031001. https://doi.org/10.1063/1.4953104

Chiapparino A, Maeda K, Turei D, Saez-Rodriguez J, Gavin AC (2016) The orchestra of lipid-transfer proteins at the crossroads between metabolism and signaling. Prog Lipid Res 61:30–39. https://doi.org/10.1016/j.plipres.2015.10.004

Chishti AH, Kim AC, Marfatia SM, Lutchman M, Hanspal M, Jindal H, Liu SC, Low PS, Rouleau GA, Mohandas N, Chasis JA, Conboy JG, Gascard P, Takakuwa Y, Huang SC, Benz EJ Jr, Bretscher A, Fehon RG, Gusella JF, Ramesh V, Solomon F, Marchesi VT, Tsukita S, Tsukita S, Hoover KB et al (1998) The FERM domain: a unique module involved in the linkage of cytoplasmic proteins to the membrane. Trends Biochem Sci 23(8):281–282

Cho W, Stahelin RV (2005) Membrane-protein interactions in cell signaling and membrane trafficking. Annu Rev Biophys Biomol Struct 34:119–151. https://doi.org/10.1146/annurev.biophys.33.110502.133337

Cho W, Stahelin RV (2006) Membrane binding and subcellular targeting of C2 domains. Biochim Biophys Acta 1761(8):838–849. https://doi.org/10.1016/j.bbalip.2006.06.014

Choi S, Thapa N, Tan X, Hedman AC, Anderson RA (2015) PIP kinases define PI4,5P(2)signaling specificity by association with effectors. Biochim Biophys Acta 1851(6):711–723. https://doi.org/10.1016/j.bbalip.2015.01.009

Choi S, Hedman AC, Sayedyahossein S, Thapa N, Sacks DB, Anderson RA (2016) Agonist-stimulated phosphatidylinositol-3,4,5-trisphosphate generation by scaffolded phosphoinositide kinases. Nat Cell Biol 18(12):1324–1335. https://doi.org/10.1038/ncb3441

Choy CH, Han BK, Botelho RJ (2017) Phosphoinositide diversity, distribution, and effector function: stepping out of the box. BioEssays 39(12). https://doi.org/10.1002/bies.201700121

Chung SH, Song WJ, Kim K, Bednarski JJ, Chen J, Prestwich GD, Holz RW (1998) The C2 domains of Rabphilin3A specifically bind phosphatidylinositol 4,5-bisphosphate containing vesicles in a Ca2+−dependent manner. In vitro characteristics and possible significance. J Biol Chem 273(17):10240–10248

Chung J, Torta F, Masai K, Lucast L, Czapla H, Tanner LB, Narayanaswamy P, Wenk MR, Nakatsu F, De Camilli P (2015) Intracellular transport. PI4P/phosphatidylserine countertransport at ORP5- and ORP8-mediated ER-plasma membrane contacts. Science 349(6246):428–432. https://doi.org/10.1126/science.aab1370

Cicchetti G, Maurer P, Wagener P, Kocks C (1999) Actin and phosphoinositide binding by the ActA protein of the bacterial pathogen Listeria monocytogenes. J Biol Chem 274(47):33616–33626

Clague MJ, Lorenzo O (2005) The myotubularin family of lipid phosphatases. Traffic 6(12):1063–1069. https://doi.org/10.1111/j.1600-0854.2005.00338.x

Cleves A, McGee T, Bankaitis V (1991) Phospholipid transfer proteins: a biological debut. Trends Cell Biol 1(1):30–34

Cockcroft S (1999) Mammalian phosphatidylinositol transfer proteins: emerging roles in signal transduction and vesicular traffic. Chem Phys Lipids 98(1–2):23–33

Cockcroft S (2012) The diverse functions of phosphatidylinositol transfer proteins. Curr Top Microbiol Immunol 362:185–208. https://doi.org/10.1007/978-94-007-5025-8_9

Cockcroft S, Carvou N (2007) Biochemical and biological functions of class I phosphatidylinositol transfer proteins. Biochim Biophys Acta 1771(6):677–691. https://doi.org/10.1016/j.bbalip.2007.03.009

Corbalan-Garcia S, Gomez-Fernandez JC (2014a) Classical protein kinases C are regulated by concerted interaction with lipids: the importance of phosphatidylinositol-4,5-bisphosphate. Biophys Rev 6(1):3–14. https://doi.org/10.1007/s12551-013-0125-z

Corbalan-Garcia S, Gomez-Fernandez JC (2014b) Signaling through C2 domains: more than one lipid target. Biochim Biophys Acta 1838(6):1536–1547. https://doi.org/10.1016/j.bbamem.2014.01.008

Coudevylle N, Montaville P, Leonov A, Zweckstetter M, Becker S (2008) Structural determinants for Ca2+ and phosphatidylinositol 4,5-bisphosphate binding by the C2A domain of rabphilin-3A. J Biol Chem 283 (51):35918–35928. https://doi.org/10.1074/jbc.M804094200

Cozier GE, Carlton J, Bouyoucef D, Cullen PJ (2004) Membrane targeting by pleckstrin homology domains. Curr Top Microbiol Immunol 282:49–88

Crowder MK, Seacrist CD, Blind RD (2017) Phospholipid regulation of the nuclear receptor superfamily. Adv Biol Regul 63:6–14. https://doi.org/10.1016/j.jbior.2016.10.006

Cullen PJ, Cozier GE, Banting G, Mellor H (2001) Modular phosphoinositide-binding domains--their role in signalling and membrane trafficking. Curr Biol 11 (21):R882–R893

Daste F, Walrant A, Holst MR, Gadsby JR, Mason J, Lee JE, Brook D, Mettlen M, Larsson E, Lee SF, Lundmark R, Gallop JL (2017) Control of actin polymerization via the coincidence of phosphoinositides and high membrane curvature. J Cell Biol 216 (11):3745–3765. https://doi.org/10.1083/jcb.201704061

Davletov B, Perisic O, Williams RL (1998) Calcium-dependent membrane penetration is a hallmark of the C2 domain of cytosolic phospholipase A2 whereas the C2A domain of synaptotagmin binds membranes electrostatically. J Biol Chem 273(30):19093–19096. https://doi.org/10.1074/jbc.273.30.19093

De Camilli P, Chen H, Hyman J, Panepucci E, Bateman A, Brunger AT (2002) The ENTH domain. FEBS Lett 513(1):11–18

De Matteis M, Godi A, Corda D (2002) Phosphoinositides and the golgi complex. Curr Opin Cell Biol 14 (4):434–447

de Saint-Jean M, Delfosse V, Douguet D, Chicanne G, Payrastre B, Bourguet W, Antonny B, Drin G (2011) Osh4p exchanges sterols for phosphatidylinositol 4-phosphate between lipid bilayers. J Cell Biol 195 (6):965–978. https://doi.org/10.1083/jcb.201104062

de Vries KJ, Heinrichs AA, Cunningham E, Brunink F, Westerman J, Somerharju PJ, Cockcroft S, Wirtz KW, Snoek GT (1995) An isoform of the phosphatidylinositol-transfer protein transfers sphingomyelin and is associated with the Golgi system. Biochem J 310(Pt 2):643–649

De Vries KJ, Westerman J, Bastiaens PI, Jovin TM, Wirtz KW, Snoek GT (1996) Fluorescently labeled phosphatidylinositol transfer protein isoforms (alpha and beta), microinjected into fetal bovine heart endothelial cells, are targeted to distinct intracellular sites.

Exp Cell Res 227(1):33–39. https://doi.org/10.1006/excr.1996.0246

Del Campo CM, Mishra AK, Wang YH, Roy CR, Janmey PA, Lambright DG (2014) Structural basis for PI(4)P-specific membrane recruitment of the Legionella pneumophila effector DrrA/SidM. Structure 22 (3):397–408. https://doi.org/10.1016/j.str.2013.12.018

Dhe-Paganon S, Ottinger EA, Nolte RT, Eck MJ, Shoelson SE (1999) Crystal structure of the pleckstrin homology-phosphotyrosine binding (PH-PTB) targeting region of insulin receptor substrate 1. Proc Natl Acad Sci U S A 96(15):8378–8383

Di Lello P, Nguyen BD, Jones TN, Potempa K, Kobor MS, Legault P, Omichinski JG (2005) NMR structure of the amino-terminal domain from the Tfb1 subunit of TFIIH and characterization of its phosphoinositide and VP16 binding sites. Biochemistry 44(21):7678–7686. https://doi.org/10.1021/bi050099s

DiNitto JP, Lambright DG (2006) Membrane and juxtamembrane targeting by PH and PTB domains. Biochim Biophys Acta 1761(8):850–867. https://doi.org/10.1016/j.bbalip.2006.04.008

DiNitto JP, Delprato A, Gabe Lee MT, Cronin TC, Huang S, Guilherme A, Czech MP, Lambright DG (2007) Structural basis and mechanism of autoregulation in 3-phosphoinositide-dependent Grp1 family Arf GTPase exchange factors. Mol Cell 28(4):569–583. https://doi.org/10.1016/j.molcel.2007.09.017

Diraviyam K, Stahelin RV, Cho W, Murray D (2003) Computer modeling of the membrane interaction of FYVE domains. J Mol Biol 328(3):721–736

Doerks T, Strauss M, Brendel M, Bork P (2000) GRAM, a novel domain in glucosyltransferases, myotubularins and other putative membrane-associated proteins. Trends Biochem Sci 25(10):483–485

Dornan GL, McPhail JA, Burke JE (2016) Type III phosphatidylinositol 4 kinases: structure, function, regulation, signalling and involvement in disease. Biochem Soc Trans 44(1):260–266. https://doi.org/10.1042/BST20150219

Dove SK, Piper RC, McEwen RK, Yu JW, King MC, Hughes DC, Thuring J, Holmes AB, Cooke FT, Michell RH, Parker PJ, Lemmon MA (2004) Svp1p defines a family of phosphatidylinositol 3,5-bisphosphate effectors. EMBO J 23(9):1922–1933. https://doi.org/10.1038/sj.emboj.7600203

Drin G (2014) Topological regulation of lipid balance in cells. Annu Rev Biochem 83:51–77. https://doi.org/10.1146/annurev-biochem-060713-035307

Dror RO, Dirks RM, Grossman JP, Xu H, Shaw DE (2012) Biomolecular simulation: a computational microscope for molecular biology. Annu Rev Biophys 41:429–452. https://doi.org/10.1146/annurev-biophys-042910-155245

Du X, Kumar J, Ferguson C, Schulz TA, Ong YS, Hong W, Prinz WA, Parton RG, Brown AJ, Yang H (2011) A role for oxysterol-binding protein-related protein 5 in endosomal cholesterol trafficking. J Cell Biol 192 (1):121–135. https://doi.org/10.1083/jcb.201004142

Dumas JJ, Merithew E, Sudharshan E, Rajamani D, Hayes S, Lawe D, Corvera S, Lambright DG (2001)

Multivalent endosome targeting by homodimeric EEA1. Mol Cell 8(5):947–958

Duning K, Wennmann DO, Bokemeyer A, Reissner C, Wersching H, Thomas C, Buschert J, Guske K, Franzke V, Floel A, Lohmann H, Knecht S, Brand SM, Poter M, Rescher U, Missler M, Seelheim P, Propper C, Boeckers TM, Makuch L, Huganir R, Weide T, Brand E, Pavenstadt H, Kremerskothen J (2013) Common exonic missense variants in the C2 domain of the human KIBRA protein modify lipid binding and cognitive performance. Transl Psychiatry 3:e272. https://doi.org/10.1038/tp.2013.49

Dyson JM, Fedele CG, Davies EM, Becanovic J, Mitchell CA (2012) Phosphoinositide phosphatases: just as important as the kinases. Subcell Biochem 58:215–279. https://doi.org/10.1007/978-94-007-3012-0_7

Ebner M, Lucic I, Leonard TA, Yudushkin I (2017) PI (3,4,5)P3 engagement restricts Akt activity to cellular membranes. Mol Cell 65(3):416–431 e416. https://doi.org/10.1016/j.molcel.2016.12.028

Edwards SD, Keep NH (2001) The 2.7 A crystal structure of the activated FERM domain of moesin: an analysis of structural changes on activation. Biochemistry 40 (24):7061–7068

Egea-Jimenez AL, Gallardo R, Garcia-Pino A, Ivarsson Y, Wawrzyniak AM, Kashyap R, Loris R, Schymkowitz J, Rousseau F, Zimmermann P (2016) Frizzled 7 and PIP2 binding by syntenin PDZ2 domain supports Frizzled 7 trafficking and signalling. Nat Commun 7:12101. https://doi.org/10.1038/ncomms12101

Eisenberg-Bord M, Shai N, Schuldiner M, Bohnert M (2016) A tether is a tether is a tether: tethering at membrane contact sites. Dev Cell 39(4):395–409. https://doi.org/10.1016/j.devcel.2016.10.022

Ellson CD, Gobert-Gosse S, Anderson KE, Davidson K, Erdjument-Bromage H, Tempst P, Thuring JW, Cooper MA, Lim ZY, Holmes AB, Gaffney PR, Coadwell J, Chilvers ER, Hawkins PT, Stephens LR (2001) PtdIns(3)P regulates the neutrophil oxidase complex by binding to the PX domain of p40(phox). Nat Cell Biol 3(7):679–682. https://doi.org/10.1038/35083076

Emptage RP, Lemmon MA, Ferguson KM (2017a) Molecular determinants of KA1 domain-mediated autoinhibition and phospholipid activation of MARK1 kinase. Biochem J 474(3):385–398. https://doi.org/10.1042/BCJ20160792

Emptage RP, Schoenberger MJ, Ferguson KM, Marmorstein R (2017b) Intramolecular autoinhibition of checkpoint kinase 1 is mediated by conserved basic motifs of the C-terminal kinase associated-1 domain. J Biol Chem 292(46):19024–19033. https://doi.org/10.1074/jbc.M117.811265

Endo M, Shirouzu M, Yokoyama S (2003) The Cdc42 binding and scaffolding activities of the fission yeast adaptor protein Scd2. J Biol Chem 278(2):843–852. https://doi.org/10.1074/jbc.M209714200

Erneux C, Edimo WE, Deneubourg L, Pirson I (2011) SHIP2 multiple functions: a balance between a negative control of PtdIns(3,4,5)P(3) level, a positive control of PtdIns(3,4)P(2) production, and intrinsic docking properties. J Cell Biochem 112 (9):2203–2209. https://doi.org/10.1002/jcb.23146

Ernst A, Appleton BA, Ivarsson Y, Zhang Y, Gfeller D, Wiesmann C, Sidhu SS (2014) A structural portrait of the PDZ domain family. J Mol Biol 426 (21):3509–3519. https://doi.org/10.1016/j.jmb.2014.08.012

Falasca M, Hughes WE, Dominguez V, Sala G, Fostira F, Fang MQ, Cazzolli R, Shepherd PR, James DE, Maffucci T (2007) The role of phosphoinositide 3-kinase C2alpha in insulin signaling. J Biol Chem 282(38):28226–28236. https://doi.org/10.1074/jbc.M704357200

Ferguson KM, Lemmon MA, Schlessinger J, Sigler PB (1994) Crystal structure at 2.2 A resolution of the pleckstrin homology domain from human dynamin. Cell 79(2):199–209

Ferguson KM, Lemmon MA, Schlessinger J, Sigler PB (1995) Structure of the high affinity complex of inositol trisphosphate with a phospholipase C pleckstrin homology domain. Cell 83(6):1037–1046

Ferguson KM, Kavran JM, Sankaran VG, Fournier E, Isakoff SJ, Skolnik EY, Lemmon MA (2000) Structural basis for discrimination of 3-phosphoinositides by pleckstrin homology domains. Mol Cell 6(2):373–384

Fernandes F, Coutinho A, Prieto M, Loura LM (2015) Electrostatically driven lipid-protein interaction: answers from FRET. Biochim Biophys Acta 1848 (9):1837–1848. https://doi.org/10.1016/j.bbamem.2015.02.023

Flesch FM, Yu JW, Lemmon MA, Burger KN (2005) Membrane activity of the phospholipase C-delta1 pleckstrin homology (PH) domain. Biochem J 389 (Pt 2):435–441. https://doi.org/10.1042/BJ20041721

Ford MG, Pearse BM, Higgins MK, Vallis Y, Owen DJ, Gibson A, Hopkins CR, Evans PR, McMahon HT (2001) Simultaneous binding of PtdIns(4,5)P2 and clathrin by AP180 in the nucleation of clathrin lattices on membranes. Science 291(5506):1051–1055. https://doi.org/10.1126/science.291.5506.1051

Ford MG, Mills IG, Peter BJ, Vallis Y, Praefcke GJ, Evans PR, McMahon HT (2002) Curvature of clathrin-coated pits driven by epsin. Nature 419(6905):361–366. https://doi.org/10.1038/nature01020

Frame MC, Patel H, Serrels B, Lietha D, Eck MJ (2010) The FERM domain: organizing the structure and function of FAK. Nat Rev Mol Cell Biol 11(11):802–814. https://doi.org/10.1038/nrm2996

Franco I, Gulluni F, Campa CC, Costa C, Margaria JP, Ciraolo E, Martini M, Monteyne D, De Luca E, Germena G, Posor Y, Maffucci T, Marengo S, Haucke V, Falasca M, Perez-Morga D, Boletta A, Merlo GR, Hirsch E (2014) PI3K class II alpha controls spatially restricted endosomal PtdIns3P and Rab11 activation to promote primary cilium function. Dev Cell 28(6):647–658. https://doi.org/10.1016/j.devcel.2014.01.022

Frazier AA, Wisner MA, Malmberg NJ, Victor KG, Fanucci GE, Nalefski EA, Falke JJ, Cafiso DS (2002)

Membrane orientation and position of the C2 domain from cPLA2 by site-directed spin labeling. Biochemistry 41(20):6282–6292

Frost A, Unger VM, De Camilli P (2009) The BAR domain superfamily: membrane-molding macromolecules. Cell 137(2):191–196. https://doi.org/10.1016/j.cell.2009.04.010

Funderburk SF, Wang QJ, Yue Z (2010) The Beclin 1-VPS34 complex--at the crossroads of autophagy and beyond. Trends Cell Biol 20(6):355–362. https://doi.org/10.1016/j.tcb.2010.03.002

Gallardo R, Ivarsson Y, Schymkowitz J, Rousseau F, Zimmermann P (2010) Structural diversity of PDZ-lipid interactions. Chembiochem 11(4):456–467. https://doi.org/10.1002/cbic.200900616

Gallop JL, Jao CC, Kent HM, Butler PJ, Evans PR, Langen R, McMahon HT (2006) Mechanism of endophilin N-BAR domain-mediated membrane curvature. EMBO J 25(12):2898–2910. https://doi.org/10.1038/sj.emboj.7601174

Gambhir A, Hangyas-Mihalyne G, Zaitseva I, Cafiso DS, Wang J, Murray D, Pentyala SN, Smith SO, McLaughlin S (2004) Electrostatic sequestration of PIP2 on phospholipid membranes by basic/aromatic regions of proteins. Biophys J 86(4):2188–2207. https://doi.org/10.1016/S0006-3495(04)74278-2

Gamper N, Shapiro MS (2007) Regulation of ion transport proteins by membrane phosphoinositides. Nat Rev Neurosci 8(12):921–934. https://doi.org/10.1038/nrn2257

Garner K, Hunt AN, Koster G, Somerharju P, Groves E, Li M, Raghu P, Holic R, Cockcroft S (2012) Phosphatidylinositol transfer protein, cytoplasmic 1 (PITPNC1) binds and transfers phosphatidic acid. J Biol Chem 287(38):32263–32276. https://doi.org/10.1074/jbc.M112.375840

Gary JD, Wurmser AE, Bonangelino CJ, Weisman LS, Emr SD (1998) Fab1p is essential for PtdIns(3)P 5-kinase activity and the maintenance of vacuolar size and membrane homeostasis. J Cell Biol 143(1):65–79

Gaullier JM, Simonsen A, D'Arrigo A, Bremnes B, Stenmark H, Aasland R (1998) FYVE fingers bind PtdIns(3)P. Nature 394(6692):432–433. https://doi.org/10.1038/28767

Ghai R, Du X, Wang H, Dong J, Ferguson C, Brown AJ, Parton RG, Wu JW, Yang H (2017) ORP5 and ORP8 bind phosphatidylinositol-4, 5-biphosphate (PtdIns(4,5)P 2) and regulate its level at the plasma membrane. Nat Commun 8(1):757. https://doi.org/10.1038/s41467-017-00861-5

Gil JE, Kim E, Kim IS, Ku B, Park WS, Oh BH, Ryu SH, Cho W, Heo WD (2012) Phosphoinositides differentially regulate protrudin localization through the FYVE domain. J Biol Chem 287(49):41268–41276. https://doi.org/10.1074/jbc.M112.419127

Gillooly DJ, Morrow IC, Lindsay M, Gould R, Bryant NJ, Gaullier JM, Parton RG, Stenmark H (2000) Localization of phosphatidylinositol 3-phosphate in yeast and mammalian cells. EMBO J 19(17):4577–4588. https://doi.org/10.1093/emboj/19.17.4577

Giordano F, Saheki Y, Idevall-Hagren O, Colombo SF, Pirruccello M, Milosevic I, Gracheva EO, Bagriantsev SN, Borgese N, De Camilli P (2013) PI(4,5)P(2)-dependent and Ca(2+)-regulated ER-PM interactions mediated by the extended synaptotagmins. Cell 153 (7):1494–1509. https://doi.org/10.1016/j.cell.2013.05.026

Godi A, Pertile P, Meyers R, Marra P, Di Tullio G, Iurisci C, Luini A, Corda D, De Matteis MA (1999) ARF mediates recruitment of PtdIns-4-OH kinase-beta and stimulates synthesis of PtdIns(4,5)P2 on the Golgi complex. Nat Cell Biol 1(5):280–287. https://doi.org/10.1038/12993

Godi A, Di Campli A, Konstantakopoulos A, Di Tullio G, Alessi DR, Kular GS, Daniele T, Marra P, Lucocq JM, De Matteis MA (2004) FAPPs control Golgi-to-cell-surface membrane traffic by binding to ARF and PtdIns(4)P. Nat Cell Biol 6(5):393–404. https://doi.org/10.1038/ncb1119

Grabon A, Khan D, Bankaitis VA (2015) Phosphatidylinositol transfer proteins and instructive regulation of lipid kinase biology. Biochim Biophys Acta 1851 (6):724–735. https://doi.org/10.1016/j.bbalip.2014.12.011

Grabon A, Orlowski A, Tripathi A, Vuorio J, Javanainen M, Rog T, Lonnfors M, McDermott MI, Siebert G, Somerharju P, Vattulainen I, Bankaitis VA (2017) Dynamics and energetics of the mammalian phosphatidylinositol transfer protein phospholipid exchange cycle. J Biol Chem 292(35):14438–14455. https://doi.org/10.1074/jbc.M117.791467

Grobler JA, Essen LO, Williams RL, Hurley JH (1996) C2 domain conformational changes in phospholipase C-delta 1. Nat Struct Biol 3(9):788–795

Guerrero-Valero M, Ferrer-Orta C, Querol-Audi J, Marin-Vicente C, Fita I, Gomez-Fernandez JC, Verdaguer N, Corbalan-Garcia S (2009) Structural and mechanistic insights into the association of PKCalpha-C2 domain to PtdIns(4,5)P2. Proc Natl Acad Sci U S A 106 (16):6603–6607. https://doi.org/10.1073/pnas.0813099106

Guillen J, Ferrer-Orta C, Buxaderas M, Perez-Sanchez D, Guerrero-Valero M, Luengo-Gil G, Pous J, Guerra P, Gomez-Fernandez JC, Verdaguer N, Corbalan-Garcia S (2013) Structural insights into the Ca2+ and PI(4,5) P2 binding modes of the C2 domains of rabphilin 3A and synaptotagmin 1. Proc Natl Acad Sci U S A 110 (51):20503–20508. https://doi.org/10.1073/pnas.1316179110

Gulyas G, Radvanszki G, Matuska R, Balla A, Hunyady L, Balla T, Varnai P (2017) Plasma membrane phosphatidylinositol 4-phosphate and 4,5-bisphosphate determine the distribution and function of K-Ras4B but not H-Ras proteins. J Biol Chem 292(46):18862–18877. https://doi.org/10.1074/jbc.M117.806679

Ham H, Sreelatha A, Orth K (2011) Manipulation of host membranes by bacterial effectors. Nat Rev Microbiol 9 (9):635–646. https://doi.org/10.1038/nrmicro2602

Hamada K, Shimizu T, Matsui T, Tsukita S, Hakoshima T (2000) Structural basis of the membrane-targeting and unmasking mechanisms of the radixin FERM domain. EMBO J 19(17):4449–4462. https://doi.org/10.1093/emboj/19.17.4449

Hammond GR, Balla T (2015) Polyphosphoinositide binding domains: key to inositol lipid biology. Biochim Biophys Acta 1851(6):746–758. https://doi.org/10.1016/j.bbalip.2015.02.013

Hammond GR, Fischer MJ, Anderson KE, Holdich J, Koteci A, Balla T, Irvine RF (2012) PI4P and PI(4,5) P2 are essential but independent lipid determinants of membrane identity. Science 337(6095):727–730. https://doi.org/10.1126/science.1222483

Hammond GR, Machner MP, Balla T (2014) A novel probe for phosphatidylinositol 4-phosphate reveals multiple pools beyond the Golgi. J Cell Biol 205 (1):113–126. https://doi.org/10.1083/jcb.201312072

Hammond GR, Takasuga S, Sasaki T, Balla T (2015) The ML1Nx2 phosphatidylinositol 3,5-bisphosphate probe shows poor selectivity in cells. PLoS One 10(10): e0139957. https://doi.org/10.1371/journal.pone.0139957

Handa Y, Suzuki M, Ohya K, Iwai H, Ishijima N, Koleske AJ, Fukui Y, Sasakawa C (2007) Shigella IpgB1 promotes bacterial entry through the ELMO-Dock180 machinery. Nat Cell Biol 9(1):121–128. https://doi.org/10.1038/ncb1526

Harlan JE, Hajduk PJ, Yoon HS, Fesik SW (1994) Pleckstrin homology domains bind to phosphatidylinositol-4,5-bisphosphate. Nature 371 (6493):168–170. https://doi.org/10.1038/371168a0

Haslam RJ, Koide HB, Hemmings BA (1993) Pleckstrin domain homology. Nature 363(6427):309–310. https://doi.org/10.1038/363309b0

Hawkins PT, Stephens LR (2016) Emerging evidence of signalling roles for PI(3,4)P2 in Class I and II PI3K-regulated pathways. Biochem Soc Trans 44 (1):307–314. https://doi.org/10.1042/BST20150248

Hawkins PT, Jackson TR, Stephens LR (1992) Platelet-derived growth factor stimulates synthesis of PtdIns (3,4,5)P3 by activating a PtdIns(4,5)P2 3-OH kinase. Nature 358(6382):157–159. https://doi.org/10.1038/358157a0

Hayakawa A, Hayes SJ, Lawe DC, Sudharshan E, Tuft R, Fogarty K, Lambright D, Corvera S (2004) Structural basis for endosomal targeting by FYVE domains. J Biol Chem 279(7):5958–5966. https://doi.org/10.1074/jbc.M310503200

He J, Haney RM, Vora M, Verkhusha VV, Stahelin RV, Kutateladze TG (2008) Molecular mechanism of membrane targeting by the GRP1 PH domain. J Lipid Res 49(8):1807–1815. https://doi.org/10.1194/jlr.M800150-JLR200

He J, Vora M, Haney RM, Filonov GS, Musselman CA, Burd CG, Kutateladze AG, Verkhusha VV, Stahelin RV, Kutateladze TG (2009) Membrane insertion of the FYVE domain is modulated by pH. Proteins 76 (4):852–860. https://doi.org/10.1002/prot.22392

Hedger G, Sansom MSP (2016) Lipid interaction sites on channels, transporters and receptors: recent insights from molecular dynamics simulations. Biochim Biophys Acta 1858(10):2390–2400. https://doi.org/10.1016/j.bbamem.2016.02.037

Heo WD, Inoue T, Park WS, Kim ML, Park BO, Wandless TJ, Meyer T (2006) PI(3,4,5)P3 and PI(4,5)P2 lipids target proteins with polybasic clusters to the plasma membrane. Science 314(5804):1458–1461. https://doi.org/10.1126/science.1134389

Hertig S, Latorraca NR, Dror RO (2016) Revealing atomic-level mechanisms of protein allostery with molecular dynamics simulations. PLoS Comput Biol 12(6):e1004746. https://doi.org/10.1371/journal.pcbi.1004746

Hilgemann DW, Ball R (1996) Regulation of cardiac Na+, Ca2+ exchange and KATP potassium channels by PIP2. Science 273(5277):956–959

Hilgemann DW, Feng S, Nasuhoglu C (2001) The complex and intriguing lives of PIP2 with ion channels and transporters. Sci STKE 2001(111):re19. https://doi.org/10.1126/stke.2001.111.re19

Hill K, Krugmann S, Andrews SR, Coadwell WJ, Finan P, Welch HC, Hawkins PT, Stephens LR (2005) Regulation of P-Rex1 by phosphatidylinositol (3,4,5)-trisphosphate and Gbetagamma subunits. J Biol Chem 280(6):4166–4173. https://doi.org/10.1074/jbc.M411262200

Hille B, Dickson EJ, Kruse M, Vivas O, Suh BC (2015) Phosphoinositides regulate ion channels. Biochim Biophys Acta 1851(6):844–856. https://doi.org/10.1016/j.bbalip.2014.09.010

Hirano S, Suzuki N, Slagsvold T, Kawasaki M, Trambaiolo D, Kato R, Stenmark H, Wakatsuki S (2006) Structural basis of ubiquitin recognition by mammalian Eap45 GLUE domain. Nat Struct Mol Biol 13(11):1031–1032. https://doi.org/10.1038/nsmb1163

Hiroaki H, Ago T, Ito T, Sumimoto H, Kohda D (2001) Solution structure of the PX domain, a target of the SH3 domain. Nat Struct Biol 8(6):526–530. https://doi.org/10.1038/88591

Hnia K, Vaccari I, Bolino A, Laporte J (2012) Myotubularin phosphoinositide phosphatases: cellular functions and disease pathophysiology. Trends Mol Med 18(6):317–327. https://doi.org/10.1016/j.molmed.2012.04.004

Holthuis JC, Menon AK (2014) Lipid landscapes and pipelines in membrane homeostasis. Nature 510 (7503):48–57. https://doi.org/10.1038/nature13474

Honigmann A, van den Bogaart G, Iraheta E, Risselada HJ, Milovanovic D, Mueller V, Mullar S, Diederichsen U, Fasshauer D, Grubmuller H, Hell SW, Eggeling C, Kuhnel K, Jahn R (2013) Phosphatidylinositol 4,5-bisphosphate clusters act as molecular beacons for vesicle recruitment. Nat Struct Mol Biol 20 (6):679–686. https://doi.org/10.1038/nsmb.2570

Hospital A, Goni JR, Orozco M, Gelpi JL (2015) Molecular dynamics simulations: advances and applications. Adv Appl Bioinforma Chem 8:37–47. https://doi.org/10.2147/AABC.S70333

Hsu F, Mao Y (2015) The structure of phosphoinositide phosphatases: Insights into substrate specificity and catalysis. Biochim Biophys Acta 1851(6):698–710. https://doi.org/10.1016/j.bbalip.2014.09.015

Hsuan J, Cockcroft S (2001) The PITP family of phosphatidylinositol transfer proteins. Genome Biol 2 (9):REVIEWS3011

Huang CL, Feng S, Hilgemann DW (1998) Direct activation of inward rectifier potassium channels by PIP2 and its stabilization by Gbetagamma. Nature 391 (6669):803–806. https://doi.org/10.1038/35882

Huang Y, Shah V, Liu T, Keshvara L (2005) Signaling through disabled 1 requires phosphoinositide binding. Biochem Biophys Res Commun 331(4):1460–1468. https://doi.org/10.1016/j.bbrc.2005.04.064

Hurley JH (2008) ESCRT complexes and the biogenesis of multivesicular bodies. Curr Opin Cell Biol 20(1):4–11. https://doi.org/10.1016/j.ceb.2007.12.002

Hyman J, Chen H, Di Fiore PP, De Camilli P, Brunger AT (2000) Epsin 1 undergoes nucleocytosolic shuttling and its eps15 interactor NH(2)-terminal homology (ENTH) domain, structurally similar to Armadillo and HEAT repeats, interacts with the transcription factor promyelocytic leukemia Zn(2)+ finger protein (PLZF). J Cell Biol 149(3):537–546

Hyvonen M, Macias MJ, Nilges M, Oschkinat H, Saraste M, Wilmanns M (1995) Structure of the binding site for inositol phosphates in a PH domain. EMBO J 14(19):4676–4685

Iijima M, Huang YE, Luo HR, Vazquez F, Devreotes PN (2004) Novel mechanism of PTEN regulation by its phosphatidylinositol 4,5-bisphosphate binding motif is critical for chemotaxis. J Biol Chem 279 (16):16606–16613. https://doi.org/10.1074/jbc.M312098200

Ile KE, Schaaf G, Bankaitis VA (2006) Phosphatidylinositol transfer proteins and cellular nanoreactors for lipid signaling. Nat Chem Biol 2(11):576–583. https://doi.org/10.1038/nchembio835

Im YJ, Hurley JH (2008) Integrated structural model and membrane targeting mechanism of the human ESCRT-II complex. Dev Cell 14(6):902–913. https://doi.org/10.1016/j.devcel.2008.04.004

Im YJ, Raychaudhuri S, Prinz WA, Hurley JH (2005) Structural mechanism for sterol sensing and transport by OSBP-related proteins. Nature 437(7055):154–158. https://doi.org/10.1038/nature03923

Ingmundson A, Delprato A, Lambright DG, Roy CR (2007) Legionella pneumophila proteins that regulate Rab1 membrane cycling. Nature 450(7168):365–369. https://doi.org/10.1038/nature06336

Inoue H, Baba T, Sato S, Ohtsuki R, Takemori A, Watanabe T, Tagaya M, Tani K (2012) Roles of

SAM and DDHD domains in mammalian intracellular phospholipase A1 KIAA0725p. Biochim Biophys Acta 1823(4):930–939. https://doi.org/10.1016/j.bbamcr.2012.02.002

Irvine RF (2003) Nuclear lipid signalling. Nat Rev Mol Cell Biol 4(5):349–360. https://doi.org/10.1038/nrm1100

Isakoff SJ, Cardozo T, Andreev J, Li Z, Ferguson KM, Abagyan R, Lemmon MA, Aronheim A, Skolnik EY (1998) Identification and analysis of PH domain-containing targets of phosphatidylinositol 3-kinase using a novel in vivo assay in yeast. EMBO J 17 (18):5374–5387. https://doi.org/10.1093/emboj/17.18.5374

Itoh T, De Camilli P (2006) BAR, F-BAR (EFC) and ENTH/ANTH domains in the regulation of membrane-cytosol interfaces and membrane curvature. Biochim Biophys Acta 1761(8):897–912. https://doi.org/10.1016/j.bbalip.2006.06.015

Itoh T, Koshiba S, Kigawa T, Kikuchi A, Yokoyama S, Takenawa T (2001) Role of the ENTH domain in phosphatidylinositol-4,5-bisphosphate binding and endocytosis. Science 291(5506):1047–1051. https://doi.org/10.1126/science.291.5506.1047

Ivarsson Y (2012) Plasticity of PDZ domains in ligand recognition and signaling. FEBS Lett 586 (17):2638–2647. https://doi.org/10.1016/j.febslet.2012.04.015

Ivarsson Y, Wawrzyniak AM, Wuytens G, Kosloff M, Vermeiren E, Raport M, Zimmermann P (2011) Cooperative phosphoinositide and peptide binding by PSD-95/discs large/ZO-1 (PDZ) domain of polychaetoid, Drosophila zonulin. J Biol Chem 286 (52):44669–44678. https://doi.org/10.1074/jbc.M111.285734

Ivarsson Y, Wawrzyniak AM, Kashyap R, Polanowska J, Betzi S, Lembo F, Vermeiren E, Chiheb D, Lenfant N, Morelli X, Borg JP, Reboul J, Zimmermann P (2013) Prevalence, specificity and determinants of lipid-interacting PDZ domains from an in-cell screen and in vitro binding experiments. PLoS One 8(2):e54581. https://doi.org/10.1371/journal.pone.0054581

Jackson CL, Walch L, Verbavatz JM (2016) Lipids and their trafficking: an integral part of cellular organization. Dev Cell 39(2):139–153. https://doi.org/10.1016/j.devcel.2016.09.030

Jain A, Holthuis JCM (2017) Membrane contact sites, ancient and central hubs of cellular lipid logistics. Biochim Biophys Acta 1864(9):1450–1458. https://doi.org/10.1016/j.bbamcr.2017.05.017

Jank T, Bohmer KE, Tzivelekidis T, Schwan C, Belyi Y, Aktories K (2012) Domain organization of Legionella effector SetA. Cell Microbiol 14(6):852–868. https://doi.org/10.1111/j.1462-5822.2012.01761.x

Jarsch IK, Daste F, Gallop JL (2016) Membrane curvature in cell biology: an integration of molecular mechanisms. J Cell Biol 214(4):375–387. https://doi.org/10.1083/jcb.201604003

Jaworski CJ, Moreira E, Li A, Lee R, Rodriguez IR (2001) A family of 12 human genes containing oxysterol-binding domains. Genomics 78(3):185–196. https://doi.org/10.1006/geno.2001.6663

Jayasundar JJ, Ju JH, He L, Liu D, Meilleur F, Zhao J, Callaway DJ, Bu Z (2012) Open conformation of ezrin bound to phosphatidylinositol 4,5-bisphosphate and to F-actin revealed by neutron scattering. J Biol Chem 287(44):37119–37133. https://doi.org/10.1074/jbc.M112.380972

Jian X, Tang WK, Zhai P, Roy NS, Luo R, Gruschus JM, Yohe ME, Chen PW, Li Y, Byrd RA, Xia D, Randazzo PA (2015) Molecular basis for cooperative binding of anionic phospholipids to the PH domain of the Arf GAP ASAP1. Structure 23(11):1977–1988. https://doi.org/10.1016/j.str.2015.08.008

Johansson M, Bocher V, Lehto M, Chinetti G, Kuismanen E, Ehnholm C, Staels B, Olkkonen VM (2003) The two variants of oxysterol binding protein-related protein-1 display different tissue expression patterns, have different intracellular localization, and are functionally distinct. Mol Biol Cell 14(3):903–915. https://doi.org/10.1091/mbc.E02-08-0459

Joseph RE, Wales TE, Fulton DB, Engen JR, Andreotti AH (2017) Achieving a graded immune response: BTK adopts a range of active/inactive conformations dictated by multiple interdomain contacts. Structure 25 (10):1481–1494 e1484. https://doi.org/10.1016/j.str.2017.07.014

Jungmichel S, Sylvestersen KB, Choudhary C, Nguyen S, Mann M, Nielsen ML (2014) Specificity and commonality of the phosphoinositide-binding proteome analyzed by quantitative mass spectrometry. Cell Rep 6(3):578–591. https://doi.org/10.1016/j.celrep.2013.12.038

Kam JL, Miura K, Jackson TR, Gruschus J, Roller P, Stauffer S, Clark J, Aneja R, Randazzo PA (2000) Phosphoinositide-dependent activation of the ADP-ribosylation factor GTPase-activating protein ASAP1. Evidence for the pleckstrin homology domain functioning as an allosteric site. J Biol Chem 275 (13):9653–9663

Kanai F, Liu H, Field SJ, Akbary H, Matsuo T, Brown GE, Cantley LC, Yaffe MB (2001) The PX domains of p47phox and p40phox bind to lipid products of PI(3) K. Nat Cell Biol 3(7):675–678. https://doi.org/10.1038/35083070

Kaneko T, Joshi R, Feller SM, Li SS (2012) Phosphotyrosine recognition domains: the typical, the atypical and the versatile. Cell Commun Signal 10 (1):32. https://doi.org/10.1186/1478-811X-10-32

Karathanassis D, Stahelin RV, Bravo J, Perisic O, Pacold CM, Cho W, Williams RL (2002) Binding of the PX domain of p47(phox) to phosphatidylinositol 3,4-bisphosphate and phosphatidic acid is masked by an intramolecular interaction. EMBO J 21 (19):5057–5068

Kavran JM, Klein DE, Lee A, Falasca M, Isakoff SJ, Skolnik EY, Lemmon MA (1998) Specificity and promiscuity in phosphoinositide binding by pleckstrin homology domains. J Biol Chem 273 (46):30497–30508

Kay BK, Yamabhai M, Wendland B, Emr SD (1999) Identification of a novel domain shared by putative components of the endocytic and cytoskeletal machinery. Protein Sci 8(2):435–438. https://doi.org/10.1110/ps.8.2.435

Kentala H, Weber-Boyvat M, Olkkonen VM (2016) OSBP-related protein family: mediators of lipid transport and signaling at membrane contact sites. Int Rev Cell Mol Biol 321:299–340. https://doi.org/10.1016/bs.ircmb.2015.09.006

Ketel K, Krauss M, Nicot AS, Puchkov D, Wieffer M, Muller R, Subramanian D, Schultz C, Laporte J, Haucke V (2016) A phosphoinositide conversion mechanism for exit from endosomes. Nature 529 (7586):408–412. https://doi.org/10.1038/nature16516

Killian JA, von Heijne G (2000) How proteins adapt to a membrane-water interface. Trends Biochem Sci 25 (9):429–434

Kim YJ, Guzman-Hernandez ML, Balla T (2011) A highly dynamic ER-derived phosphatidylinositol-synthesizing organelle supplies phosphoinositides to cellular membranes. Dev Cell 21(5):813–824. https://doi.org/10.1016/j.devcel.2011.09.005

Kim S, Kedan A, Marom M, Gavert N, Keinan O, Selitrennik M, Laufman O, Lev S (2013a) The phosphatidylinositol-transfer protein Nir2 binds phosphatidic acid and positively regulates phosphoinositide signalling. EMBO Rep 14(10):891–899. https://doi.org/10.1038/embor.2013.113

Kim YJ, Hernandez ML, Balla T (2013b) Inositol lipid regulation of lipid transfer in specialized membrane domains. Trends Cell Biol 23(6):270–278. https://doi.org/10.1016/j.tcb.2013.01.009

Kim YJ, Guzman-Hernandez ML, Wisniewski E, Balla T (2015) Phosphatidylinositol-phosphatidic acid exchange by Nir2 at ER-PM contact sites maintains phosphoinositide signaling competence. Dev Cell 33 (5):549–561. https://doi.org/10.1016/j.devcel.2015.04.028

Kim YJ, Guzman-Hernandez ML, Wisniewski E, Echeverria N, Balla T (2016) Phosphatidylinositol and phosphatidic acid transport between the ER and plasma membrane during PLC activation requires the Nir2 protein. Biochem Soc Trans 44(1):197–201. https://doi.org/10.1042/BST20150187

Kleyn PW, Fan W, Kovats SG, Lee JJ, Pulido JC, Wu Y, Berkemeier LR, Misumi DJ, Holmgren L, Charlat O, Woolf EA, Tayber O, Brody T, Shu P, Hawkins F, Kennedy B, Baldini L, Ebeling C, Alperin GD, Deeds J, Lakey ND, Culpepper J, Chen H, Glucksmann-Kuis MA, Carlson GA, Duyk GM, Moore KJ (1996) Identification and characterization of the mouse obesity gene tubby: a member of a novel gene family. Cell 85(2):281–290

Knight JD, Falke JJ (2009) Single-molecule fluorescence studies of a PH domain: new insights into the membrane docking reaction. Biophys J 96 (2):566–582. https://doi.org/10.1016/j.bpj.2008.10.020

Kohout SC, Corbalan-Garcia S, Gomez-Fernandez JC, Falke JJ (2003) C2 domain of protein kinase C alpha: elucidation of the membrane docking surface by site-directed fluorescence and spin labeling. Biochemistry 42(5):1254–1265. https://doi.org/10.1021/bi026596f

Konermann L, Vahidi S, Sowole MA (2014) Mass spectrometry methods for studying structure and dynamics of biological macromolecules. Anal Chem 86 (1):213–232. https://doi.org/10.1021/ac4039306

Kontos CD, Stauffer TP, Yang WP, York JD, Huang L, Blanar MA, Meyer T, Peters KG (1998) Tyrosine 1101 of Tie2 is the major site of association of p85 and is required for activation of phosphatidylinositol 3-kinase and Akt. Mol Cell Biol 18(7):4131–4140

Krick R, Tolstrup J, Appelles A, Henke S, Thumm M (2006) The relevance of the phosphatidylinositolphosphat-binding motif FRRGT of Atg18 and Atg21 for the Cvt pathway and autophagy. FEBS Lett 580(19):4632–4638. https://doi.org/10.1016/j.febslet.2006.07.041

Krick R, Busse RA, Scacioc A, Stephan M, Janshoff A, Thumm M, Kuhnel K (2012) Structural and functional characterization of the two phosphoinositide binding sites of PROPPINs, a beta-propeller protein family. Proc Natl Acad Sci U S A 109(30):E2042–E2049. https://doi.org/10.1073/pnas.1205128109

Kular GS, Chaudhary A, Prestwich G, Swigart P, Wetzker R, Cockcroft S (2002) Co-operation of phosphatidylinositol transfer protein with phosphoinositide 3-kinase gamma in vitro. Adv Enzym Regul 42:53–61

Kulkarni S, Das S, Funk CD, Murray D, Cho W (2002) Molecular basis of the specific subcellular localization of the C2-like domain of 5-lipoxygenase. J Biol Chem 277(15):13167–13174. https://doi.org/10.1074/jbc.M112393200

Kutateladze TG (2006) Phosphatidylinositol 3-phosphate recognition and membrane docking by the FYVE domain. Biochim Biophys Acta 1761(8):868–877. https://doi.org/10.1016/j.bbalip.2006.03.011

Kutateladze TG (2010) Translation of the phosphoinositide code by PI effectors. Nat Chem Biol 6 (7):507–513. https://doi.org/10.1038/nchembio.390

Kutateladze T, Overduin M (2001) Structural mechanism of endosome docking by the FYVE domain. Science 291(5509):1793–1796. https://doi.org/10.1126/science.291.5509.1793

Kutateladze TG, Ogburn KD, Watson WT, de Beer T, Emr SD, Burd CG, Overduin M (1999) Phosphatidylinositol 3-phosphate recognition by the FYVE domain. Mol Cell 3(6):805–811

Kutateladze TG, Capelluto DG, Ferguson CG, Cheever ML, Kutateladze AG, Prestwich GD, Overduin M (2004) Multivalent mechanism of membrane insertion by the FYVE domain. J Biol Chem 279(4):3050–3057. https://doi.org/10.1074/jbc.M309007200

Kweon DH, Shin YK, Shin JY, Lee JH, Lee JB, Seo JH, Kim YS (2006) Membrane topology of helix 0 of the Epsin N-terminal homology domain. Mol Cells 21 (3):428–435

Laganowsky A, Reading E, Allison TM, Ulmschneider MB, Degiacomi MT, Baldwin AJ, Robinson CV (2014) Membrane proteins bind lipids selectively to modulate their structure and function. Nature 510 (7503):172–175. https://doi.org/10.1038/nature13419

Lai CL, Landgraf KE, Voth GA, Falke JJ (2010) Membrane docking geometry and target lipid stoichiometry of membrane-bound PKCalpha C2 domain: a combined molecular dynamics and experimental study. J Mol Biol 402(2):301–310. https://doi.org/10.1016/j.jmb.2010.07.037

Lai CL, Jao CC, Lyman E, Gallop JL, Peter BJ, McMahon HT, Langen R, Voth GA (2012) Membrane binding and self-association of the epsin N-terminal homology domain. J Mol Biol 423(5):800–817. https://doi.org/10.1016/j.jmb.2012.08.010

Larijani B, Allen-Baume V, Morgan CP, Li M, Cockcroft S (2003) EGF regulation of PITP dynamics is blocked by inhibitors of phospholipase C and of the Ras-MAP kinase pathway. Curr Biol 13(1):78–84

Lawe DC, Patki V, Heller-Harrison R, Lambright D, Corvera S (2000) The FYVE domain of early endosome antigen 1 is required for both phosphatidylinositol 3-phosphate and Rab5 binding. Critical role of this dual interaction for endosomal localization. J Biol Chem 275(5):3699–3705

Lee AG (2003) Lipid-protein interactions in biological membranes: a structural perspective. Biochim Biophys Acta 1612(1):1–40

Lee SA, Eyeson R, Cheever ML, Geng J, Verkhusha VV, Burd C, Overduin M, Kutateladze TG (2005) Targeting of the FYVE domain to endosomal membranes is regulated by a histidine switch. Proc Natl Acad Sci U S A 102(37):13052–13057. https://doi.org/10.1073/pnas.0503900102

Legendre-Guillemin V, Wasiak S, Hussain NK, Angers A, McPherson PS (2004) ENTH/ANTH proteins and clathrin-mediated membrane budding. J Cell Sci 117 (Pt 1):9–18. https://doi.org/10.1242/jcs.00928

Lehto M, Laitinen S, Chinetti G, Johansson M, Ehnholm C, Staels B, Ikonen E, Olkkonen VM (2001) The OSBP-related protein family in humans. J Lipid Res 42(8):1203–1213

Lemmon MA (2003) Phosphoinositide recognition domains. Traffic 4(4):201–213

Lemmon MA (2004) Pleckstrin homology domains: not just for phosphoinositides. Biochem Soc Trans 32. (Pt 5:707–711. https://doi.org/10.1042/BST0320707

Lemmon MA (2007) Pleckstrin homology (PH) domains and phosphoinositides. Biochem Soc Symp 74:81–93. https://doi.org/10.1042/BSS0740081

Lemmon MA (2008) Membrane recognition by phospholipid-binding domains. Nat Rev Mol Cell Biol 9(2):99–111. https://doi.org/10.1038/nrm2328

Lemmon MA, Ferguson KM (2001) Molecular determinants in pleckstrin homology domains that allow specific recognition of phosphoinositides. Biochem Soc Trans 29(Pt 4):377–384

Lemmon MA, Ferguson KM, O'Brien R, Sigler PB, Schlessinger J (1995) Specific and high-affinity binding of inositol phosphates to an isolated pleckstrin homology domain. Proc Natl Acad Sci U S A 92 (23):10472–10476

Lemmon MA, Ferguson KM, Abrams CS (2002) Pleckstrin homology domains and the cytoskeleton. FEBS Lett 513(1):71–76

Lenoir M, Kufareva I, Abagyan R, Overduin M (2015) Membrane and protein interactions of the pleckstrin homology domain superfamily. Membranes (Basel) 5 (4):646–663. https://doi.org/10.3390/membranes5040646

Lev S (2010) Non-vesicular lipid transport by lipid-transfer proteins and beyond. Nat Rev Mol Cell Biol 11(10):739–750. https://doi.org/10.1038/nrm2971

Levine TP, Munro S (1998) The pleckstrin homology domain of oxysterol-binding protein recognises a determinant specific to Golgi membranes. Curr Biol 8 (13):729–739

Li H, Marshall AJ (2015) Phosphatidylinositol (3,4) bisphosphate-specific phosphatases and effector proteins: a distinct branch of PI3K signaling. Cell Signal 27(9):1789–1798. https://doi.org/10.1016/j.cellsig.2015.05.013

Li X, Routt SM, Xie Z, Cui X, Fang M, Kearns MA, Bard M, Kirsch DR, Bankaitis VA (2000) Identification of a novel family of nonclassic yeast phosphatidylinositol transfer proteins whose function modulates phospholipase D activity and Sec14p-independent cell growth. Mol Biol Cell 11(6):1989–2005

Li H, Tremblay JM, Yarbrough LR, Helmkamp GM Jr (2002) Both isoforms of mammalian phosphatidylinositol transfer protein are capable of binding and transporting sphingomyelin. Biochim Biophys Acta 1580(1):67–76

Li J, Mao X, Dong LQ, Liu F, Tong L (2007) Crystal structures of the BAR-PH and PTB domains of human APPL1. Structure 15(5):525–533. https://doi.org/10.1016/j.str.2007.03.011

Liang B, Tamm LK (2016) NMR as a tool to investigate the structure, dynamics and function of membrane proteins. Nat Struct Mol Biol 23(6):468–474. https://doi.org/10.1038/nsmb.3226

Lietzke SE, Bose S, Cronin T, Klarlund J, Chawla A, Czech MP, Lambright DG (2000) Structural basis of 3-phosphoinositide recognition by pleckstrin homology domains. Mol Cell 6(2):385–394

Lindahl E, Sansom MS (2008) Membrane proteins: molecular dynamics simulations. Curr Opin Struct Biol 18 (4):425–431. https://doi.org/10.1016/j.sbi.2008.02.003

Liu X, Ridgway ND (2014) Characterization of the sterol and phosphatidylinositol 4-phosphate binding properties of Golgi-associated OSBP-related protein 9 (ORP9). PLoS One 9(9):e108368. https://doi.org/10.1371/journal.pone.0108368

Liu Y, Kahn RA, Prestegard JH (2014) Interaction of Fapp1 with Arf1 and PI4P at a membrane surface: an example of coincidence detection. Structure 22 (3):421–430. https://doi.org/10.1016/j.str.2013.12.011

Lo WT, Vujicic Zagar A, Gerth F, Lehmann M, Puchkov D, Krylova O, Freund C, Scapozza L, Vadas O, Haucke V (2017) A coincidence detection mechanism controls PX-BAR domain-mediated endocytic membrane remodeling via an allosteric structural switch. Dev Cell 43(4):522–529.e524. https://doi.org/10.1016/j.devcel.2017.10.019

Loewen CJ, Levine TP (2005) A highly conserved binding site in vesicle-associated membrane protein-associated protein (VAP) for the FFAT motif of lipid-binding proteins. J Biol Chem 280(14):14097–14104. https://doi.org/10.1074/jbc.M500147200

Lomize AL, Pogozheva ID, Lomize MA, Mosberg HI (2007) The role of hydrophobic interactions in positioning of peripheral proteins in membranes. BMC Struct Biol 7:44. https://doi.org/10.1186/1472-6807-7-44

Lorenzo O, Urbe S, Clague MJ (2005) Analysis of phosphoinositide binding domain properties within the myotubularin-related protein MTMR3. J Cell Sci 118(Pt 9):2005–2012. https://doi.org/10.1242/jcs.02325

Lumb CN, Sansom MS (2012) Finding a needle in a haystack: the role of electrostatics in target lipid recognition by PH domains. PLoS Comput Biol 8(7): e1002617. https://doi.org/10.1371/journal.pcbi.1002617

Lumb CN, He J, Xue Y, Stansfeld PJ, Stahelin RV, Kutateladze TG, Sansom MS (2011) Biophysical and computational studies of membrane penetration by the GRP1 pleckstrin homology domain. Structure 19 (9):1338–1346. https://doi.org/10.1016/j.str.2011.04.010

Luo X, Wasilko DJ, Liu Y, Sun J, Wu X, Luo ZQ, Mao Y (2015) Structure of the legionella virulence factor, SidC reveals a unique PI(4)P-specific binding domain essential for its targeting to the bacterial phagosome. PLoS Pathog 11(6):e1004965. https://doi.org/10.1371/journal.ppat.1004965

Machner MP, Isberg RR (2006) Targeting of host Rab GTPase function by the intravacuolar pathogen legionella pneumophila. Dev Cell 11(1):47–56. https://doi.org/10.1016/j.devcel.2006.05.013

Macias MJ, Musacchio A, Ponstingl H, Nilges M, Saraste M, Oschkinat H (1994) Structure of the pleckstrin homology domain from beta-spectrin. Nature 369(6482):675–677. https://doi.org/10.1038/369675a0

Maffucci T, Falasca M (2001) Specificity in pleckstrin homology (PH) domain membrane targeting: a role for a phosphoinositide-protein co-operative mechanism. FEBS Lett 506(3):173–179

Malaby AW, van den Berg B, Lambright DG (2013) Structural basis for membrane recruitment and allosteric activation of cytohesin family Arf GTPase exchange factors. Proc Natl Acad Sci U S A 110 (35):14213–14218. https://doi.org/10.1073/pnas.1301883110

Malek M, Kielkowska A, Chessa T, Anderson KE, Barneda D, Pir P, Nakanishi H, Eguchi S, Koizumi A, Sasaki J, Juvin V, Kiselev VY, Niewczas I, Gray A, Valayer A, Spensberger D,

Imbert M, Felisbino S, Habuchi T, Beinke S, Cosulich S, Le Novere N, Sasaki T, Clark J, Hawkins PT, Stephens LR (2017) PTEN regulates PI(3,4)P2 signaling downstream of class I PI3K. Mol Cell 68 (3):566–580.e510. https://doi.org/10.1016/j.molcel.2017.09.024

Malkova S, Stahelin RV, Pingali SV, Cho W, Schlossman ML (2006) Orientation and penetration depth of monolayer-bound p40phox-PX. Biochemistry 45 (45):13566–13575. https://doi.org/10.1021/bi0611331

Manna D, Albanese A, Park WS, Cho W (2007) Mechanistic basis of differential cellular responses of phosphatidylinositol 3,4-bisphosphate- and phosphatidylinositol 3,4,5-trisphosphate-binding pleckstrin homology domains. J Biol Chem 282 (44):32093–32105. https://doi.org/10.1074/jbc.M703517200

Manna D, Bhardwaj N, Vora MS, Stahelin RV, Lu H, Cho W (2008) Differential roles of phosphatidylserine, PtdIns(4,5)P2, and PtdIns(3,4,5)P3 in plasma membrane targeting of C2 domains. Molecular dynamics simulation, membrane binding, and cell translocation studies of the PKCalpha C2 domain. J Biol Chem 283 (38):26047–26058. https://doi.org/10.1074/jbc.M802617200

Mao Y, Nickitenko A, Duan X, Lloyd TE, Wu MN, Bellen H, Quiocho FA (2000) Crystal structure of the VHS and FYVE tandem domains of Hrs, a protein involved in membrane trafficking and signal transduction. Cell 100(4):447–456

Mao Y, Chen J, Maynard JA, Zhang B, Quiocho FA (2001) A novel all helix fold of the AP180 amino-terminal domain for phosphoinositide binding and clathrin assembly in synaptic vesicle endocytosis. Cell 104(3):433–440

Marat AL, Haucke V (2016) Phosphatidylinositol 3-phosphates-at the interface between cell signalling and membrane traffic. EMBO J 35(6):561–579. https://doi.org/10.15252/embj.201593564

Marat AL, Wallroth A, Lo WT, Muller R, Norata GD, Falasca M, Schultz C, Haucke V (2017) mTORC1 activity repression by late endosomal phosphatidylinositol 3,4-bisphosphate. Science 356 (6341):968–972. https://doi.org/10.1126/science.aaf8310

Marsh D (2008) Protein modulation of lipids, and vice-versa, in membranes. Biochim Biophys Acta 1778 (7–8):1545–1575. https://doi.org/10.1016/j.bbamem.2008.01.015

Martelli AM, Ognibene A, Buontempo F, Fini M, Bressanin D, Goto K, McCubrey JA, Cocco L, Evangelisti C (2011) Nuclear phosphoinositides and their roles in cell biology and disease. Crit Rev Biochem Mol Biol 46(5):436–457. https://doi.org/10.3109/10409238.2011.609530

Masuda M, Takeda S, Sone M, Ohki T, Mori H, Kamioka Y, Mochizuki N (2006) Endophilin BAR domain drives membrane curvature by two newly identified structure-based mechanisms. EMBO J 25 (12):2889–2897. https://doi.org/10.1038/sj.emboj.7601176

Mattila PK, Pykäläinen A, Saarikangas J, Paavilainen VO, Vihinen H, Jokitalo E, Lappalainen P (2007) Missing-in-metastasis and IRSp53 deform PI(4,5)P2-rich membranes by an inverse BAR domain–like mechanism. J Cell Biol 176(7):953–964. https://doi.org/10.1083/jcb.200609176

Mayer BJ, Ren R, Clark KL, Baltimore D (1993) A putative modular domain present in diverse signaling proteins. Cell 73(4):629–630

McCrea HJ, De Camilli P (2009) Mutations in phosphoinositide metabolizing enzymes and human disease. Physiology (Bethesda) 24:8–16. https://doi.org/10.1152/physiol.00035.2008

McLaughlin S, Murray D (2005) Plasma membrane phosphoinositide organization by protein electrostatics. Nature 438(7068):605–611. https://doi.org/10.1038/nature04398

McMahon HT, Gallop JL (2005) Membrane curvature and mechanisms of dynamic cell membrane remodelling. Nature 438(7068):590–596. https://doi.org/10.1038/nature04396

Mesmin B, Bigay J, Moser von Filseck J, Lacas-Gervais S, Drin G, Antonny B (2013) A four-step cycle driven by PI(4)P hydrolysis directs sterol/PI(4)P exchange by the ER-Golgi tether OSBP. Cell 155(4):830–843. https://doi.org/10.1016/j.cell.2013.09.056

Michell RH, Heath VL, Lemmon MA, Dove SK (2006) Phosphatidylinositol 3,5-bisphosphate: metabolism and cellular functions. Trends Biochem Sci 31 (1):52–63. https://doi.org/10.1016/j.tibs.2005.11.013

Millard TH, Bompard G, Heung MY, Dafforn TR, Scott DJ, Machesky LM, Futterer K (2005) Structural basis of filopodia formation induced by the IRSp53/MIM homology domain of human IRSp53. EMBO J 24 (2):240–250. https://doi.org/10.1038/sj.emboj.7600535

Miller SE, Sahlender DA, Graham SC, Honing S, Robinson MS, Peden AA, Owen DJ (2011) The molecular basis for the endocytosis of small R-SNAREs by the clathrin adaptor CALM. Cell 147(5):1118–1131. https://doi.org/10.1016/j.cell.2011.10.038

Miller SE, Mathiasen S, Bright NA, Pierre F, Kelly BT, Kladt N, Schauss A, Merrifield CJ, Stamou D, Honing S, Owen DJ (2015) CALM regulates clathrin-coated vesicle size and maturation by directly sensing and driving membrane curvature. Dev Cell 33 (2):163–175. https://doi.org/10.1016/j.devcel.2015.03.002

Mim C, Unger VM (2012) Membrane curvature and its generation by BAR proteins. Trends Biochem Sci 37 (12):526–533. https://doi.org/10.1016/j.tibs.2012.09.001

Mim C, Cui H, Gawronski-Salerno JA, Frost A, Lyman E, Voth GA, Unger VM (2012) Structural basis of membrane bending by the N-BAR protein endophilin. Cell 149(1):137–145. https://doi.org/10.1016/j.cell.2012.01.048

Misawa H, Ohtsubo M, Copeland NG, Gilbert DJ, Jenkins NA, Yoshimura A (1998) Cloning and characterization of a novel class II phosphoinositide 3-kinase containing C2 domain. Biochem Biophys Res

Commun 244(2):531–539. https://doi.org/10.1006/bbrc.1998.8294

Misra S, Hurley JH (1999) Crystal structure of a phosphatidylinositol 3-phosphate-specific membrane-targeting motif, the FYVE domain of Vps27p. Cell 97(5):657–666

Moleirinho S, Tilston-Lunel A, Angus L, Gunn-Moore F, Reynolds PA (2013) The expanding family of FERM proteins. Biochem J 452(2):183–193. https://doi.org/10.1042/BJ20121642

Monje-Galvan V, Klauda JB (2016) Peripheral membrane proteins: tying the knot between experiment and computation. Biochim Biophys Acta 1858(7 Pt B):1584–1593. https://doi.org/10.1016/j.bbamem.2016.02.018

Montaville P, Coudevylle N, Radhakrishnan A, Leonov A, Zweckstetter M, Becker S (2008) The PIP2 binding mode of the C2 domains of rabphilin-3A. Protein Sci 17(6):1025–1034. https://doi.org/10.1110/ps.073326608

Morales KA, Yang Y, Cole TR, Igumenova TI (2016) Dynamic response of the C2 domain of protein kinase calpha to Ca(2+) binding. Biophys J 111(8):1655–1667. https://doi.org/10.1016/j.bpj.2016.09.008

Moravcevic K, Mendrola JM, Schmitz KR, Wang YH, Slochower D, Janmey PA, Lemmon MA (2010) Kinase associated-1 domains drive MARK/PAR1 kinases to membrane targets by binding acidic phospholipids. Cell 143(6):966–977. https://doi.org/10.1016/j.cell.2010.11.028

Moravcevic K, Oxley CL, Lemmon MA (2012) Conditional peripheral membrane proteins: facing up to limited specificity. Structure 20(1):15–27. https://doi.org/10.1016/j.str.2011.11.012

Moravcevic K, Alvarado D, Schmitz KR, Kenniston JA, Mendrola JM, Ferguson KM, Lemmon MA (2015) Comparison of Saccharomyces cerevisiae F-BAR domain structures reveals a conserved inositol phosphate binding site. Structure 23(2):352–363. https://doi.org/10.1016/j.str.2014.12.009

Mortier E, Wuytens G, Leenaerts I, Hannes F, Heung MY, Degeest G, David G, Zimmermann P (2005) Nuclear speckles and nucleoli targeting by PIP2-PDZ domain interactions. EMBO J 24(14):2556–2565. https://doi.org/10.1038/sj.emboj.7600722

Moser von Filseck J, Copic A, Delfosse V, Vanni S, Jackson CL, Bourguet W, Drin G (2015a) Intracellular transport. Phosphatidylserine transport by ORP/Osh proteins is driven by phosphatidylinositol 4-phosphate. Science 349(6246):432–436. https://doi.org/10.1126/science.aab1346

Moser von Filseck J, Vanni S, Mesmin B, Antonny B, Drin G (2015b) A phosphatidylinositol-4-phosphate powered exchange mechanism to create a lipid gradient between membranes. Nat Commun 6:6671. https://doi.org/10.1038/ncomms7671

Mu Y, Cai P, Hu S, Ma S, Gao Y (2014) Characterization of diverse internal binding specificities of PDZ domains by yeast two-hybrid screening of a special peptide library. PLoS One 9(2):e88286. https://doi.org/10.1371/journal.pone.0088286

Muallem S, Chung WY, Jha A, Ahuja M (2017) Lipids at membrane contact sites: cell signaling and ion transport. EMBO Rep 18(11):1893–1904. https://doi.org/10.15252/embr.201744331

Mukhopadhyay S, Jackson PK (2011) The tubby family proteins. Genome Biol 12(6):225. https://doi.org/10.1186/gb-2011-12-6-225

Mulgrew-Nesbitt A, Diraviyam K, Wang J, Singh S, Murray P, Li Z, Rogers L, Mirkovic N, Murray D (2006) The role of electrostatics in protein-membrane interactions. Biochim Biophys Acta 1761(8):812–826. https://doi.org/10.1016/j.bbalip.2006.07.002

Murphy SE, Levine TP (2016) VAP, a versatile access point for the endoplasmic reticulum: review and analysis of FFAT-like motifs in the VAPome. Biochim Biophys Acta 1861(8 Pt B):952–961. https://doi.org/10.1016/j.bbalip.2016.02.009

Murray D, Honig B (2002) Electrostatic control of the membrane targeting of C2 domains. Mol Cell 9(1):145–154

Musacchio A, Gibson T, Rice P, Thompson J, Saraste M (1993) The PH domain: a common piece in the structural patchwork of signalling proteins. Trends Biochem Sci 18(9):343–348

Nakatsu F, Baskin JM, Chung J, Tanner LB, Shui G, Lee SY, Pirruccello M, Hao M, Ingolia NT, Wenk MR, De Camilli P (2012) PtdIns4P synthesis by PI4KIIIalpha at the plasma membrane and its impact on plasma membrane identity. J Cell Biol 199(6):1003–1016. https://doi.org/10.1083/jcb.201206095

Nalefski EA, McDonagh T, Somers W, Seehra J, Falke JJ, Clark JD (1998) Independent folding and ligand specificity of the C2 calcium-dependent lipid binding domain of cytosolic phospholipase A2. J Biol Chem 273(3):1365–1372

Naughton FB, Kalli AC, Sansom MS (2016) Association of peripheral membrane proteins with membranes: free energy of binding of GRP1 PH domain with phosphatidylinositol phosphate-containing model bilayers. J Phys Chem Lett 7(7):1219–1224. https://doi.org/10.1021/acs.jpclett.6b00153

Nawrotek A, Zeghouf M, Cherfils J (2016) Allosteric regulation of Arf GTPases and their GEFs at the membrane interface. Small GTPases 7(4):283–296. https://doi.org/10.1080/21541248.2016.1215778

Noben-Trauth K, Naggert JK, North MA, Nishina PM (1996) A candidate gene for the mouse mutation tubby. Nature 380(6574):534–538. https://doi.org/10.1038/380534a0

Nourry C, Grant SG, Borg JP (2003) PDZ domain proteins: plug and play! Sci STKE 2003(179):RE7. https://doi.org/10.1126/stke.2003.179.re7

Obara K, Sekito T, Niimi K, Ohsumi Y (2008) The Atg18-Atg2 complex is recruited to autophagic membranes via phosphatidylinositol 3-phosphate and exerts an essential function. J Biol Chem 283

(35):23972–23980. https://doi.org/10.1074/jbc. M803180200

Ocaka L, Spalluto C, Wilson DI, Hunt DM, Halford S (2005) Chromosomal localization, genomic organization and evolution of the genes encoding human phosphatidylinositol transfer protein membrane-associated (PITPNM) 1, 2 and 3. Cytogenet Genome Res 108(4):293–302. https://doi.org/10.1159/000081519

Ono Y, Fujii T, Igarashi K, Kuno T, Tanaka C, Kikkawa U, Nishizuka Y (1989) Phorbol ester binding to protein kinase C requires a cysteine-rich zinc-finger-like sequence. Proc Natl Acad Sci U S A 86 (13):4868–4871

Ooms LM, Binge LC, Davies EM, Rahman P, Conway JR, Gurung R, Ferguson DT, Papa A, Fedele CG, Vieusseux JL, Chai RC, Koentgen F, Price JT, Tiganis T, Timpson P, McLean CA, Mitchell CA (2015) The inositol polyphosphate 5-phosphatase PIPP regulates AKT1-dependent breast cancer growth and metastasis. Cancer Cell 28(2):155–169. https://doi.org/10.1016/j.ccell.2015.07.003

Osada S, Mizuno K, Saido TC, Akita Y, Suzuki K, Kuroki T, Ohno S (1990) A phorbol ester receptor/protein kinase, nPKC eta, a new member of the protein kinase C family predominantly expressed in lung and skin. J Biol Chem 265(36):22434–22440

Pang X, Fan J, Zhang Y, Zhang K, Gao B, Ma J, Li J, Deng Y, Zhou Q, Egelman EH, Hsu VW, Sun F (2014) A PH domain in ACAP1 possesses key features of the BAR domain in promoting membrane curvature. Dev Cell 31(1):73–86. https://doi.org/10.1016/j.devcel.2014.08.020

Park WS, Heo WD, Whalen JH, O'Rourke NA, Bryan HM, Meyer T, Teruel MN (2008) Comprehensive identification of PIP3-regulated PH domains from C. elegans to H. sapiens by model prediction and live imaging. Mol Cell 30(3):381–392. https://doi.org/10.1016/j.molcel.2008.04.008

Parkinson GN, Vines D, Driscoll PC, Djordjevic S (2008) Crystal structures of PI3K-C2alpha PX domain indicate conformational change associated with ligand binding. BMC Struct Biol 8:13. https://doi.org/10.1186/1472-6807-8-13

Pasenkiewicz-Gierula M, Baczynski K, Markiewicz M, Murzyn K (2016) Computer modelling studies of the bilayer/water interface. Biochim Biophys Acta 1858 (10):2305–2321. https://doi.org/10.1016/j.bbamem.2016.01.024

Payrastre B, Gaits-Iacovoni F, Sansonetti P, Tronchere H (2012) Phosphoinositides and cellular pathogens. Subcell Biochem 59:363–388. https://doi.org/10.1007/978-94-007-3015-1_12

Pearson MA, Reczek D, Bretscher A, Karplus PA (2000) Structure of the ERM protein moesin reveals the FERM domain fold masked by an extended actin binding tail domain. Cell 101(3):259–270

Pendaries C, Tronchere H, Plantavid M, Payrastre B (2003) Phosphoinositide signaling disorders in human diseases. FEBS Lett 546(1):25–31

Perisic O, Fong S, Lynch DE, Bycroft M, Williams RL (1998) Crystal structure of a calcium-phospholipid binding domain from cytosolic phospholipase A2. J Biol Chem 273(3):1596–1604

Peter BJ, Kent HM, Mills IG, Vallis Y, Butler PJ, Evans PR, McMahon HT (2004) BAR domains as sensors of membrane curvature: the amphiphysin BAR structure. Science 303(5657):495–499. https://doi.org/10.1126/science.1092586

Phillips SE, Ile KE, Boukhelifa M, Huijbregts RP, Bankaitis VA (2006) Specific and nonspecific membrane-binding determinants cooperate in targeting phosphatidylinositol transfer protein beta-isoform to the mammalian trans-Golgi network. Mol Biol Cell 17(6):2498–2512. https://doi.org/10.1091/mbc.E06-01-0089

Pizarro-Cerda J, Kuhbacher A, Cossart P (2015) Phosphoinositides and host-pathogen interactions. Biochim Biophys Acta 1851(6):911–918. https://doi.org/10.1016/j.bbalip.2014.09.011

Pogozheva ID, Tristram-Nagle S, Mosberg HI, Lomize AL (2013) Structural adaptations of proteins to different biological membranes. Biochim Biophys Acta 1828(11):2592–2608. https://doi.org/10.1016/j.bbamem.2013.06.023

Ponting CP (1996) Novel domains in NADPH oxidase subunits, sorting nexins, and PtdIns 3-kinases: binding partners of SH3 domains? Protein Sci 5 (11):2353–2357. https://doi.org/10.1002/pro.5560051122

Posor Y, Eichhorn-Gruenig M, Puchkov D, Schoneberg J, Ullrich A, Lampe A, Muller R, Zarbakhsh S, Gulluni F, Hirsch E, Krauss M, Schultz C, Schmoranzer J, Noe F, Haucke V (2013) Spatiotemporal control of endocytosis by phosphatidylinositol-3,4-bisphosphate. Nature 499(7457):233–237. https://doi.org/10.1038/nature12360

Prehoda KE, Lee DJ, Lim WA (1999) Structure of the enabled/VASP homology 1 domain-peptide complex: a key component in the spatial control of actin assembly. Cell 97(4):471–480

Premkumar L, Bobkov AA, Patel M, Jaroszewski L, Bankston LA, Stec B, Vuori K, Cote JF, Liddington RC (2010) Structural basis of membrane targeting by the Dock180 family of Rho family guanine exchange factors (Rho-GEFs). J Biol Chem 285 (17):13211–13222. https://doi.org/10.1074/jbc.M110.102517

Prinz WA (2010) Lipid trafficking sans vesicles: where, why, how? Cell 143(6):870–874. https://doi.org/10.1016/j.cell.2010.11.031

Psachoulia E, Sansom MS (2008) Interactions of the pleckstrin homology domain with phosphatidylinositol phosphate and membranes: characterization via molecular dynamics simulations. Biochemistry 47 (14):4211–4220. https://doi.org/10.1021/bi702319k

Pykalainen A, Boczkowska M, Zhao H, Saarikangas J, Rebowski G, Jansen M, Hakanen J, Koskela EV, Peranen J, Vihinen H, Jokitalo E, Salminen M, Ikonen E, Dominguez R, Lappalainen P (2011)

Pinkbar is an epithelial-specific BAR domain protein that generates planar membrane structures. Nat Struct Mol Biol 18(8):902–907. https://doi.org/10.1038/nsmb.2079

Pylypenko O, Lundmark R, Rasmuson E, Carlsson SR, Rak A (2007) The PX-BAR membrane-remodeling unit of sorting nexin 9. EMBO J 26(22):4788–4800. https://doi.org/10.1038/sj.emboj.7601889

Qualmann B, Koch D, Kessels MM (2011) Let's go bananas: revisiting the endocytic BAR code. EMBO J 30(17):3501–3515. https://doi.org/10.1038/emboj.2011.266

Ragaz C, Pietsch H, Urwyler S, Tiaden A, Weber SS, Hilbi H (2008) The Legionella pneumophila phosphatidylinositol-4 phosphate-binding type IV substrate SidC recruits endoplasmic reticulum vesicles to a replication-permissive vacuole. Cell Microbiol 10 (12):2416–2433. https://doi.org/10.1111/j.1462-5822.2008.01219.x

Rameh LE, Arvidsson A, Carraway KL 3rd, Couvillon AD, Rathbun G, Crompton A, VanRenterghem B, Czech MP, Ravichandran KS, Burakoff SJ, Wang DS, Chen CS, Cantley LC (1997) A comparative analysis of the phosphoinositide binding specificity of pleckstrin homology domains. J Biol Chem 272 (35):22059–22066

Ramel D, Lagarrigue F, Pons V, Mounier J, Dupuis-Coronas S, Chicanne G, Sansonetti PJ, Gaits-Iacovoni-F, Tronchere H, Payrastre B (2011) Shigella flexneri infection generates the lipid PI5P to alter endocytosis and prevent termination of EGFR signaling. Sci Signal 4(191):ra61. https://doi.org/10.1126/scisignal.2001619

Ravichandran KS, Zhou MM, Pratt JC, Harlan JE, Walk SF, Fesik SW, Burakoff SJ (1997) Evidence for a requirement for both phospholipid and phosphotyrosine binding via the Shc phosphotyrosine-binding domain in vivo. Mol Cell Biol 17(9):5540–5549

Raychaudhuri S, Im YJ, Hurley JH, Prinz WA (2006) Nonvesicular sterol movement from plasma membrane to ER requires oxysterol-binding protein-related proteins and phosphoinositides. J Cell Biol 173 (1):107–119. https://doi.org/10.1083/jcb.200510084

Redfern RE, Redfern D, Furgason ML, Munson M, Ross AH, Gericke A (2008) PTEN phosphatase selectively binds phosphoinositides and undergoes structural changes. Biochemistry 47(7):2162–2171. https://doi.org/10.1021/bi702114w

Renard HF, Simunovic M, Lemiere J, Boucrot E, Garcia-Castillo MD, Arumugam S, Chambon V, Lamaze C, Wunder C, Kenworthy AK, Schmidt AA, McMahon HT, Sykes C, Bassereau P, Johannes L (2015) Endophilin-A2 functions in membrane scission in clathrin-independent endocytosis. Nature 517 (7535):493–496. https://doi.org/10.1038/nature14064

Reue K (2009) The lipin family: mutations and metabolism. Curr Opin Lipidol 20(3):165–170. https://doi.org/10.1097/MOL.0b013e32832adee5

Rizo J, Sudhof TC (1998) C2-domains, structure and function of a universal Ca2+−binding domain. J Biol Chem 273(26):15879–15882

Roderick SL, Chan WW, Agate DS, Olsen LR, Vetting MW, Rajashankar KR, Cohen DE (2002) Structure of human phosphatidylcholine transfer protein in complex with its ligand. Nat Struct Biol 9(7):507–511. https://doi.org/10.1038/nsb812

Rogaski B, Klauda JB (2012) Membrane-binding mechanism of a peripheral membrane protein through microsecond molecular dynamics simulations. J Mol Biol 423(5):847–861. https://doi.org/10.1016/j.jmb.2012.08.015

Rohacs T, Lopes CM, Jin T, Ramdya PP, Molnar Z, Logothetis DE (2003) Specificity of activation by phosphoinositides determines lipid regulation of Kir channels. Proc Natl Acad Sci U S A 100(2):745–750. https://doi.org/10.1073/pnas.0236364100

Romanowski MJ, Soccio RE, Breslow JL, Burley SK (2002) Crystal structure of the Mus musculus cholesterol-regulated START protein 4 (StarD4) containing a StAR-related lipid transfer domain. Proc Natl Acad Sci U S A 99(10):6949–6954. https://doi.org/10.1073/pnas.052140699

Roy NS, Yohe ME, Randazzo PA, Gruschus JM (2016) Allosteric properties of PH domains in Arf regulatory proteins. Cell Logist 6(2):e1181700. https://doi.org/10.1080/21592799.2016.1181700

Ryan MM, Temple BR, Phillips SE, Bankaitis VA (2007) Conformational dynamics of the major yeast phosphatidylinositol transfer protein sec14p: insight into the mechanisms of phospholipid exchange and diseases of sec14p-like protein deficiencies. Mol Biol Cell 18(5):1928–1942. https://doi.org/10.1091/mbc.E06-11-1024

Saarikangas J, Zhao H, Pykalainen A, Laurinmaki P, Mattila PK, Kinnunen PK, Butcher SJ, Lappalainen P (2009) Molecular mechanisms of membrane deformation by I-BAR domain proteins. Curr Biol 19 (2):95–107. https://doi.org/10.1016/j.cub.2008.12.029

Saarikangas J, Zhao H, Lappalainen P (2010) Regulation of the actin cytoskeleton-plasma membrane interplay by phosphoinositides. Physiol Rev 90(1):259–289. https://doi.org/10.1152/physrev.00036.2009

Saksena S, Sun J, Chu T, Emr SD (2007) ESCRTing proteins in the endocytic pathway. Trends Biochem Sci 32(12):561–573. https://doi.org/10.1016/j.tibs.2007.09.010

Salama SR, Cleves AE, Malehorn DE, Whitters EA, Bankaitis VA (1990) Cloning and characterization of Kluyveromyces lactis SEC14, a gene whose product stimulates Golgi secretory function in Saccharomyces cerevisiae. J Bacteriol 172(8):4510–4521

Salomon D, Guo Y, Kinch LN, Grishin NV, Gardner KH, Orth K (2013) Effectors of animal and plant pathogens use a common domain to bind host phosphoinositides. Nat Commun 4:2973. https://doi.org/10.1038/ncomms3973

Salzer U, Kostan J, Djinovic-Carugo K (2017) Deciphering the BAR code of membrane modulators. Cell Mol Life Sci 74(13):2413–2438. https://doi.org/10.1007/s00018-017-2478-0

Sanchez-Bautista S, Marin-Vicente C, Gomez-Fernandez JC, Corbalan-Garcia S (2006) The C2 domain of PKCalpha is a Ca2+ −dependent PtdIns(4,5)P2 sensing domain: a new insight into an old pathway. J Mol Biol 362(5):901–914. https://doi.org/10.1016/j.jmb.2006.07.093

Santagata S, Boggon TJ, Baird CL, Gomez CA, Zhao J, Shan WS, Myszka DG, Shapiro L (2001) G-protein signaling through tubby proteins. Science 292 (5524):2041–2050. https://doi.org/10.1126/science.1061233

Saras J, Heldin CH (1996) PDZ domains bind carboxy-terminal sequences of target proteins. Trends Biochem Sci 21(12):455–458

Sarkes D, Rameh LE (2010) A novel HPLC-based approach makes possible the spatial characterization of cellular PtdIns5P and other phosphoinositides. Biochem J 428(3):375–384. https://doi.org/10.1042/BJ20100129

Sasaki T, Takasuga S, Sasaki J, Kofuji S, Eguchi S, Yamazaki M, Suzuki A (2009) Mammalian phosphoinositide kinases and phosphatases. Prog Lipid Res 48 (6):307–343. https://doi.org/10.1016/j.plipres.2009.06.001

Sato TK, Overduin M, Emr SD (2001) Location, location, location: membrane targeting directed by PX domains. Science 294(5548):1881–1885. https://doi.org/10.1126/science.1065763

Sbrissa D, Ikonomov OC, Filios C, Delvecchio K, Shisheva A (2012) Functional dissociation between PIKfyve-synthesized PtdIns5P and PtdIns(3,5)P2 by means of the PIKfyve inhibitor YM201636. Am J Physiol Cell Physiol 303(4):C436–C446. https://doi.org/10.1152/ajpcell.00105.2012

Scacioc A, Schmidt C, Hofmann T, Urlaub H, Kuhnel K, Perez-Lara A (2017) Structure based biophysical characterization of the PROPPIN Atg18 shows Atg18 oligomerization upon membrane binding. Sci Rep 7 (1):14008. https://doi.org/10.1038/s41598-017-14337-5

Schaaf G, Ortlund EA, Tyeryar KR, Mousley CJ, Ile KE, Garrett TA, Ren J, Woolls MJ, Raetz CR, Redinbo MR, Bankaitis VA (2008) Functional anatomy of phospholipid binding and regulation of phosphoinositide homeostasis by proteins of the sec14 superfamily. Mol Cell 29(2):191–206. https://doi.org/10.1016/j.molcel.2007.11.026

Schaaf G, Dynowski M, Mousley CJ, Shah SD, Yuan P, Winklbauer EM, de Campos MK, Trettin K, Quinones MC, Smirnova TI, Yanagisawa LL, Ortlund EA, Bankaitis VA (2011) Resurrection of a functional phosphatidylinositol transfer protein from a pseudo-Sec14 scaffold by directed evolution. Mol Biol Cell 22 (6):892–905. https://doi.org/10.1091/mbc.E10-11-0903

Scheffzek K, Welti S (2012) Pleckstrin homology (PH) like domains – versatile modules in protein-interaction platforms. FEBS Lett 586 (17):2662–2673. https://doi.org/10.1016/j.febslet.2012.06.006

Schink KO, Tan KW, Stenmark H (2016) Phosphoinositides in control of membrane dynamics. Annu Rev Cell Dev Biol 32:143–171. https://doi.org/10.1146/annurev-cellbio-111315-125349

Schoebel S, Blankenfeldt W, Goody RS, Itzen A (2010) High-affinity binding of phosphatidylinositol 4-phosphate by Legionella pneumophila DrrA. EMBO Rep 11(8):598–604. https://doi.org/10.1038/embor.2010.97

Schoneberg J, Lehmann M, Ullrich A, Posor Y, Lo WT, Lichtner G, Schmoranzer J, Haucke V, Noe F (2017) Lipid-mediated PX-BAR domain recruitment couples local membrane constriction to endocytic vesicle fission. Nat Commun 8:15873. https://doi.org/10.1038/ncomms15873

Schouten A, Agianian B, Westerman J, Kroon J, Wirtz KW, Gros P (2002) Structure of apo-phosphatidylinositol transfer protein alpha provides insight into membrane association. EMBO J 21(9):2117–2121. https://doi.org/10.1093/emboj/21.9.2117

Schulz TA, Choi MG, Raychaudhuri S, Mears JA, Ghirlando R, Hinshaw JE, Prinz WA (2009) Lipid-regulated sterol transfer between closely apposed membranes by oxysterol-binding protein homologues. J Cell Biol 187(6):889–903. https://doi.org/10.1083/jcb.200905007

Seet LF, Hong W (2006) The Phox (PX) domain proteins and membrane traffic. Biochim Biophys Acta 1761 (8):878–896. https://doi.org/10.1016/j.bbalip.2006.04.011

Segui B, Allen-Baume V, Cockcroft S (2002) Phosphatidylinositol transfer protein beta displays minimal sphingomyelin transfer activity and is not required for biosynthesis and trafficking of sphingomyelin. Biochem J 366(Pt 1):23–34. https://doi.org/10.1042/BJ20020317

Senju Y, Kalimeri M, Koskela EV, Somerharju P, Zhao H, Vattulainen I, Lappalainen P (2017) Mechanistic principles underlying regulation of the actin cytoskeleton by phosphoinositides. Proc Natl Acad Sci U S A 114(43):E8977–E8986. https://doi.org/10.1073/pnas.1705032114

Servant G, Weiner OD, Herzmark P, Balla T, Sedat JW, Bourne HR (2000) Polarization of chemoattractant receptor signaling during neutrophil chemotaxis. Science 287(5455):1037–1040

Sha B, Phillips SE, Bankaitis VA, Luo M (1998) Crystal structure of the Saccharomyces cerevisiae phosphatidylinositol-transfer protein. Nature 391 (6666):506–510. https://doi.org/10.1038/35179

Shadan S, Holic R, Carvou N, Ee P, Li M, Murray-Rust J, Cockcroft S (2008) Dynamics of lipid transfer by phosphatidylinositol transfer proteins in cells. Traffic 9(10):1743–1756. https://doi.org/10.1111/j.1600-0854.2008.00794.x

Shah ZH, Jones DR, Sommer L, Foulger R, Bultsma Y, D'Santos C, Divecha N (2013) Nuclear phosphoinositides and their impact on nuclear functions. FEBS J 280(24):6295–6310. https://doi.org/10.1111/febs.12543

Shao X, Davletov BA, Sutton RB, Sudhof TC, Rizo J (1996) Bipartite Ca2+−binding motif in C2 domains of synaptotagmin and protein kinase C. Science 273 (5272):248–251

Shimada A, Niwa H, Tsujita K, Suetsugu S, Nitta K, Hanawa-Suetsugu K, Akasaka R, Nishino Y, Toyama M, Chen L, Liu ZJ, Wang BC, Yamamoto M, Terada T, Miyazawa A, Tanaka A, Sugano S, Shirouzu M, Nagayama K, Takenawa T, Yokoyama S (2007) Curved EFC/F-BAR-domain dimers are joined end to end into a filament for membrane invagination in endocytosis. Cell 129(4):761–772. https://doi.org/10.1016/j.cell.2007.03.040

Sidhu G, Li W, Laryngakis N, Bishai E, Balla T, Southwick F (2005) Phosphoinositide 3-kinase is required for intracellular Listeria monocytogenes actin-based motility and filopod formation. J Biol Chem 280(12):11379–11386. https://doi.org/10.1074/jbc.M414533200

Silkov A, Yoon Y, Lee H, Gokhale N, Adu-Gyamfi E, Stahelin RV, Cho W, Murray D (2011) Genome-wide structural analysis reveals novel membrane binding properties of AP180 N-terminal homology (ANTH) domains. J Biol Chem 286(39):34155–34163. https://doi.org/10.1074/jbc.M111.265611

Simonsen A, Lippe R, Christoforidis S, Gaullier JM, Brech A, Callaghan J, Toh BH, Murphy C, Zerial M, Stenmark H (1998) EEA1 links PI(3)K function to Rab5 regulation of endosome fusion. Nature 394 (6692):494–498. https://doi.org/10.1038/28879

Simunovic M, Voth GA, Callan-Jones A, Bassereau P (2015) When physics takes over: BAR proteins and membrane curvature. Trends Cell Biol 25 (12):780–792. https://doi.org/10.1016/j.tcb.2015.09.005

Skoble J, Portnoy DA, Welch MD (2000) Three regions within ActA promote Arp2/3 complex-mediated actin nucleation and Listeria monocytogenes motility. J Cell Biol 150(3):527–538

Skruzny M, Desfosses A, Prinz S, Dodonova SO, Gieras A, Uetrecht C, Jakobi AJ, Abella M, Hagen WJ, Schulz J, Meijers R, Rybin V, Briggs JA, Sachse C, Kaksonen M (2015) An organized co-assembly of clathrin adaptors is essential for endocytosis. Dev Cell 33(2):150–162. https://doi.org/10.1016/j.devcel.2015.02.023

Slagsvold T, Aasland R, Hirano S, Bache KG, Raiborg C, Trambaiolo D, Wakatsuki S, Stenmark H (2005) Eap45 in mammalian ESCRT-II binds ubiquitin via a phosphoinositide-interacting GLUE domain. J Biol Chem 280(20):19600–19606. https://doi.org/10.1074/jbc.M501510200

Smirnova TI, Chadwick TG, MacArthur R, Poluektov O, Song L, Ryan MM, Schaaf G, Bankaitis VA (2006) The chemistry of phospholipid binding by the Saccharomyces cerevisiae phosphatidylinositol transfer protein Sec14p as determined by EPR spectroscopy. J Biol Chem 281(46):34897–34908. https://doi.org/10.1074/jbc.M603054200

Smirnova TI, Chadwick TG, Voinov MA, Poluektov O, van Tol J, Ozarowski A, Schaaf G, Ryan MM, Bankaitis VA (2007) Local polarity and hydrogen bonding inside the Sec14p phospholipid-binding cavity: high-field multi-frequency electron paramagnetic resonance studies. Biophys J 92(10):3686–3695. https://doi.org/10.1529/biophysj.106.097899

Smith WJ, Nassar N, Bretscher A, Cerione RA, Karplus PA (2003) Structure of the active N-terminal domain of Ezrin. Conformational and mobility changes identify keystone interactions. J Biol Chem 278(7):4949–4956. https://doi.org/10.1074/jbc.M210601200

Snoek GT, Berrie CP, Geijtenbeek TB, van der Helm HA, Cadee JA, Iurisci C, Corda D, Wirtz KW (1999) Overexpression of phosphatidylinositol transfer protein alpha in NIH3T3 cells activates a phospholipase A. J Biol Chem 274(50):35393–35399

Sohn M, Ivanova P, Brown HA, Toth DJ, Varnai P, Kim YJ, Balla T (2016) Lenz-Majewski mutations in PTDSS1 affect phosphatidylinositol 4-phosphate metabolism at ER-PM and ER-Golgi junctions. Proc Natl Acad Sci U S A 113(16):4314–4319. https://doi.org/10.1073/pnas.1525719113

Song X, Xu W, Zhang A, Huang G, Liang X, Virbasius JV, Czech MP, Zhou GW (2001) Phox homology domains specifically bind phosphatidylinositol phosphates. Biochemistry 40(30):8940–8944

Sot B, Behrmann E, Raunser S, Wittinghofer A (2013) Ras GTPase activating (RasGAP) activity of the dual specificity GAP protein Rasal requires colocalization and C2 domain binding to lipid membranes. Proc Natl Acad Sci U S A 110(1):111–116. https://doi.org/10.1073/pnas.1201658110

Stahelin RV, Long F, Diraviyam K, Bruzik KS, Murray D, Cho W (2002) Phosphatidylinositol 3-phosphate induces the membrane penetration of the FYVE domains of Vps27p and Hrs. J Biol Chem 277 (29):26379–26388. https://doi.org/10.1074/jbc.M201106200

Stahelin RV, Burian A, Bruzik KS, Murray D, Cho W (2003a) Membrane binding mechanisms of the PX domains of NADPH oxidase p40phox and p47phox. J Biol Chem 278(16):14469–14479. https://doi.org/10.1074/jbc.M212579200

Stahelin RV, Long F, Peter BJ, Murray D, De Camilli P, McMahon HT, Cho W (2003b) Contrasting membrane interaction mechanisms of AP180 N-terminal homology (ANTH) and epsin N-terminal homology (ENTH) domains. J Biol Chem 278(31):28993–28999. https://doi.org/10.1074/jbc.M302865200

Stahelin RV, Rafter JD, Das S, Cho W (2003c) The molecular basis of differential subcellular localization of C2 domains of protein kinase C-alpha and group IVa cytosolic phospholipase A2. J Biol Chem 278 (14):12452–12460. https://doi.org/10.1074/jbc.M212864200

Stahelin RV, Ananthanarayanan B, Blatner NR, Singh S, Bruzik KS, Murray D, Cho W (2004) Mechanism of membrane binding of the phospholipase D1 PX domain. J Biol Chem 279(52):54918–54926. https://doi.org/10.1074/jbc.M407798200

Stahelin RV, Karathanassis D, Bruzik KS, Waterfield MD, Bravo J, Williams RL, Cho W (2006) Structural and membrane binding analysis of the Phox homology domain of phosphoinositide 3-kinase-C2alpha. J Biol Chem 281(51):39396–39406. https://doi.org/10.1074/jbc.M607079200

Stahelin RV, Karathanassis D, Murray D, Williams RL, Cho W (2007) Structural and membrane binding analysis of the Phox homology domain of Bem1p: basis of phosphatidylinositol 4-phosphate specificity. J Biol Chem 282(35):25737–25747. https://doi.org/10.1074/jbc.M702861200

Stahelin RV, Scott JL, Frick CT (2014) Cellular and molecular interactions of phosphoinositides and peripheral proteins. Chem Phys Lipids 182:3–18. https://doi.org/10.1016/j.chemphyslip.2014.02.002

Stauffer TP, Ahn S, Meyer T (1998) Receptor-induced transient reduction in plasma membrane PtdIns(4,5) P2 concentration monitored in living cells. Curr Biol 8(6):343–346

Stefan CJ, Manford AG, Baird D, Yamada-Hanff J, Mao Y, Emr SD (2011) Osh proteins regulate phosphoinositide metabolism at ER-plasma membrane contact sites. Cell 144(3):389–401. https://doi.org/10.1016/j.cell.2010.12.034

Steffen P, Schafer DA, David V, Gouin E, Cooper JA, Cossart P (2000) Listeria monocytogenes ActA protein interacts with phosphatidylinositol 4,5-bisphosphate in vitro. Cell Motil Cytoskeleton 45(1):58–66. https://doi.org/10.1002/(SICI)1097-0169(200001)45:1<58::AID-CM6>3.0.CO;2-Y

Stenmark H, Aasland R, Toh BH, D'Arrigo A (1996) Endosomal localization of the autoantigen EEA1 is mediated by a zinc-binding FYVE finger. J Biol Chem 271(39):24048–24054

Stolt PC, Jeon H, Song HK, Herz J, Eck MJ, Blacklow SC (2003) Origins of peptide selectivity and phosphoinositide binding revealed by structures of disabled-1 PTB domain complexes. Structure 11(5):569–579

Stolt PC, Vardar D, Blacklow SC (2004) The dual-function disabled-1 PTB domain exhibits site independence in binding phosphoinositide and peptide ligands. Biochemistry 43(34):10979–10987. https://doi.org/10.1021/bi0490921

Stolt PC, Chen Y, Liu P, Bock HH, Blacklow SC, Herz J (2005) Phosphoinositide binding by the disabled-1 PTB domain is necessary for membrane localization and Reelin signal transduction. J Biol Chem 280 (10):9671–9677. https://doi.org/10.1074/jbc.M413356200

Stromhaug PE, Reggiori F, Guan J, Wang CW, Klionsky DJ (2004) Atg21 is a phosphoinositide binding protein required for efficient lipidation and localization of Atg8 during uptake of aminopeptidase I by selective autophagy. Mol Biol Cell 15(8):3553–3566. https://doi.org/10.1091/mbc.E04-02-0147

Sutton RB, Davletov BA, Berghuis AM, Sudhof TC, Sprang SR (1995) Structure of the first C2 domain of synaptotagmin I: a novel Ca2+/phospholipid-binding fold. Cell 80(6):929–938

Szentpetery Z, Balla A, Kim YJ, Lemmon MA, Balla T (2009) Live cell imaging with protein domains capable of recognizing phosphatidylinositol 4,5-bisphosphate; a comparative study. BMC Cell Biol 10:67. https://doi.org/10.1186/1471-2121-10-67

Takeuchi H, Kanematsu T, Misumi Y, Sakane F, Konishi H, Kikkawa U, Watanabe Y, Katan M, Hirata M (1997) Distinct specificity in the binding of inositol phosphates by pleckstrin homology domains of pleckstrin, RAC-protein kinase, diacylglycerol kinase and a new 130 kDa protein. Biochim Biophys Acta 1359(3):275–285

Takeuchi H, Matsuda M, Yamamoto T, Kanematsu T, Kikkawa U, Yagisawa H, Watanabe Y, Hirata M (1998) PTB domain of insulin receptor substrate-1 binds inositol compounds. Biochem J 334 (Pt 1):211–218

Tan X, Thapa N, Choi S, Anderson RA (2015) Emerging roles of PtdIns(4,5)P2--beyond the plasma membrane. J Cell Sci 128(22):4047–4056. https://doi.org/10.1242/jcs.175208

Tani K, Kogure T, Inoue H (2012) The intracellular phospholipase A1 protein family. Biomol Concepts 3 (5):471–478. https://doi.org/10.1515/bmc-2012-0014

Teo H, Gill DJ, Sun J, Perisic O, Veprintsev DB, Vallis Y, Emr SD, Williams RL (2006) ESCRT-I core and ESCRT-II GLUE domain structures reveal role for GLUE in linking to ESCRT-I and membranes. Cell 125(1):99–111. https://doi.org/10.1016/j.cell.2006.01.047

Thapa N, Tan X, Choi S, Lambert PF, Rapraeger AC, Anderson RA (2016) The hidden conundrum of phosphoinositide signaling in cancer. Trends Cancer 2(7):378–390. https://doi.org/10.1016/j.trecan.2016.05.009

Tilley SJ, Skippen A, Murray-Rust J, Swigart PM, Stewart A, Morgan CP, Cockcroft S, McDonald NQ (2004) Structure-function analysis of human [corrected] phosphatidylinositol transfer protein alpha bound to phosphatidylinositol. Structure 12 (2):317–326. https://doi.org/10.1016/j.str.2004.01.013

Toker A, Cantley LC (1997) Signalling through the lipid products of phosphoinositide-3-OH kinase. Nature 387 (6634):673–676. https://doi.org/10.1038/42648

Tong J, Yang H, Yang H, Eom SH, Im YJ (2013) Structure of Osh3 reveals a conserved mode of phosphoinositide binding in oxysterol-binding proteins. Structure 21 (7):1203–1213. https://doi.org/10.1016/j.str.2013.05.007

Tong J, Manik MK, Yang H, Im YJ (2016) Structural insights into nonvesicular lipid transport by the oxysterol binding protein homologue family. Biochim

Biophys Acta 1861(8 Pt B):928–939. https://doi.org/10.1016/j.bbalip.2016.01.008

Tsujishita Y, Hurley JH (2000) Structure and lipid transport mechanism of a StAR-related domain. Nat Struct Biol 7(5):408–414. https://doi.org/10.1038/75192

Tsujita K, Itoh T, Ijuin T, Yamamoto A, Shisheva A, Laporte J, Takenawa T (2004) Myotubularin regulates the function of the late endosome through the gram domain-phosphatidylinositol 3,5-bisphosphate interaction. J Biol Chem 279(14):13817–13824. https://doi.org/10.1074/jbc.m312294200

Tuzi S, Uekama N, Okada M, Yamaguchi S, Saito H, Yagisawa H (2003) Structure and dynamics of the phospholipase C-delta1 pleckstrin homology domain located at the lipid bilayer surface. J Biol Chem 278(30):28019–28025. https://doi.org/10.1074/jbc.M300101200

Tyson GH, Halavaty AS, Kim H, Geissler B, Agard M, Satchell KJ, Cho W, Anderson WF, Hauser AR (2015) A novel phosphatidylinositol 4,5-bisphosphate binding domain mediates plasma membrane localization of ExoU and other patatin-like phospholipases. J Biol Chem 290(5):2919–2937. https://doi.org/10.1074/jbc.M114.611251

Uchida Y, Hasegawa J, Chinnapen D, Inoue T, Okazaki S, Kato R, Wakatsuki S, Misaki R, Koike M, Uchiyama Y, Iemura S, Natsume T, Kuwahara R, Nakagawa T, Nishikawa K, Mukai K, Miyoshi E, Taniguchi N, Sheff D, Lencer WI, Taguchi T, Arai H (2011) Intracellular phosphatidylserine is essential for retrograde membrane traffic through endosomes. Proc Natl Acad Sci U S A 108(38):15846–15851. https://doi.org/10.1073/pnas.1109101108

Vadas O, Burke JE (2015) Probing the dynamic regulation of peripheral membrane proteins using hydrogen deuterium exchange-MS (HDX-MS). Biochem Soc Trans 43(5):773–786. https://doi.org/10.1042/BST20150065

van den Bogaart G, Meyenberg K, Diederichsen U, Jahn R (2012) Phosphatidylinositol 4,5-bisphosphate increases Ca2+ affinity of synaptotagmin-1 by 40-fold. J Biol Chem 287(20):16447–16453. https://doi.org/10.1074/jbc.M112.343418

van Meer G, Voelker DR, Feigenson GW (2008) Membrane lipids: where they are and how they behave. Nat Rev Mol Cell Biol 9(2):112–124. https://doi.org/10.1038/nrm2330

Van Paridon PA, Somerharju P, Wirtz KW (1987) Phosphatidylinositol-transfer protein and cellular phosphatidylinositol metabolism. Biochem Soc Trans 15(3):321–323

Vance JE (2015) Phospholipid synthesis and transport in mammalian cells. Traffic 16(1):1–18. https://doi.org/10.1111/tra.12230

Vanhaesebroeck B, Guillermet-Guibert J, Graupera M, Bilanges B (2010) The emerging mechanisms of isoform-specific PI3K signalling. Nat Rev Mol Cell Biol 11(5):329–341. https://doi.org/10.1038/nrm2882

Varnai P, Balla T (1998) Visualization of phosphoinositides that bind pleckstrin homology domains:

calcium- and agonist-induced dynamic changes and relationship to myo-[3H]inositol-labeled phosphoinositide pools. J Cell Biol 143(2):501–510

Varnai P, Rother KI, Balla T (1999) Phosphatidylinositol 3-kinase-dependent membrane association of the Bruton's tyrosine kinase pleckstrin homology domain visualized in single living cells. J Biol Chem 274(16):10983–10989

Varnai P, Gulyas G, Toth DJ, Sohn M, Sengupta N, Balla T (2017) Quantifying lipid changes in various membrane compartments using lipid binding protein domains. Cell Calcium 64:72–82. https://doi.org/10.1016/j.ceca.2016.12.008

Verdaguer N, Corbalan-Garcia S, Ochoa WF, Fita I, Gomez-Fernandez JC (1999) Ca(2+) bridges the C2 membrane-binding domain of protein kinase Calpha directly to phosphatidylserine. EMBO J 18(22):6329–6338. https://doi.org/10.1093/emboj/18.22.6329

Vetter IR, Nowak C, Nishimoto T, Kuhlmann J, Wittinghofer A (1999) Structure of a Ran-binding domain complexed with Ran bound to a GTP analogue: implications for nuclear transport. Nature 398(6722):39–46. https://doi.org/10.1038/17969

Vihtelic TS, Hyde DR, O'Tousa JE (1991) Isolation and characterization of the Drosophila retinal degeneration B (rdgB) gene. Genetics 127(4):761–768

Vihtelic TS, Goebl M, Milligan S, O'Tousa JE, Hyde DR (1993) Localization of Drosophila retinal degeneration B, a membrane-associated phosphatidylinositol transfer protein. J Cell Biol 122(5):1013–1022

Vollert CS, Uetz P (2004) The phox homology (PX) domain protein interaction network in yeast. Mol Cell Proteomics 3(11):1053–1064. https://doi.org/10.1074/mcp.M400081-MCP200

Vonkova I, Saliba AE, Deghou S, Anand K, Ceschia S, Doerks T, Galih A, Kugler KG, Maeda K, Rybin V, van Noort V, Ellenberg J, Bork P, Gavin AC (2015) Lipid cooperativity as a general membrane-recruitment principle for PH domains. Cell Rep 12(9):1519–1530. https://doi.org/10.1016/j.celrep.2015.07.054

Vordtriede PB, Doan CN, Tremblay JM, Helmkamp GM Jr, Yoder MD (2005) Structure of PITPbeta in complex with phosphatidylcholine: comparison of structure and lipid transfer to other PITP isoforms. Biochemistry 44(45):14760–14771. https://doi.org/10.1021/bi051191r

Walker SM, Leslie NR, Perera NM, Batty IH, Downes CP (2004) The tumour-suppressor function of PTEN requires an N-terminal lipid-binding motif. Biochem J 379(Pt 2):301–307. https://doi.org/10.1042/BJ20031839

Wang J, Gambhir A, Hangyas-Mihalyne G, Murray D, Golebiewska U, McLaughlin S (2002) Lateral sequestration of phosphatidylinositol 4,5-bisphosphate by the basic effector domain of myristoylated alanine-rich C kinase substrate is due to nonspecific electrostatic interactions. J Biol Chem 277(37):34401–34412. https://doi.org/10.1074/jbc.M203954200

Wang YJ, Wang J, Sun HQ, Martinez M, Sun YX, Macia E, Kirchhausen T, Albanesi JP, Roth MG, Yin HL (2003) Phosphatidylinositol 4 phosphate regulates targeting of clathrin adaptor AP-1 complexes to the Golgi. Cell 114(3):299–310

Wang J, Gambhir A, McLaughlin S, Murray D (2004) A computational model for the electrostatic sequestration of PI(4,5)P2 by membrane-adsorbed basic peptides. Biophys J 86(4):1969–1986. https://doi.org/10.1016/S0006-3495(04)74260-5

Watt SA, Kular G, Fleming IN, Downes CP, Lucocq JM (2002) Subcellular localization of phosphatidylinositol 4,5-bisphosphate using the pleckstrin homology domain of phospholipase C delta1. Biochem J 363 (Pt 3):657–666

Watton SJ, Downward J (1999) Akt/PKB localisation and $3'$ phosphoinositide generation at sites of epithelial cell-matrix and cell-cell interaction. Curr Biol 9 (8):433–436

Wawrzyniak AM, Kashyap R, Zimmermann P (2013) Phosphoinositides and PDZ domain scaffolds. Adv Exp Med Biol 991:41–57. https://doi.org/10.1007/978-94-007-6331-9_4

Weber SS, Ragaz C, Hilbi H (2009) The inositol polyphosphate 5-phosphatase OCRL1 restricts intracellular growth of Legionella, localizes to the replicative vacuole and binds to the bacterial effector LpnE. Cell Microbiol 11(3):442–460. https://doi.org/10.1111/j.1462-5822.2008.01266.x

Weber-Boyvat M, Kentala H, Peranen J, Olkkonen VM (2015) Ligand-dependent localization and function of ORP-VAP complexes at membrane contact sites. Cell Mol Life Sci 72(10):1967–1987. https://doi.org/10.1007/s00018-014-1786-x

Wei Y, Stec B, Redfield AG, Weerapana E, Roberts MF (2015) Phospholipid-binding sites of phosphatase and tensin homolog (PTEN): exploring the mechanism of phosphatidylinositol 4,5-bisphosphate activation. J Biol Chem 290(3):1592–1606. https://doi.org/10.1074/jbc.M114.588590

Weigele BA, Orchard RC, Jimenez A, Cox GW, Alto NM (2017) A systematic exploration of the interactions between bacterial effector proteins and host cell membranes. Nat Commun 8(1):532. https://doi.org/10.1038/s41467-017-00700-7

Weixel KM, Blumental-Perry A, Watkins SC, Aridor M, Weisz OA (2005) Distinct Golgi populations of phosphatidylinositol 4-phosphate regulated by phosphatidylinositol 4-kinases. J Biol Chem 280 (11):10501–10508. https://doi.org/10.1074/jbc.M414304200

Welch MD, Rosenblatt J, Skoble J, Portnoy DA, Mitchison TJ (1998) Interaction of human Arp2/3 complex and the Listeria monocytogenes ActA protein in actin filament nucleation. Science 281 (5373):105–108

Welch HC, Coadwell WJ, Ellson CD, Ferguson GJ, Andrews SR, Erdjument-Bromage H, Tempst P, Hawkins PT, Stephens LR (2002) P-Rex1, a PtdIns (3,4,5)P3- and Gbetagamma-regulated guanine-nucleotide exchange factor for Rac. Cell 108(6):809–821

Whited AM, Johs A (2015) The interactions of peripheral membrane proteins with biological membranes. Chem Phys Lipids 192:51–59. https://doi.org/10.1016/j.chemphyslip.2015.07.015

Whorton MR, MacKinnon R (2011) Crystal structure of the mammalian GIRK2 K+ channel and gating regulation by G proteins, PIP2, and sodium. Cell 147 (1):199–208. https://doi.org/10.1016/j.cell.2011.07.046

Wientjes FB, Reeves EP, Soskic V, Furthmayr H, Segal AW (2001) The NADPH oxidase components p47 (phox) and p40(phox) bind to moesin through their PX domain. Biochem Biophys Res Commun 289 (2):382–388. https://doi.org/10.1006/bbrc.2001.5982

Williams RL, Urbe S (2007) The emerging shape of the ESCRT machinery. Nat Rev Mol Cell Biol 8 (5):355–368. https://doi.org/10.1038/nrm2162

Wirtz KW (1991) Phospholipid transfer proteins. Annu Rev Biochem 60:73–99. https://doi.org/10.1146/annurev.bi.60.070191.000445

Wirtz KW (1997) Phospholipid transfer proteins revisited. Biochem J 324(Pt 2):353–360

Wishart MJ, Taylor GS, Dixon JE (2001) Phoxy lipids: revealing PX domains as phosphoinositide binding modules. Cell 105(7):817–820

Wong LH, Levine TP (2016) Lipid transfer proteins do their thing anchored at membrane contact sites... but what is their thing? Biochem Soc Trans 44 (2):517–527. https://doi.org/10.1042/BST20150275

Wong LH, Copic A, Levine TP (2017) Advances on the transfer of lipids by lipid transfer proteins. Trends Biochem Sci 42(7):516–530. https://doi.org/10.1016/j.tibs.2017.05.001

Wu H, Feng W, Chen J, Chan LN, Huang S, Zhang M (2007) PDZ domains of Par-3 as potential phosphoinositide signaling integrators. Mol Cell 28 (5):886–898. https://doi.org/10.1016/j.molcel.2007.10.028

Wyckoff GJ, Solidar A, Yoden MD (2010) Phosphatidylinositol transfer proteins: sequence motifs in structural and evolutionary analyses. J Biomed Sci Eng 3 (1):65–77. https://doi.org/10.4236/jbise.2010.31010

Wymann MP, Schneiter R (2008) Lipid signalling in disease. Nat Rev Mol Cell Biol 9(2):162–176. https://doi.org/10.1038/nrm2335

Xing Y, Liu D, Zhang R, Joachimiak A, Songyang Z, Xu W (2004) Structural basis of membrane targeting by the Phox homology domain of cytokine-independent survival kinase (CISK-PX). J Biol Chem 279 (29):30662–30669. https://doi.org/10.1074/jbc.M404107200

Xu J, Liu D, Gill G, Songyang Z (2001) Regulation of cytokine-independent survival kinase (CISK) by the Phox homology domain and phosphoinositides. J Cell Biol 154(4):699–705. https://doi.org/10.1083/jcb.200105089

Xu C, Watras J, Loew LM (2003) Kinetic analysis of receptor-activated phosphoinositide turnover. J Cell

Biol 161(4):779–791. https://doi.org/10.1083/jcb. 200301070

Xu M, Arnaud L, Cooper JA (2005) Both the phosphoi-nositide and receptor binding activities of Dab1 are required for Reelin-stimulated Dab1 tyrosine phos-phorylation. Brain Res Mol Brain Res 139 (2):300–305. https://doi.org/10.1016/j.molbrainres. 2005.06.001

Xu Q, Bateman A, Finn RD, Abdubek P, Astakhova T, Axelrod HL, Bakolitsa C, Carlton D, Chen C, Chiu HJ, Chiu M, Clayton T, Das D, Deller MC, Duan L, Ellrott K, Ernst D, Farr CL, Feuerhelm J, Grant JC, Grzechnik A, Han GW, Jaroszewski L, Jin KK, Klock HE, Knuth MW, Kozbial P, Krishna SS, Kumar A, Marciano D, McMullan D, Miller MD, Morse AT, Nigoghossian E, Nopakun A, Okach L, Puckett C, Reyes R, Rife CL, Sefcovic N, Tien HJ, Trame CB, van den Bedem H, Weekes D, Wooten T, Hodgson KO, Wooley J, Elsliger MA, Deacon AM, Godzik A, Lesley SA, Wilson IA (2010) Bacterial pleckstrin homology domains: a prokaryotic origin for the PH domain. J Mol Biol 396(1):31–46. https://doi.org/10. 1016/j.jmb.2009.11.006

Yadav S, Garner K, Georgiev P, Li M, Gomez-Espinosa E, Panda A, Mathre S, Okkenhaug H, Cockcroft S, Raghu P (2015) RDGBalpha, a PtdIns-PtdOH transfer protein, regulates G-protein-coupled PtdIns(4,5)P2 signalling during Drosophila phototransduction. J Cell Sci 128 (17):3330–3344. https://doi.org/10.1242/jcs.173476

Yamamoto E, Kalli AC, Yasuoka K, Sansom MSP (2016) Interactions of pleckstrin homology domains with membranes: adding back the bilayer via high-throughput molecular dynamics. Structure 24(8):1421–1431. https://doi.org/10.1016/j.str.2016.06.002

Yamamoto E, Akimoto T, Kalli AC, Yasuoka K, Sansom MS (2017) Dynamic interactions between a membrane binding protein and lipids induce fluctuating diffusiv-ity. Sci Adv 3(1):e1601871. https://doi.org/10.1126/ sciadv.1601871

Yan D, Olkkonen VM (2008) Characteristics of oxysterol binding proteins. Int Rev Cytol 265:253–285. https:// doi.org/10.1016/S0074-7696(07)65007-4

Yau WM, Wimley WC, Gawrisch K, White SH (1998) The preference of tryptophan for membrane interfaces. Biochemistry 37(42):14713–14718. https://doi.org/10. 1021/bi980809c

Yeung T, Terebiznik M, Yu L, Silvius J, Abidi WM, Philips M, Levine T, Kapus A, Grinstein S (2006) Receptor activation alters inner surface potential dur-ing phagocytosis. Science 313(5785):347–351. https:// doi.org/10.1126/science.1129551

Yoder MD, Thomas LM, Tremblay JM, Oliver RL, Yarbrough LR, Helmkamp GM Jr (2001) Structure of a multifunctional protein. Mammalian phosphatidy-linositol transfer protein complexed with phosphatidyl-choline. J Biol Chem 276(12):9246–9252. https://doi. org/10.1074/jbc.M010131200

Yoon Y, Tong J, Lee PJ, Albanese A, Bhardwaj N, Kallberg M, Digman MA, Lu H, Gratton E, Shin YK,

Cho W (2010) Molecular basis of the potent membrane-remodeling activity of the epsin 1 N-terminal homology domain. J Biol Chem 285(1):531–540. https://doi.org/ 10.1074/jbc.M109.068015

Yoon Y, Zhang X, Cho W (2012) Phosphatidylinositol 4,5-bisphosphate (PtdIns(4,5)P2) specifically induces membrane penetration and deformation by Bin/amphiphysin/Rvs (BAR) domains. J Biol Chem 287(41):34078–34090. https://doi.org/10.1074/jbc. M112.372789

Yu JW, Lemmon MA (2001) All phox homology (PX) domains from Saccharomyces cerevisiae specifi-cally recognize phosphatidylinositol 3-phosphate. J Biol Chem 276(47):44179–44184. https://doi.org/10. 1074/jbc.M108811200

Yu JW, Mendrola JM, Audhya A, Singh S, Keleti D, DeWald DB, Murray D, Emr SD, Lemmon MA (2004) Genome-wide analysis of membrane targeting by S. cerevisiae pleckstrin homology domains. Mol Cell 13(5):677–688

Yun M, Keshvara L, Park CG, Zhang YM, Dickerson JB, Zheng J, Rock CO, Curran T, Park HW (2003) Crystal structures of the Dab homology domains of mouse disabled 1 and 2. J Biol Chem 278(38):36572–36581. https://doi.org/10.1074/jbc.M304384200

Zalevsky J, Grigorova I, Mullins RD (2001) Activation of the Arp2/3 complex by the Listeria acta protein. Acta binds two actin monomers and three subunits of the Arp2/3 complex. J Biol Chem 276(5):3468–3475. https://doi.org/10.1074/jbc.M006407200

Zhao L, Ma Y, Seemann J, Huang LJ (2010) A regulating role of the JAK2 FERM domain in hyperactivation of JAK2(V617F). Biochem J 426(1):91–98. https://doi. org/10.1042/BJ20090615

Zhao H, Michelot A, Koskela EV, Tkach V, Stamou D, Drubin DG, Lappalainen P (2013) Membrane-sculpting BAR domains generate stable lipid microdomains. Cell Rep 4(6):1213–1223. https://doi. org/10.1016/j.celrep.2013.08.024

Zhou MM, Fesik SW (1995) Structure and function of the phosphotyrosine binding (PTB) domain. Prog Biophys Mol Biol 64(2–3):221–235

Zhou MM, Ravichandran KS, Olejniczak EF, Petros AM, Meadows RP, Sattler M, Harlan JE, Wade WS, Burakoff SJ, Fesik SW (1995) Structure and ligand recognition of the phosphotyrosine binding domain of Shc. Nature 378 (6557):584–592. https://doi.org/10.1038/378584a0

Zhou CZ, Li de La Sierra-Gallay I, Quevillon-Cheruel S, Collinet B, Minard P, Blondeau K, Henckes G, Aufrere R, Leulliot N, Graille M, Sorel I, Savarin P, de la Torre F, Poupon A, Janin J, van Tilbeurgh H (2003) Crystal structure of the yeast Phox homology (PX) domain protein Grd19p complexed to phosphati-dylinositol-3-phosphate. J Biol Chem 278 (50):50371–50376. https://doi.org/10.1074/jbc. M304392200

Zhu G, Chen J, Liu J, Brunzelle JS, Huang B, Wakeham N, Terzyan S, Li X, Rao Z, Li G, Zhang XC (2007) Structure of the APPL1 BAR-PH domain and characterization of its interaction with Rab5. EMBO J 26(14):3484–3493. https://doi.org/10.1038/sj.emboj.7601771

Zhu Y, Hu L, Zhou Y, Yao Q, Liu L, Shao F (2010) Structural mechanism of host Rab1 activation by the bifunctional Legionella type IV effector SidM/DrrA. Proc Natl Acad Sci U S A 107(10):4699–4704. https://doi.org/10.1073/pnas.0914231107

Zimmerberg J, Kozlov MM (2006) How proteins produce cellular membrane curvature. Nat Rev Mol Cell Biol 7 (1):9–19. https://doi.org/10.1038/nrm1784

Zimmermann P (2006) The prevalence and significance of PDZ domain-phosphoinositide interactions. Biochim Biophys Acta 1761(8):947–956. https://doi.org/10.1016/j.bbalip.2006.04.003

Zimmermann P, Meerschaert K, Reekmans G, Leenaerts I, Small JV, Vandekerckhove J, David G, Gettemans J (2002) PIP(2)-PDZ domain binding controls the association of syntenin with the plasma membrane. Mol Cell 9(6):1215–1225

Zolov SN, Bridges D, Zhang Y, Lee WW, Riehle E, Verma R, Lenk GM, Converso-Baran K, Weide T, Albin RL, Saltiel AR, Meisler MH, Russell MW, Weisman LS (2012) In vivo, Pikfyve generates PI(3,5) P2, which serves as both a signaling lipid and the major precursor for PI5P. Proc Natl Acad Sci U S A 109 (43):17472–17477. https://doi.org/10.1073/pnas.1203106109

Adv Exp Med Biol - Protein Reviews (2019) 20: 139–157
https://doi.org/10.1007/5584_2018_295
© Springer Nature Singapore Pte Ltd. 2018
Published online: 30 November 2018

Physiological Functions of Phosphoinositide-Modifying Enzymes and Their Interacting Proteins in Arabidopsis

Tomoko Hirano and Masa H. Sato

Abstract

The integrity of cellular membranes is maintained not only by structural phospholipids such as phosphatidylcholine and phosphatidylethanolamine, but also by regulatory phospholipids, phosphatidylinositol phosphates (phosphoinositides). Although phosphoinositides constitute minor membrane phospholipids, they exert a wide variety of regulatory functions in all eukaryotic cells. They act as key markers of membrane surfaces that determine the biological integrity of cellular compartments to recruit various phosphoinositide-binding proteins. This review focuses on recent progress on the significance of phosphoinositides, their modifying enzymes, and phosphoinositide-binding proteins in Arabidopsis.

Keywords

Arabidopsis · Biological membranes · Organelles · Phosphoinositides · PI-binding proteins

T. Hirano and M. H. Sato (✉)
Graduate School of Life and Environmental Sciences, Kyoto Prefectural University, Kyoto, Japan
e-mail: mhsato@kpu.ac.jp

1 Introduction

Cells and sub-cellular compartments are surrounded by biological membranes to create specialized areas for controlled biological reactions. The integrity of cellular membranes is maintained through the presence of particular membrane-associated proteins that mediate various processes, including transport of molecules, signal transduction and regulating membrane regions. Biological membranes are mainly composed of structural glycerophospholipids (i.e. phosphatidylcholine and phosphatidylethanolamine), although phosphatidylinositol and its derivatives constitute minor membrane glycerophospholipids that exert a wide variety of regulatory functions in all eukaryotic cells (Balla 2013).

The hydroxyl groups of the D3, D4, and D5 positions of the inositol head group of phosphatidylinositol can be phosphorylated by specific lipid kinases to generate phosphoinositides, including three phosphatidylinositol (PI) monophosphates, two PI bisphosphates, and one PI trisphosphate (Di Paolo and De Camilli 2006). These phosphoinositides can be interconverted by various phosphoinositide kinases and phosphatases to modify the phosphorylation states of the inositol head group; alternatively, PIs are hydrolyzed by phospholipase Cs (PLCs) to generate diacylglycerol (DAG) and a soluble inositol group is then transferred into the cytosol (Balla 2013). However, only up to five phosphoinositides are generated due to a lack of PI(3,4)P$_2$ and PI(3,4,5)P$_3$ in plants (Meijer and Munnik 2003; Meijer et al. 2001) (Fig. 1). The existence of phosphoinositides in plants was described a decade ago; however, the roles of phosphoinositides, their modifying enzymes and effector proteins of phosphoinositides have been reported recently (Gerth et al. 2017). In this review, we focus on the recent progress of physiological functions of phosphoinositides, as well as their modifying enzymes and effector proteins in Arabidopsis.

2   Structures of Phosphoinositide-Modifying Enzymes

In the Arabidopsis genome, there are 20 PI-kinases, 27 PI phosphatases and nine PLCs (Fig. 2). We summarize the structure and physiological function of each phosphoinositide-modifying enzyme in the following sections.

2.1   Phosphatidylinositol 3-Kinase

Phosphatidylinositol 3-kinase, or Vacuolar Protein Sorting 34 (VPS34) produces phosphatidylinositol 3-phosphate from phosphatidylinositol by phosphorylating the D3 position of the inositol ring. Arabidopsis VPS34 (AtVPS34) is similar to yeast VPS34 and is encoded in the Arabidopsis genome as a single essential gene (Mueller-Roeber and Pical 2002). AtVPS34 has two conserved domains: a catalytic domain with the ATP binding site near the C terminus and a calcium-dependent lipid-binding domain near the N terminus.

2.2   Phosphatidylinositol 4-Kinase

PI can alternatively be phosphorylated in the D4 position of the inositol ring, resulting in the formation of phosphatidylinositol 4-phosphate (PI4P). PI4P is produced from phosphatidylinositol via phosphatidylinositol-4-kinases (PI4Ks). PI4P is the most abundant isomer, representing 80% of the total plant phosphatidylinositol phosphate pool (Meijer and Munnik 2003). In plants, PI4P

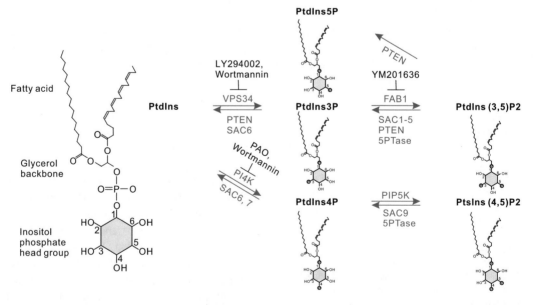

Fig. 1 The plant phosphoinositides bio-synthesis pathway. The structure of PIs and PIs-producing enzymes, and their inhibitors are shown in the PIs bio-synthesis pathways

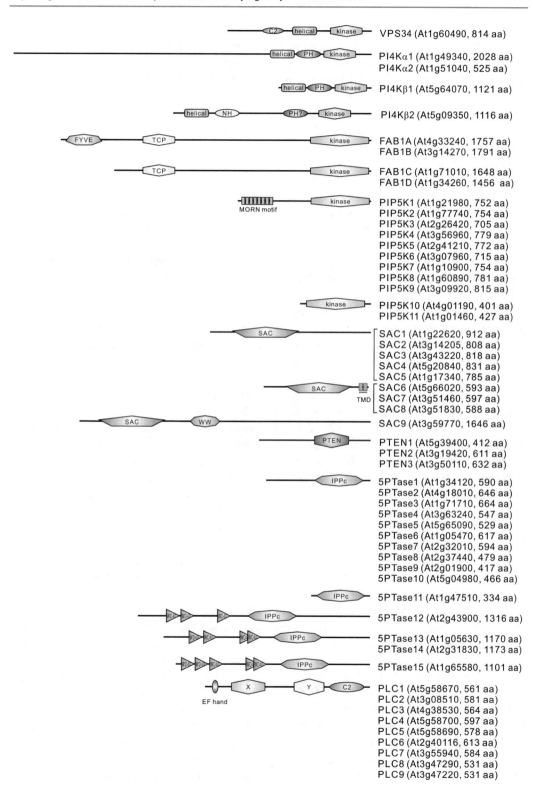

Fig. 2 Protein structure of PI-kinases and PI-phosphatases in Arabidopsis. Location of conserved domains are shown in respective proteins on the families of PI3-kinase, PI4-kinase, PI3P -kinase, PI4P 5-kinase, PTEN, SAC, and 5PTase

is important for controlling cell membrane identity and signaling to confer a negatively charged electrostatic field on the inner surface of the plasma membrane (Simon et al. 2016). The Arabidopsis genome encodes at least three sub-families of PI 4-kinases, namely PI4Kα, PI4Kβ and PI4Kγ (Mueller-Roeber and Pical 2002). PI4Kα and PI4Kβ have two isoforms each (PI4Kα1, PI4Kα2 and PI4Kβ1, PI4Kβ2, respectively) in the genome. PI4Kα1, PI4Kβ1, and PI4Kβ2 but not PI4Kα1 have been reported to have complete PI4K activity. PI4Kα contains the unique lipid kinase domain, a pleckstrin homology (PH) as well as a catalytic domain. The PH domain has the highest affinity for PI4P, the product of the reaction. However, Stevenson et al. (1998) found that while PI4Kβ1 does not have a PH domain, PI4Kβ1 was stimulated two-fold by PI4P. PI4Kα1 is associated primarily with membranes in the perinuclear region, whereas PI4Kβ1 is associated with small vesicles throughout the cytosol (Stevenson-Paulik et al. 2003).

PI4Kγs are characterized as a new family of protein kinases in Arabidopsis. Although eight genes encoding putative PI4Kγs (PI4Kγ1-γ8) have been identified in the Arabidopsis genome, no functional PI4K activity has yet been demonstrated (Galvão et al. 2008; Mueller-Roeber and Pical 2002).

2.3 Phosphatidylinositol 3-Phosphate 5-Kinases

PI3P can serve as a substrate for PI3P 5-kinases, which further phosphorylate the inositol ring in the D5 position to produce phosphatidylinositol 3,5-bisphosphate [$PI(3,5)P_2$]. Formation of aploid and binucleate cell (FAB1)/PIKfyve localizes to endolysosomes and vacuoles and plays an essential role in endosomal membrane trafficking, including vacuolar sorting, endocytosis of membrane proteins, ion transport, cytoskeleton dynamics and retrograde transport in mammalian and yeast cells (Efe et al. 2005; Ikonomov et al. 2011; Shisheva 2008). The Arabidopsis genome encodes four FAB1s, which have been termed FAB1A-FAB1D, based on their similarities to

the yeast enzyme FAB1 (Whitley et al. 2009). Yeast FAB1 and mammalian PIKfyve have a conserved N-terminal FYVE (Fab1p, YOTB, Vac1p, and EEA1)-finger domain responsible for PI3P binding to localize to the endosomes, the central part of the Cpn_TCP1 (HSP chaperonin T-complex protein 1) homology domain and the C-terminal kinase catalytic domain (Shisheva 2008).

In Arabidopsis, only FAB1A and FAB1B possess a conserved FYVE domain and the *fab1afab1b* double mutant reveals a male gametophyte-lethal phenotype, suggesting functional redundancy between FAB1A and FAB1B in Arabidopsis (Whitley et al. 2009). Studies have shown that FAB1A partially co-localizes with the late endosomes and multivesicular bodies (MVBs), which are precursors of vacuoles; knockdown of *FAB1A* and *FAB1B* alter the maturation state and localization of the late endosomes (Hirano et al. 2011, 2015). The functions of FYVE domain-lacking unconventional FAB1C and FAB1D remain unknown, although knockouts of *FAB1B* and *FAB1C* are known to decrease the rate of ABA-induced stomatal closure (Bak et al. 2013). Moreover, FAB1B and FAB1D are known to have complementary roles in regulating membrane recycling, vacuolar pH and homeostatic control of ROS in pollen tube growth (Serrazina et al. 2014).

2.4 Phosphatidylinositol 4-Phosphate 5-Kinases

Phosphatidylinositol 4,5-bisphosphate functions as a site-specific signal on membranes to promote various physiological processes that control cytoskeletal reorganization and membrane trafficking. $PI(4,5)P_2$ is generated by phosphorylation of the D5 position of the inositol ring of PI4P via phosphatidylinositol 4-phosphate 5-kinase (PIP5K). In the Arabidopsis genome, 11 isoforms of PIP5K (PIP5K1-PIP5K11) have been identified and these are categorized into two sub-families (sub-family A; PIP5K10 and PIP5K11, sub-family B; PIP5K1-PIP5K9) based on the existence of the N-terminal membrane occupation and recognition nexus (MORN)-repeat domain (Mueller-Roeber and Pical 2002).

2.5 Suppressor of Actin Phosphatases (SAC), Phosphatase and Tensin Homolog Deleted on Chromosome 10 (PTEN) and Phosphoinositide 5-Phosphatases (5PTases)

PIs can be dephosphorylated by three different kinds of phosphatases: suppressor of actin (SAC) phosphatases, the phosphatase and tensin homolog deleted on chromosome 10 (PTEN) and phosphoinositide 5-phosphatases (5PTases). In the Arabidopsis genome, these enzymes are encoded by a family of nine genes for SAC (SAC1-SAC9; Zhong and Ye 2003), three members for PTEN (PTEN1-PTEN3; Gupta 2002) and a family of 15 genes for 5PTase (At5PTase1-At5PTases15; Berdy et al. 2001). These multiple phosphatases can convert PIP_2 to PIPs or PIPs to PI. For instance, SAC1, PTENs, and 5PTase11 are likely to dephosphorylate PI $(3,5)P_2$ to PI3P (Ercetin et al. 2008; Zhong et al. 2005). Additionally, 5PTase6 and 5PTase7, also called CVP2 and CVL1, respectively, display activities against $PI(4,5)P_2$, resulting in the formation of PI4P (Carland and Nelson 2009). Furthermore, SAC7/ROOT HAIR DEFECTIVE 4 (RHD4) dephosphorylates PI4P to PI (Thole et al. 2008).

2.6 Phosphoinositide-Specific Phospholipase Cs

PLC enzymes can hydrolyze glycerophospholipids between glycerol and phosphate, resulting in the formation of DAG and the phosphorylated head group. In the case of phosphoinositide-specific PLCs (PI-PLCs), the reaction generates soluble inositol polyphosphates (IPPs) in addition to DAG (Fig. 1); the hydrolysis of PtdIns4P yields DAG and inositol 1,4-bisphosphate (IP_2) or inositol 1,4,5-trisphosphate (IP_3). In plants, two types of PLCs exist: the phosphoinositide-specific PLCs (PI-PLCs) and phosphatidylcholine-cleaving PLCs (PC-PLCs). Arabidopsis PI-PLCs are most similar in sequence to the zeta sub-family of mammalian PI-PLCs and the Arabidopsis genome encodes nine PI-PLC isoforms (PLC1-PLC9; Pokotylo et al. 2014; Tasma et al. 2008).

3 Sub-Cellular Localization of Plant Phosphoinositides in Arabidopsis

PIs are enriched in membranes of particular organelles and function as markers to assemble specific functional proteins onto membranes that determine organelle identity. Therefore, specific PI species predominantly localize to particular organelles or specific membrane domains (Balla 2007). The intracellular localization of PIs *in vivo* has been investigated via immunodetection by PI-specific antibodies or using biosensors that are composed of fluorescently tagged PI-binding domains for labeling specific PI species (Balla 2013).

In Arabidopsis, fluorescently tagged 2xFYVE domain labels of PI3P localize to vacuolar membranes, late endosomes and cell plates of dividing cells (Simon et al. 2014; Vermeer et al. 2006). The PH domain of the Four-phosphate-adaptor protein 1 (FAPP1)-based PI4P probe reveals how PI4P resides in the Golgi apparatus, trans-Golgi network, late endosome, plasma membrane (PM) and outer plastid envelope (Kim et al. 2001). Expression of a green or yellow fluorescent protein (GFP or YFP) fused to the PH domain of human PLCδ1 for visualization of PI $(4,5)P_2$ localized to the PM (Van Leeuwen et al. 2007; Simon et al. 2014) reveals its polarized localization in growing pollen tubes (Ischebeck et al. 2008), root hair tips (Braun et al. 1999; Kusano et al. 2008; Van Leeuwen et al. 2007; Stenzel et al. 2008) and root cells (Tejos et al. 2014). Using immunogold electron microscopy, $PI(4,5)P_2$ has also been shown to be enriched in microdomains on the PM (Furt et al. 2010).

Recently, a novel type of $PI(3,5)P_2$ probe has been developed in Arabidopsis (Hirano et al. 2017a) based on a tandem repeat of the cytosolic phosphoinositide-interacting domain (ML1N) of the mammalian lysosomal transient receptor potential cation channel (Li et al. 2013). Using this probe, it is possible to visualize $PI(3,5)P_2$ localized to the limited membranes of the FAB1- and sorting nexin1 (SNX1)-positive late endosomes (Hirano et al. 2017a).

The localization of PIs is basically shared with the localization of their producing enzymes. For example, PI 4-kinases reside in the trans-Golgi network (Preuss et al. 2006; Thole et al. 2008), PI4P 5-kinases reside in the PM (Ischebeck et al. 2013; Kusano et al. 2008; Lou et al. 2007; Sousa et al. 2008; Stenzel et al. 2008; Tejos et al. 2014), while PI3P 5-kinase, FAB1 resides in late endosomes (Hirano et al. 2017b) (Fig. 3).

4   Physiological Function of PI-Modifying Enzymes

4.1   Controlling Exocytosis and Endocytosis

Plant PIs not only influence the secretory pathway, but also control processes involved in the recycling of membrane and/or membrane-integral cargo. *pi4kβ1pi4kβ2* knockout mutant plants produce secretory vesicles with highly variable sizes (Kang et al. 2011), indicating that PI4Kβ1/2 regulate membrane trafficking from the trans-Golgi network (TGN) to the PM and the vacuoles. The Rab-E sub-class of the Rab family of small GTPases is related to mammalian Rab 8 and is involved in the secretory pathway from the Golgi to the PM. Arabidopsis PIP5K2 interacts with all five members of the GTP-bound active form of Rab-Es with the MORN domain, suggesting that Rab-Es and PIP5K2 and/or its product, $PI(4,5)P_2$, coordinately control membrane trafficking from the TGN to the PM (Camacho et al. 2009).

Using a combination of fluorescent reporter proteins and the $PI(3,5)P_2$-specific inhibitor YM202636, Jefferies et al. (2008) reported that $PI(3,5)P_2$ affects various membrane trafficking

events, mostly in the post-Golgi routes. Treatment with YM201636 effectively reduced $PI(3,5)P_2$ concentration not only in the wild type but also in *FAB1A*-overexpressing Arabidopsis plants. In particular, reduced $PI(3,5)P_2$ levels caused abnormal membrane dynamics of PM proteins, AUX1 and BOR1, with different trafficking patterns. Secretion and morphological characteristics of late endosomes and vacuoles were also affected by decreased $PI(3,5)P_2$ production. These pleiotropic defects in the post-Golgi trafficking events were caused by inhibition of $PI(3,5)P_2$ production. This effect is probably mediated by the inhibition of maturation of FAB1-positive late endosomes, thereby impairing late endosome function. In conclusion, in Arabidopsis, late endosomes are involved in multiple post-Golgi membrane trafficking routes, including not only vacuolar trafficking and endocytosis, but also secretion (Hirano et al. 2017a).

4.2   Controlling Polarization of the PM Proteins

In multicellular organisms, establishment of cell polarity by asymmetric distribution of PM membrane proteins such as receptors, transporters, and pumps, is essential for development and response to environmental stimuli. In this process, phosphoinositides and their modifying enzymes are crucial for the establishment of polarity of various functional PM proteins controlling membrane trafficking events (Balla 2013; Di Paolo and De Camilli 2006). In Arabidopsis, polarized localization of PIN auxin efflux carrier proteins is mediated by the combined action of several phosphoinositide signaling pathways.

Phosphatidylinositol-4-monophosphate 5-kinases (PIP5Ks) produce $PI(4,5)P_2$ and have been implicated in polarized membrane trafficking in Arabidopsis. PIP5K1 and PIP5K2 are expressed ubiquitously in Arabidopsis and the *pip5k1pip5k2* double mutant with reduced $PI(4,5)P_2$ levels showed dwarf phenotypes (Ischebeck et al. 2013). These phenotypes are attributed to the auxin distribution defect caused by mislocalization of auxin efflux carriers PIN1

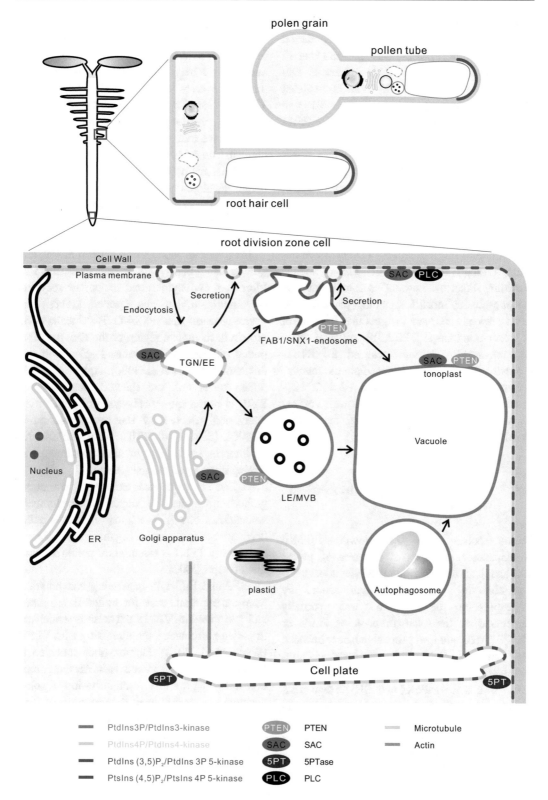

Fig. 3 Organelle localization map of plant phosphoinositides. PIs localizes to the specific organelle membranes. PIP5K/ PI(4,5)P$_2$ localizes to the root-hair tip and upper-bottom side of the plasma membrane in the root differentiation zone

and PIN2. In the *pip5k1pip5k2* double mutant, clathrin light chain is also displayed in an altered localization pattern. These data indicate that PI $(4,5)P_2$ produced by PIP5K1/2 influences PIN polarization by controlling clathrin-mediated membrane trafficking (Ischebeck et al. 2013); in addition, controlling PIN2 and PIN3 by PIP5K2 influences lateral root formation and root gravitropic responses (Mei et al. 2012).

Bipolar localization of PIP5K1 and PIP5K2 and their product $PI(4,5)P_2$ is required for apical and basal localization of PIN proteins, which are essential for embryonic and post-embryonic patterning in roots (Tejos et al. 2014). This reveals a critical role of $PIP5K2/PI(4,5)P_2$ in root development through regulation of PIN proteins, providing direct evidence of cross-talk between the phosphatidylinositol signaling pathway and auxin response and new insights into the control of polar auxin transport. $PI(3,5)P_2$ and its producing enzyme FAB1, are localized in SNX1-positive late endosomes that mediate the maturation process of late endosomes. FAB1/SNX1-positive endosomes are associated with cortical microtubules, thereby controlling basal PIN polarity of roots (Hirano et al. 2015).

4.3 Pollen Development and Pollen Tube Growth

During pollen tube tip growth, PIP5K6 accumulates at the apex of the growing pollen tube and regulates clathrin-dependent endocytosis. The abnormal PM accumulation caused by overexpression of PIP5K6 was partially suppressed by the overexpression of PLC2 or PI4Kβ1, indicating a proper balance between PI4P and $PI(4,5)P_2$, which regulate clathrin-dependent endocytosis at the pollen tube tip (Zhao et al. 2010). PIP5K4 and PIP5K5 also accumulate at the pollen tube tip and are involved in pollen tube growth in Arabidopsis and *Nicotiana tabacum*. Overexpression of either PIP5K4 or PIP5K5 induces multiple tip-branching pollen tubes and causes massive deposition of pectin accompanied by PM deposition (Ischebeck et al.

2008). These reports show that $PI(4,5)P_2$ regulates balanced membrane trafficking and apical deposition of pectin during pollen tube growth. PIP5K1 and PIP5K2 have been shown to be expressed during the early stages of pollen development. The *pip5k1pip5k2* double mutant could not produce any viable pollen, indicating that these PIP5Ks have redundant functions in pollen development. The double mutant pollens have morphological defects of vacuoles in the later stages of pollen development (Ugalde et al. 2016). These results show that PIP5K1/2-mediated $PI(4,5)P_2$ signaling is an important cue for vacuole formation during male gametophyte development.

In the *pip5k4* mutant, pollen germination, tube growth and polarity are significantly impaired. Moreover, endocytosis and membrane recycling are inhibited in its pollen tubes and overexpression of PIP5K4-GFP delocalized the protein to the apical region of the PM, perturbed pollen tube growth and caused apical cell wall thickening (Sousa et al. 2008). Apical growth of pollen tubes and the specific localization of SYP124 at the sub-apical region of pollen tubes were also affected by the overexpression of PIP5K4. (Silva et al. 2010). Thus, PIP5K4 has an important role in regulating membrane trafficking during pollen tube growth.

PTEN1 is expressed exclusively in pollen grains during the late stages of development. Knockdown of *PTEN1* expression causes pollen death after mitosis, indicating that PI phosphatase activity of PTEN1 is essential for pollen development (Gupta 2002).

PI3P and $PI(3,5)P_2$-generating enzymes also have an important role for pollen development and germination; VPS15 plays an essential role in pollen germination by interacting with VPS34 (Wang et al. 2012). Homozygous mutation of FAB1A and FAB1B affects vacuolar reorganization following the first mitotic division of pollen development, resulting in the cessation of large abnormal vacuoles at the tricellular stage. This demonstrates the crucial role of FAB1 and PI $(3,5)P_2$ for successful pollen development (Whitley et al. 2009).

4.4   Root Hair Elongation

Phosphoinositide signaling is an important factor for regulating tip growth of root hairs. In particular, in the root cortex, epidermal cells and root hairs, the Arabidopsis *PIP5K3* gene was identified and found to be expressed. On membranes of root hairs, the product of PIP5K3, $PI(4,5)P_2$, functions as a site-specific signal to promote cytoskeletal reorganization and membrane trafficking. Localization of $PI(4,5)P_2$ to apices of growing root hairs suggests that PI$(4,5)P_2$ plays an important role in tip growth (Braun et al. 1999; Van Leeuwen et al. 2007). Fluorescence-tagged PIP5K3 localized to the periphery of the apical region of root hair cells and a T-DNA insertion mutant of *PIP5K3* displays significantly shorter root hairs than the wild type (Stenzel et al. 2008). In contrast, overexpression of *PIP5K3* causes longer root hairs and multiple protruding sites on a single trichoblast, and YFP-fused PIP5K3 complements the short root hair phenotype and localizes to the PM, cytoplasmic space of elongating root hair apices, growing root hair bulges and notably, to sites about to form root hair bulges (Kusano et al. 2008). *PIP5K3* and *PIP5K4* genes respond to phosphate deficiency-dependent root-hair elongation, indicating that a phosphate deficiency signal is transferred to the regulatory pathway for root hair elongation via *PIP5K* genes (Wada et al. 2015). In addition, PIP5K3 regulates planar polarity of root epidermal cells with Rho-of-plant (ROP)2 and ROP6, as well as dynamin-related proteins (DRP)1A and DRP2B, which accumulate in a sterol-enriched polar membrane domain during root hair initiation. A AGCVIII kinase, D6 protein kinase (D6PK) is a modulator of this process (Stanislas et al. 2015). In roots, PI4Kα1 is selectively recruited to the TGN by the action of RabA4b, providing a link between Ca^{2+} and $PI(4,5)P_2$-dependent signals during the polarized secretion of cell wall components in tip-growing root hair cells (Preuss et al. 2006). In addition, *RHD4* encodes a SAC1-type phosphatidylinositol-4-phosphate phosphatase, which localizes to membranes at the tips of growing root hairs and regulates proper root hair development, and RHD4 shows a preference for PI4P *in vitro* and *rhd4–1* roots accumulate higher levels of PI4P (Thole et al. 2008). These data suggest that PI4Kβ1 and RHD4 cooperatively regulate PI4P levels at the root hair tip region, thereby controlling RAB4A4b-dependent tip growth of root hairs.

4.5   Stress Response

Exogenous stresses, including salt and osmotic stresses, control the expression of enzymes that generate or degrade PIs, to regulate the total amounts of PIs. This suggests that PIs might take part in not only endogenous developmental processes but also in the adaptive responses of plants to a changing environment. Salinity and hyperosmotic stresses induce a rapid synthesis of $PI(4,5)P_2$ (Pical et al. 1999). Rapid accumulation of $PI(4,5)P_2$ and IP_3 correlate with calcium mobilization in salt-stressed Arabidopsis (Dewald et al. 2001). In salt stress conditions, $PI(4,5)P_2$ is associated with clathrin-coated vesicles (CCVs), suggesting that $PI(4,5)P_2$ acts in stress-induced formation of CCVs (König et al. 2008). Heat stress induces a rapid accumulation of phosphatidic acid (PA) and $PI(4,5)P_2$ by activation of phospholipase D and PIP kinase to regulate ion channels and the cytoskeleton (Mishkind et al. 2009). Thus, PI(4,5) P_2, IP_3 and PA have important roles in tolerance against various environmental stresses. Several PI-modifying enzymes are reportedly involved in stress-induced PI changes in Arabidopsis. The synthesis of P4P by PI4Ks occurs in parallel with PI-PLC activation in response to cold stress (Delage et al. 2012).

The *PLC1* gene is expressed at very low levels under normal growth conditions, but is induced to a significant extent under various environmental stresses, such as hydration, salinity and low temperature (Hirayama et al. 1995). *PLC2* is an essential gene in Arabidopsis and is required not only for reproductive development

(Di Fino et al. 2017; Li et al. 2015) and seedling growth, but also for the endoplasmic reticulum stress response (Kanehara et al. 2015). Mutants of *PLC3* and *PLC9* additively exhibit a serious thermosensitive phenotype and heat-induced elevation of Ca^{2+} is impaired in *plc3* and *plc9* mutants. These results indicate that *PLC3* and *PLC9* function in the signaling pathway leading to thermotolerance (Gao et al. 2014b; Zheng et al. 2012). *plc4* mutant seedlings exhibit increased tolerance against salt stress, whereas transgenic seedlings with overexpression of *PLC4* show a salt-hypersensitive phenotype, indicating that *PLC4* is a negative regulator of salt tolerance (Xia et al. 2017). Exposure of *pikIIIβ1pikβ2* double mutant to cold stress results in significantly decreased PA, the product of the PI-PLC pathway (Delage et al. 2012).

PI(4,5)P$_2$ signaling is terminated by the action of inositol polyphosphate phosphatases and PI phosphatases such as actin mutation (SAC) domain phosphatases. The expression of the *SAC6* gene, which dephosphorylates PI3P and PI4P, is highly induced by salinity stress (Zhong and Ye 2003). *SAC9* is expressed constitutively (Zhong and Ye 2003) and mutations in *SAC9* lead to over-accumulation of PI(4,5)P$_2$ under normal growth conditions, suggesting that SAC9 phosphatase is involved in modulating phosphoinositide signals during the stress response (Williams 2005).

PI3P and PI(3,5)P$_2$ also have crucial roles for stress responses in Arabidopsis. Osmotic stress induces PI(3,5)P$_2$ not only in animal cells but also in plant cells (Dove et al. 1997). PI(3,5)P$_2$ also accumulates under oxidative stress conditions (Hirano et al. 2017b). During drought or salinity stress conditions, proline metabolism is also regulated by a class-III phosphatidylinositol 3-kinase (PI3K), VPS34 (Leprince et al. 2014).

4.6 Stomatal Opening and Closing

The pharmacological inhibitors wortmannin (WM) and LY294002 have been shown to inhibit PI3K and PI4K activities in guard cells and promote stomatal opening induced by white light or the circadian clock (Jung et al. 2002). WM and LY294002 also inhibit stomatal closure mediated by abscisic acid (ABA) to inhibit ABA-induced ROS generation (Park et al. 2003). ABA-induced changes in organization of actin filaments are inhibited by both WM and LY295002 or by expressing PI3P- and PI4P-binding domains (Choi et al. 2008). Fragmented vacuoles in guard cells after ABA-induced stomatal closure are fused into forming large central vacuoles following WM treatment (Zheng et al. 2014). These results suggest that PI3P and PI4P regulate stomatal closure to control ROS generation, actin dynamics and vacuolar morphology in guard cells. In contrast, PI(4,5)P$_2$ produced by PIP5K4 is also important for stomatal opening to inhibit anion channels of guard cells (Lee et al. 2007b).

Vacuolar acidification is necessary for ABA-induced stomatal closure. An inhibitor of FAB1 PI3P 5-kinase, YM201636, inhibits vacuolar convolution, which induces delayed stomatal closure in response to ABA. These results suggest that rapid ABA-induced stomatal closure requires PI(3,5)P$_2$ for regulating vacuolar acidification and convolution (Bak et al. 2013). For vacuolar acidification, PI(3,5)P$_2$ does not affect the activity of vacuolar H^+-pyrophosphatase or H^+-ATPase, but inhibits an inwardly rectifying conductance mediated by chloride channel a (CLC-a) (Carpaneto et al. 2017).

4.7 Autophagy

SH3P2 associates with the PI3K complex and interacts with ATG8s to function as a novel regulator for autophagosome biogenesis in Arabidopsis (Zhuang et al. 2013, 2017). In addition, ATG9 regulates autophagosome progression from the endoplasmic reticulum to control the trafficking of ATG18a in a PI3P-dependent manner (Zhuang et al. 2017); moreover, ATG18a has also been isolated as a PI(3,5)P$_2$-binding protein *in vitro* (Oxley et al. 2013).

4.8   Stem Cells

PIP5K1 is reportedly strongly expressed in pro-cambial cells in vascular tissues of leaves, flowers and roots, suggesting the involvement of PIP5K and PI(4,5)P$_2$ in the control of cell proliferation (Elge et al. 2001). PI4Ps are related to the CLE/WOX signaling pathways and play an important role in promoting stem cell specification. In addition, the Arabidopsis protein phosphatase type-2C proteins POLTERGEIST (POL) and POLTERGEIST LIKE 1 (PLL1) are essential for maintaining both the root and shoot stem cells. POL is catalytically activated by PI4P *in vitro*, suggesting that phospholipids play an important role in promoting stem cell specification (Gagne and Clark 2010).

4.9   Plastid Division

The existence of PI kinase activity was reported in chloroplast envelope membranes and PI3P as well as PI4P were identified as its products (Bovet et al. 2001). Arabidopsis dynamin-like 2 (ADL2) protein that binds specifically to PI4P targets the plastid envelope membranes, demonstrating that ADL2 may be involved in vesicle formation at the chloroplast envelope membrane (Cheong et al. 2013). Loss-of-function of PI4Kα1 and PI4Kβ2 decreases the PI4P levels of chloroplasts, resulting in chloroplast division, and PI4P binds to PLASTID DIVISION1 (PDV1) and PDV2 proteins that recruit DYNAMIN-RELATED PROTEIN5B (DRP5B). These findings suggest that PI4P is a regulator of chloroplast division in a PDV1- and DRP5B-dependent manner (Okazaki et al. 2015).

4.10   Pathogen Response

Salicylic acid (SA) has a central role in defense against pathogen attack. In Arabidopsis suspension cells, addition of SA into the medium induces a rapid decrease in the PI pool parallel to an increase in the PI4P and PI(4,5)P$_2$ contents.

These changes are inhibited by type-III PI 4-kinase inhibitors, suggesting that PI4Ks are possibly involved in this process (Krinke et al. 2007). Subsequently, PI4Kβ1, PI4Kβ2 and PLANT U-BOX 13 (PUB13) have been identified as RabA4b effector proteins, which play important roles during SA-mediated plant defense signaling (Antignani et al. 2015).

EDR2 acts as a negative regulator of cell death, specifically the cell death elicited by pathogen attack and mediated by the SA defense pathway. In addition, PI4P may have a role in limiting cell death via its effect on EDR2 (Vorwerk et al. 2007).

4.11   Regulation of Cytoskeleton and Organelle Morphology

Membrane trafficking and polarization of cells are tightly linked to the dynamics of the actin cytoskeleton, which is required for the directional movement of vesicles and organelles (Wang and Hussey 2015). PI(4,5)P$_2$ is involved in regulating the dynamic assembly of the actin cytoskeleton through interaction with the actin-binding protein, CP1, in pollen tube development (Huang et al. 2006). The FRAGILE FIBER3 (*FRA3*) gene encodes 5PTase15, that exhibits phosphatase activity toward PI(4,5)P$_2$ and PI(3,4,5)P$_3$; *FRA3* also plays an essential role in secondary cell wall synthesis in fiber cells and xylem vessels to regulate actin organization (Zhong et al. 2004). The *fragile fiber 7* mutant (*fra7*) causes alternations in cell morphogenesis, cell wall synthesis and actin organization. *FRA7* encodes another type of PI phosphatase, SAC1, that localizes to the Golgi apparatus and exhibits highly specific phosphatase activity toward PI(3,5)P$_2$ (Zhong et al. 2005).

In contrast, SAC2-SAC5 localize to the tonoplast along with PI3P and gain- and loss-of-function mutants of these SAC phosphatases affect PI(3,5)P$_2$ content, thereby influencing vacuolar trafficking and morphogenesis (Nováková et al. 2014). Similarly, treatment with a PI(3,5)P$_2$ synthesis inhibitor induces morphological changes in the central vacuole (Hirano et al. 2017a) and pollen carrying mutant alleles of both *FAB1A* and *FAB1B* cause a severe defect in vacuolar reorganization following the first mitotic division of development (Whitley et al. 2009).

These results suggest that the regulation of PI $(3,5)P_2$ content is crucial for morphogenesis of the central vacuole. Intriguingly, neither FAB1 nor $PI(3,5)P_2$ have ever been observed in the central vacuolar membrane in Arabidopsis (Hirano et al. 2015, 2017b), although these molecules localize to vacuolar and lysosomal membranes in yeast and mammals (McCartney et al. 2014; Shisheva 2008). The non-existence of $PI(3,5)P_2$ in the vascular membrane suggests that $PI(3,5)P_2$ transported to the central vacuolar membrane is rapidly erased by vacuole-localized SAC phosphoinositide phosphatases for controlling vacuole size and morphology in plants (Nováková et al. 2014).

SNX1-positive endosomes are associated with cortical microtubules via the microtubule-associated protein CLASP to link microtubules and auxin transport (Ambrose et al. 2013). FAB1 or $PI(3,5)P_2$ mediates the maturation process of SNX1/FAB1-positive late endosomes to establish endosome-cortical microtubule interaction for controlling PIN protein trafficking, and a loss-of *FAB1* function causes the release of SNX1/FAB1-positive endosomes from cortical microtubules and disrupts proper cortical microtubule organization and morphology of the endosomes (Hirano et al. 2015). These results support the idea that $PI(3,5)P_2$ regulates the organization of cortical microtubules and late endosomes to connect with each other beneath the PM in Arabidopsis.

5 Phosphoinositide-Binding Proteins

PIs function as scaffolds or ligands of various effector proteins, which may be positively or negatively regulated in plant physiological activity. In this section, we introduce previously reported effectors and binding proteins of PIs (Table 1).

5.1 Non-specific Phosphoinositide- and Phosphoinositide Kinase-Binding Proteins

Several regulatory proteins of cell polarity have been isolated as non-specific PI-binding proteins. For instance, an AGCVIIIa kinase, PINOID (PID), plays a fundamental role in the asymmetrical localization of membrane proteins during polar auxin transport to control the polarity of PIN proteins (Armengot et al. 2016), which binds to phosphoinositides non-selectively (Zegzouti et al. 2006). AGC kinases are regulated via 3-phosphoinositide-dependent protein kinase-1 (PDK1) (Garcia et al. 2012), which interacts via its PH domain with PA, PI3P, $PI(3,4,5)P_3$ and PI $(3,4)P_2$ and to a lesser extent with $PI(4,5)P_2$ and PI4P *in vitro* (Deak et al. 1999). Also, novel Sec14-like proteins, Patellin1 and Patellin2, which are involved in cell-plate expansion, have been characterized as PI-binding proteins with less binding specificity to various PIs (Peterman et al. 2004; Suzuki et al. 2016).

The Arabidopsis stem cell factors, POL and PLL1, are integral components of the CLE/WOX signaling pathways and directly bind to multiple lipids including cardiolipin, phosphatidylserine, PI3P, PI4P and PI5P (Gagne and Clark 2010).

5.2 Phosphatidylinositol 3-Phosphate

In Arabidopsis, various proteins having the PH domain, FYVE domain, PX domain and N-terminal ENTH domain have been reported to specifically bind to PI3P. A PH domain-containing protein, AtPH1, binds to PI3P *in vivo* and acts in the late endosome compartment to control the localization of the metal transporter NRAMP1 (Agorio et al. 2017). The Arabidopsis dynamin-like 6 (ADL6)/DRP2a shows high-affinity binding to PI3P via its PH domain and is involved in vesicle trafficking from the TGN to the central vacuole (Lee et al. 2002). A FYVE domain protein 1 (FYVE1)/endosomal sorting 1 (FREE1), which is essential for vacuole

Table 1 Phosphoinositide-binding proteins and phosphoinositide kinases in Arabidopsis

Interacting protein	Gene ID of interacting protein	Interacting domain	Interacting PIs/PIs-kinase	References
PDK1	AT5G04510	PH	PI3P, PI(4,5)P$_2$	Deak et al. (1999), Anthony et al. (2004)
AtPH1	AT2G29700	PH	PI3P	Agorio et al. (2017)
FYVE1	AT1G20110	FYVE	PI3P	Kolb et al. (2015)
EpsinR2	AT2G43160	ENTH	PI3P	Lee et al. 2007a
SNX1	AT5G06140	PX	PI3P, PI(3,5)P$_2$, FAB1	Hirano et al. (2015)
SNX2b	AT5G07120	PX	PI3P	Phan et al. (2008)
DRP2A	AT1G10290	PH	PI3P, PI4P, PI(4,5)P$_2$	Lee et al. (2002)
ADL2	AT4G33650	PH	PI4P	Kim et al. (2001)
ATG18a	AT3G62770		PI3P, PI(3,5)P$_2$	Oxley et al. (2013)
Patellin1	AT1G72150		Pi(3,5)P$_2$	Oxley et al. (2013)
Patellin2	AT1G22530		PI3P, PI4P, PI(3,5)P$_2$, PI(4,5)P$_2$	Suzuki et al. (2016)
SH3P1	AT1G31440	BAR	PI4P, PI(4,5)P$_2$	Lam et al. (2001)
SH3P2	AT4G34660	BAR	PI3P, PI3K, PI(4,5)P$_2$, PI(3,4,5)P$_3$	Zhuang et al. (2013)
ROF1	AT3G25230		PI3K, PI(3,5)P$_2$	Karali et al. (2012)
VPS15	AT4G29380		PI3K	Wang et al. (2012)
POL	AT2G46920		PI4P	Gagne and Clark (2010)
PLL1	AT2G35350		PI4P	Gagne and Clark (2010)
SFC/VAN3	AT5G13300		PI4P	Koizumi et al. (2005)
AGD1	AT5G61980		PI4P	Koizumi et al. (2005)
TPR-like	AT1G55890		Pi(3,5)P$_2$	Oxley et al. (2013)
PCaP1/MAP25	AT4G20260		PI(3,5)P$_2$, PI(4,5)P$_2$, PI(3,4,5)P$_3$	Nagasaki et al. (2008)
PCaP2/MAP18	AT5G44610		PI(3,5)P$_2$, PI(4,5)P$_2$, PI(3,4,5)P$_3$	Kato et al. (2013)
TLP3	AT2G47900		PI(3,4)P$_2$, PI(3,5)P$_2$, PI(4,5)P$_2$	Bao et al. (2014)
TLP9	AT3G06380		PI(3,4)P$_2$, PI(3,5)P$_2$, PI(4,5)P$_2$	Bao et al. (2014)
AtCP	AT3G05520		Pi(4,5)P$_2$	Huang et al. (2006)
AtSFH1	AT4G34580		Pi(4,5)P$_2$	Ghosh et al. (2015)
RAB-E1D	AT5G03520	MORN	PIP5K2	Camacho et al. (2009)

biogenesis (Kolb et al. 2015), binds to PI3P and ubiquitin and specifically interacts with Vps23 via PTAP-like tetrapeptide motifs to be incorporated into the ESCRT-I complex (Gao et al. 2014a). FREE1 also regulates IRT1-dependent metal transport and homeostasis (Barberon et al. 2014). The Arabidopsis SH3 domain-containing protein (SH3P) family, which has a BAR (Bin-Amphiphysin-Rvs) domain, is composed of three isoforms, SH3P1-SH3P3 (Lam et al. 2001). SH3 domain-containing protein 1 (SH3P1) binds to PA, PI4P, and PI(4,5)P$_2$ and is possibly involved in clathrin-mediated vesicle trafficking along with the actin cytoskeleton (Lam et al. 2001). Alternatively, SH3P2 binds to phosphatidylinositol 3-phosphate (PtdIns3P) and functions downstream of the phosphatidylinositol 3-kinase (PI3K) complex (Zhuang et al. 2013). It also functions as a novel regulator for autophagosome biogenesis (Zhuang et al. 2013), as a ubiquitin-binding protein along with ESCRT-I and the deubiquitinating enzyme AMSH3 (Nagel et al. 2017). The BAR domain of SH3P2 was recently reported to also bind to PI(4,5)P$_2$ and PI(3,4,5)P$_2$ so as to be involved in membrane tabulation during cell-plate formation (Ahn et al. 2017). The PX domain of AtSNX2b binds to PI3P *in vitro* and this association is required for the localization of GFP-AtSNX2b to TGN, the pre-vacuolar compartment and endosomes (Phan et al. 2008). A member of the Epsin family of proteins having the epsin N-terminal homology (ENTH) domain, EpsinR2, binds to PI3P with its ENTH domain and plays an important role in protein trafficking through interaction with δ-adaptin, VTI12, and clathrin (Lee et al. 2007a).

5.3 Phosphatidylinositol 4-Phosphate

Arabidopsis dynamin-like 2a (ADL2) is a PI4P-binding protein that is involved in mitochondrial and chloroplast division (Arimura et al. 2004; Kim et al. 2001; Logan et al. 2004). Integral outer envelope membrane proteins PDV1 and PDV2 are components of the plastid division machinery and specifically bind to PI4P at the plastid envelope membrane (Okazaki et al. 2015). A class 1 ADP-ribosylation factor GTPase-activating protein (ARF-GAP), AGD1, functions in controlling signaling pathways with phosphoinositides and actin to control root hair development (Yoo et al. 2012). AGD1 has a PI-binding PH domain (Vernoud et al. 2003), suggesting PI4P-binding activity. Another class 1 ARF-GAP, VAN3/AGD3/SFC, is involved in leaf vascular network formation and specifically binds to PI4P, necessary for its TGN (Koizumi et al. 2005).

5.4 Phosphatidylinositol 5-Phosphate

UHRF1 is a multi-domain protein crucially linking histone H3 modification sites and DNA methylation. UHRF1 preferentially binds to PI5P with the polybasic region (PBR), resulting in a conformational rearrangement of the tandem Tudor domain to bind H3K9me3 (Gelato et al. 2014).

5.5 Phosphatidylinositol 3,5-Bisphosphate

$PI(3,5)P_2$ increases in response to osmotic stress. A plant immunophilin ROF1, containing a FK506 Binding Domain (FKBP) and tricopeptide repeat domains (TPR) plays an important role in the osmotic/salt stress responses of germinating seeds. ROF1 interacts directly with PI3P and PI $(3,5)P_2$ (Karali et al. 2012). Hydrophilic cation-binding proteins, PCAP1, and PCAP2, which are involved in stomatal closure (Nagata et al. 2016) and root hair morphogenesis (Kato et al. 2013), were found to preferentially interact with $PI(3,5)$ P_2 and $PI(3,4,5)P_3$. Moreover, these interactions were found to be inhibited by association with calmodulin in a Ca^{2+}-dependent manner (Kato et al. 2013; Nagasaki et al. 2008). An endosomal protein, SNX1, which is involved in the auxin pathway, binds not only to PI3P but also $PI(3,5)$ P_2 (Hirano et al. 2015).

5.6 Phosphatidylinositol 4,5-Bisphosphate

Phospholipase Ds (PLDs) hydrolyze membrane lipids to generate PA and a free head group. This activity is widespread in plants and plays multiple regulatory roles in diverse plant processes, including ABA signaling, programmed cell death, root hair patterning, root growth, freezing tolerance and other stress responses (Qin and Wang 2002). Phospholipase D β and γ have a conserved Ca^{2+}-dependent phospholipid-binding (C2) domain that binds to $PI(4,5)P_2$ for their activity (Pappan et al. 1997; Qin et al. 1997). A Sec14-nodulin protein AtSfh1 interacts with $PI(4,5)P_2$ in the C-terminal nodulin domain, playing an essential aspect of the polarity signaling program in root hairs (Ghosh et al. 2015). Tubby and Tubby-like proteins (TLPs) play essential roles in the development and function of mammal neuronal cells. Arabidopsis TLPs, TLP3, and TLP9 are PM-tethered PIP2- $(PI(3,4)P_2, PI(3,5)P_2, PI(4,5)$ $P_2)$ binding proteins that function redundantly in ABA- and osmotic stress-mediated seed germination (Bao et al. 2014). The heterodimeric actin capping protein from Arabidopsis (AtCP) binds to PA as well as PI(4,5)P$_2$. The interaction with PA inhibits the actin-binding activity of CP to induce actin polymerization in response to extracellular stimuli or pollen tube tip growth. (Huang et al. 2006). A SEC14 family protein PATTELLIN2 (PATL2) regulated by MAP kinase 4 (MPK4), has the ability to bind to various phosphoinositides, particularly to PI3P and PI4P in the unphosphorylated state and to PI(4,5)P$_2$ in the phosphorylated state (Suzuki et al. 2016).

6   Conclusions and Future Prospects

As sessile organisms, plants in their natural environment are continuously threatened by various abiotic and biotic stresses. To survive such stressful conditions, plants have evolved complex regulatory networks that control their responses to changes in environmental conditions (Yamaguchi-Shinozaki and Shinozaki 2006). Recent progress in the physiological functions of PIs in plants indicate that the stress-induced dynamics of PIs and the PI signaling cascade would contribute to the control of other regulatory pathways including ROP, MAP kinases and Ca^{2+} signaling, but the links between PI signaling and other regulatory pathways have not been well explored in detail. In addition, plants have a more complex endomembrane system and membrane trafficking machinery than yeasts and animals to build up their bodies, establish cell polarity and secrete various secondary metabolites (Paul et al. 2014). PI molecules determine the organelle identity and membrane domain specificity of particular PM domains to recruit specific effector molecules on the membrane. Recently, the important roles of a number of PI-interacting proteins or PI-modifying enzymes have been identified as membrane trafficking regulators. Future studies are necessary to unravel the exact roles of PIs in the complex regulatory mechanisms for adapting to environmental stresses, the regulation of membrane trafficking and establishment of organelle identity.

Acknowledgements This work was supported by the Ministry of Education, Culture, Sports, Science and Technology of Japan, a grant-in-aid for Scientific Research (B) (16H05068 to M.H.S.).

Disclosures The authors have no conflicts of interest to declare.

References

Agorio A, Giraudat J, Bianchi M et al (2017) Phosphatidylinositol 3-phosphate–binding protein AtPH1 controls the localization of the metal transporter NRAMP1 in *Arabidopsis*. Proc Natl Acad Sci U S A 114:E3354–E3363

Ahn G, Kim H, Kim DH et al (2017) SH3 domain-containing protein 2 plays a crucial role at the step of membrane tubulation during cell plate formation in plants. Plant Cell 29:1388–1405

Ambrose C, Ruan Y, Gardiner J et al (2013) CLASP interacts with sorting nexin 1 to link microtubules and auxin transport via PIN2 recycling in *Arabidopsis thaliana*. Dev Cell 24:649–659

Antignani V, Klocko AL, Bak G et al (2015) Recruitment of PLANT U-BOX 13 and the PI4Kβ1/β2 -phosphatidylinositol-4 kinases by the small GTPase RabA4B plays important roles during salicylic acid-mediated plant defense signaling in Arabidopsis. Plant Cell 27:243–261

Anthony RG, Henriques R, Helfer A et al (2004) A protein kinase target of a PDK1 signalling pathway is involved in root hair growth in Arabidopsis. EMBO J 23:572–581

Arimura SI, Aida GP, Fujimoto M et al (2004) Arabidopsis dynamin-like protein 2a (ADL2a), like ADL2b, is involved in plant mitochondrial division. Plant Cell Physiol 45:236–242

Armengot L, Marquès-Bueno MM, Jaillais Y (2016) Regulation of polar auxin transport by protein and lipid kinases. J Exp Bot 67:4015–4037

Bak G, Lee E-J, Lee Y (2013) Rapid structural changes and acidification of guard cell vacuoles during stomatal closure require phosphatidylinositol 3,5-bisphosphate. Plant Cell 25:2202–2216

Balla T (2007) Imaging and manipulating phosphoinositides in living cells. J Physiol 582:927–937

Balla T (2013) Phosphoinositides: tiny lipids with giant impact on cell regulation. Physiol Rev 93:1019–1137

Bao Y, Song WM, Jin YL et al (2014) Characterization of Arabidopsis Tubby-like proteins and redundant function of AtTLP3 and AtTLP9 in plant response to ABA and osmotic stress. Plant Mol Biol 86:471–483

Barberon M, Dubeaux G, Kolb C et al (2014) Polarization of IRON-REGULATED TRANSPORTER 1 (IRT1) to the plant-soil interface plays crucial role in metal homeostasis. Proc Natl Acad Sci U S A 111:8293–8298

Berdy SE, Kudla J, Gruissem W et al (2001) Molecular characterization of At5PTase1, an inositol phosphatase capable of terminating inositol trisphosphate signaling. Plant Physiol 126:801–810

Bovet L, Müller MO, Siegenthaler PA (2001) Three distinct lipid kinase activities are present in spinach chloroplast envelope membranes: phosphatidylinositol phosphorylation is sensitive to wortmannin and not dependent on chloroplast ATP. Biochem Biophys Res Commun 289:269–275

Braun M, Baluska F, von Witsch M et al (1999) Redistribution of actin, profilin and phosphatidylinositol-4,5-bisphosphate in growing and maturing root hairs. Planta 209:435–443

Camacho L, Smertenko AP, Perez-Gomez J et al (2009) Arabidopsis Rab-E GTPases exhibit a novel interaction with a plasma-membrane phosphatidylinositol-4-phosphate 5-kinase. J Cell Sci 122:4383–4392

Carland F, Nelson T (2009) CVP2- and CVL1-mediated phosphoinositide signaling as a regulator of the ARF GAP SFC/VAN3 in establishment of foliar vein patterns. Plant J 59:895–907

Carpaneto A, Boccaccio A, Lagostena L et al (2017) The signaling lipid phosphatidylinositol-3,5-bisphosphate targets plant CLC-a anion/H$^+$ exchange activity. EMBO Rep 18:1100–1107

Cheong H, Kim C-Y, Jeon J-S et al (2013) *Xanthomonas oryzae* pv. oryzae type III effector XopN targets OsVOZ2 and a putative thiamine synthase as a virulence factor in rice. PLoS One 8:e73346

Choi Y, Lee Y, Jeòn BW et al (2008) Phosphatidylinositol 3- and 4-phosphate modulate actin filament reorganization in guard cells of day flower. Plant Cell Environ 31:366–377

Deak M, Casamayor A, Currie RA et al (1999) Characterisation of a plant 3-phosphoinositide-dependent protein kinase-1 homologue which contains a pleckstrin homology domain. FEBS Lett 451:220–226

Delage E, Ruelland E, Guillas I et al (2012) Arabidopsis type-III phosphatidylinositol 4-kinases β1 and β2 are upstream of the phospholipase C pathway triggered by cold exposure. Plant Cell Physiol 53:565–576

Dewald DB, Torabinejad J, Jones CA et al (2001) Rapid accumulation of phosphatidylinositol 4,5-bisphosphate and inositol 1,4,5-trisphosphate correlates with calcium mobilization in salt-stressed Arabidopsis. Plant Physiol 126:759–769

Di Fino LM, D'Ambrosio JM, Tejos R et al (2017) Arabidopsis phosphatidylinositol-phospholipase C2 (PLC2) is required for female gametogenesis and embryo development. Planta 245:717–728

Di Paolo G, De Camilli P (2006) Phosphoinositides in cell regulation and membrane dynamics. Nature 443:651–657

Dove SK, Cooke FT, Douglas MR et al (1997) Osmotic stress activates phosphatidylinositol-3,5-bisphosphate synthesis. Nature 390:187–192

Efe JA, Botelho RJ, Emr SD (2005) The Fab1 phosphatidylinositol kinase pathway in the regulation of vacuole morphology. Curr Opin Cell Biol 17:402–408

Elge S, Brearley C, Xia HJ et al (2001) An Arabidopsis inositol phospholipid kinase strongly expressed in procambial cells: synthesis of PtdIns(4,5)P$_2$ and PtdIns(3,4,5)P$_3$ in insect cells by 5-phosphorylation of precursors. Plant J 26:561–571

Ercetin ME, Ananieva EA, Safaee NM et al (2008) A phosphatidylinositol phosphate-specific myo-inositol polyphosphate 5-phosphatase required for seedling growth. Plant Mol Biol 67:375–388

Furt F, Konig S, Bessoule JJ et al (2010) Polyphosphoinositides are enriched in plant membrane rafts and form microdomains in the plasma membrane. Plant Physiol 152:2173–2187

Gagne JM, Clark SE (2010) The Arabidopsis stem cell factor POLTERGEIST is membrane localized and phospholipid stimulated. Plant Cell 22:729–743

Galvão RM, Kota U, Soderblom EJ et al (2008) Characterization of a new family of protein kinases from Arabidopsis containing phosphoinositide 3/4-kinase and ubiquitin-like domains. Biochem J 409:117–127

Gao K, Liu YL, Li B et al (2014a) Arabidopsis thaliana phosphoinositide-specific phospholipase C isoform 3 (AtPLC3) and AtPLC9 have an additive effect on thermotolerance. Plant Cell Physiol 55:1873–1883

Gao C, Luo M, Zhao Q et al (2014b) A Unique plant ESCRT component, FREE1, regulates multivesicular body protein sorting and plant growth. Curr Biol 24:2556–2563

Garcia AV, Al-Yousif M, Hirt H (2012) Role of AGC kinases in plant growth and stress responses. Cell Mol Life Sci 69:3259–3267

Gelato KA, Tauber M, Ong MS et al (2014) Accessibility of different histone H3-binding domains of UHRF1 is allosterically regulated by phosphatidylinositol 5-phosphate. Mol Cell 54:905–919

Gerth K, Lin F, Menzel W et al (2017) Guilt by association: a phenotype-based view of the plant phosphoinositide network. Annu Rev Plant Biol 68:349–374

Ghosh R, de Campos MKF, Huang J et al (2015) Sec14-nodulin proteins and the patterning of phosphoinositide landmarks for developmental control of membrane morphogenesis. Mol Biol Cell 26:1764–1781

Gupta R (2002) A tumor suppressor homolog, AtPTEN1, is essential for pollen development in Arabidopsis. Plant Cell 14:2495–2507

Hirano T, Matsuzawa T, Takegawa K et al (2011) Loss-of-function and gain-of-function mutations in FAB1A/B impair endomembrane homeostasis, conferring pleiotropic developmental abnormalities in Arabidopsis. Plant Physiol 155:797–807

Hirano T, Munnik T, Sato MH (2015) Phosphatidylinositol 3-phosphate 5-kinase, FAB1/PIKfyve mediates endosome maturation to establish endosome-cortical microtubule interaction in Arabidopsis. Plant Physiol 169:1961–1974

Hirano T, Munnik T, Sato MH (2017a) Inhibition of phosphatidylinositol 3,5-bisphosphate production has pleiotropic effects on various membrane trafficking routes in Arabidopsis. Plant Cell Physiol 58:120–129

Hirano T, Stecker K, Munnik T et al (2017b) Visualization of phosphatidylinositol 3,5-bisphosphate dynamics by a tandem ML1N-based fluorescent protein probe in Arabidopsis. Plant Cell Physiol 58:120–129

Hirayama T, Ohto C, Mizoguchi T et al (1995) A gene encoding a phosphatidylinositol-specific phospholipase C is induced by dehydration and salt stress in *Arabidopsis thaliana*. Proc Natl Acad Sci U S A 92:3903–3907

Huang S, Gao L, Blanchoin L et al (2006) Heterodimeric capping protein from Arabidopsis is regulated by phosphatidic acid. Mol Biol Cell 17:1946–1968

Ikonomov OC, Sbrissa D, Delvecchio K et al (2011) The phosphoinositide kinase PIKfyve is vital in early embryonic development: preimplantation lethality of PIKfyve−/− embryos but normality of PIKfyve+/− mice. J Biol Chem 286:13404–13413

Ischebeck T, Stenzel I, Heilmann I (2008) Type B phosphatidylinositol-4-phosphate 5-kinases mediate Arabidopsis and *Nicotiana tabacum* pollen tube growth by regulating apical pectin secretion. Plant Cell 20:3312–3330

Ischebeck T, Werner S, Krishnamoorthy P et al (2013) Phosphatidylinositol 4,5-bisphosphate influences PIN polarization by controlling clathrin-mediated membrane trafficking in Arabidopsis. Plant Cell 25:4894–4911

Jefferies HBJ, Cooke FT, Jat P et al (2008) A selective PIKfyve inhibitor blocks PtdIns(3,5)P$_2$ production and disrupts endomembrane transport and retroviral budding. EMBO Rep 9:164–170

Jung JY, Kim YW, Kwak JM et al (2002) Phosphatidylinositol 3- and 4-phosphate are required for normal stomatal movements. Plant Cell 14:2399–2412

Kanehara K, Yu CY, Cho Y et al (2015) Arabidopsis AtPLC2 is a primary phosphoinositide-specific phospholipase C in phosphoinositide metabolism and the endoplasmic reticulum stress response. PLoS Genet 11:e1005511

Kang B-H, Nielsen E, Preuss ML et al (2011) Electron tomography of RabA4b- and PI-4Kβ1-labeled trans Golgi network compartments in Arabidopsis. Traffic 12:313–329

Karali D, Oxley D, Runions J et al (2012) The *Arabidopsis thaliana* immunophilin ROF1 directly interacts with PI (3)P and PI(3,5)P$_2$ and affects germination under osmotic stress. PLoS One 7:e48241

Kato M, Aoyama T, Maeshima M (2013) The Ca^{2+}-binding protein PCaP2 located on the plasma membrane is involved in root hair development as a possible signal transducer. Plant J 74:690–700

Kim YW, Park DS, Park SC et al (2001) Arabidopsis dynamin-like 2 that binds specifically to phosphatidylinositol 4-phosphate assembles into a high-molecular weight complex in vivo and in vitro. Plant Physiol 127:1243–1255

Koizumi K, Naramoto S, Sawa S et al (2005) VAN3 ARF-GAP-mediated vesicle transport is involved in leaf vascular network formation. Development 132:1699–1711

Kolb C, Nagel M-K, Kalinowska K et al (2015) FYVE1 is essential for vacuole biogenesis and intracellular trafficking in Arabidopsis. Plant Physiol 167:1361–1373

König S, Ischebeck T, Lerche J et al (2008) Salt-stress-induced association of phosphatidylinositol 4,5-bisphosphate with clathrin-coated vesicles in plants. Biochem J 415:387–399

Krinke O, Ruelland E, Valentová O et al (2007) Phosphatidylinositol 4-kinase activation is an early response to salicylic acid in Arabidopsis suspension cells. Plant Physiol 144:1347–1359

Kusano H, Testerink C, Vermeer JEM et al (2008) The Arabidopsis phosphatidylinositol phosphate 5-kinase PIP5K3 is a key regulator of root hair tip growth. Plant Cell 20:367–380

Lam BC, Sage TL, Bianchi F et al (2001) Role of SH3 domain – containing proteins in clathrin-mediated vesicle trafficking in Arabidopsis. Plant Cell 13:2499–2512

Lee SH, Jin JB, Song J et al (2002) The intermolecular interaction between the PH domain and the C-terminal domain of Arabidopsis dynamin-like 6 determines lipid binding specificity. J Biol Chem 277:31842–31849

Lee Y, Kim YW, Jeon BW et al (2007a) Phosphatidylinositol 4,5-bisphosphate is important for stomatal opening. Plant J 52:803–816

Lee G-J, Kim H, Kang H et al (2007b) EpsinR2 interacts with clathrin, adaptor protein-3, AtVTI12, and phosphatidylinositol-3-phosphate. Implications for EpsinR2 function in protein trafficking in plant cells. Plant Physiol 143:1561–1575

Leprince A-S, Magalhaes N, De Vos D et al (2014) Involvement of phosphatidylinositol 3-kinase in the regulation of proline catabolism in *Arabidopsis thaliana*. Front Plant Sci 5:772

Li X, Wang X, Zhang X et al (2013) Genetically encoded fluorescent probe to visualize intracellular phosphatidylinositol 3,5-bisphosphate localization and dynamics. Proc Natl Acad Sci U S A 110:21165–21170

Li L, He Y, Wang Y et al (2015) Arabidopsis PLC2 is involved in auxin-modulated reproductive development. Plant J 84:504–515

Logan DC, Scott I, Tobin AK (2004) ADL2a, like ADL2b, is involved in the control of higher plant mitochondrial morphology. J Exp Bot 55:783–785

Lou Y, Gou J-Y, Xue H-W (2007) PIP5K9, an Arabidopsis phosphatidylinositol monophosphate kinase, interacts with a cytosolic invertase to negatively regulate sugar-mediated root growth. Plant Cell 19:163–181

McCartney AJ, Zhang Y, Weisman LS (2014) Phosphatidylinositol 3,5-bisphosphate: low abundance, high significance. BioEssays 36:52–64

Mei Y, Jia W-J, Chu Y-J et al (2012) Arabidopsis phosphatidylinositol monophosphate 5-kinase 2 is involved in root gravitropism through regulation of polar auxin transport by affecting the cycling of PIN proteins. Cell Res 22:581–597

Meijer HJG, Munnik T (2003) Phospholipid-based signaling in plants. Annu Rev Plant Biol 54:265–306

Meijer HJG, Berrie CP, Iurisci C et al (2001) Identification of a new polyphosphoinositide in plants, phosphatidylinositol 5-monophosphate (PtdIns5P), and its accumulation upon osmotic stress. Biochem J 498:491–498

Mishkind M, Vermeer JEM, Darwish E et al (2009) Heat stress activates phospholipase D and triggers PIP$_2$ accumulation at the plasma membrane and nucleus. Plant J 60:10–21

Mueller-Roeber B, Pical C (2002) Inositol phospholipid metabolism in Arabidopsis. Characterized and putative isoforms of inositol phospholipid kinase and phosphoinositide-specific phospholipase C. Plant Physiol 130:22–46

Nagasaki N, Tomioka R, Maeshima M (2008) A hydrophilic cation-binding protein of Arabidopsis thaliana, AtPCaP1, is localized to plasma membrane via N-myristoylation and interacts with calmodulin and the phosphatidylinositol phosphates PtdIns(3,4,5)P$_3$ and PtdIns(3,5)P$_2$. FEBS J 275:2267–2282

Nagata C, Miwa C, Tanaka N et al (2016) A novel-type phosphatidylinositol phosphate-interactive, Ca-binding protein PCaP1 in Arabidopsis thaliana: stable association with plasma membrane and partial involvement in stomata closure. J Plant Res 129:539–550

Nagel M-K, Kalinowska K, Vogel K et al (2017) Arabidopsis SH3P2 is an ubiquitin-binding protein that functions together with ESCRT-I and the deubiquitylating enzyme AMSH3. Proc Natl Acad Sci U S A 29:E7197–E7204

Nováková P, Hirsch S, Feraru E et al (2014) SAC phosphoinositide phosphatases at the tonoplast mediate vacuolar function in Arabidopsis. Proc Natl Acad Sci U S A 111:2818–2823

Okazaki K, Miyagishima S, Wada H (2015) Phosphatidylinositol 4-phosphate negatively regulates chloroplast division in Arabidopsis. Plant Cell 27:663–674

Oxley D, Ktictakis N, Farmaki T (2013) Differential isolation and identification of PI(3)P and PI(3,5)P2 binding proteins from Arabidopsis thaliana using an agarose-phosphatidylinositol-phosphate affinity chromatography. J Proteome 91:580–594

Pappan K, Qin WS, Dyer JH et al (1997) Molecular cloning and functional analysis of polyphosphoinositide-dependent phospholipase D, PLD beta, from Arabidopsis. J Biol Chem 272:7055–7061

Park K, Jung J, Park J et al (2003) A role for phosphatidylinositol 3-phosphate in abscisic acid-induced reactive oxygen species generation in guard cells. Plant Physiol 132:92–98

Paul P, Simm S, Mirus O et al (2014) The complexity of vesicle transport factors in plants examined by orthology search. PLoS One 9:e97745

Peterman TK, Ohol YM, McReynolds LJ et al (2004) Patellin1, a novel Sec14-like protein, localizes to the cell plate and binds phosphoinositides. Plant Physol 136:3080–3094

Phan NQ, Kim SJ, Bassham DC (2008) Overexpression of Arabidopsis sorting nexin AtSNX2b inhibits endocytic trafficking to the vacuole. Mol Plant 1:961–976

Pical C, Westergren T, Dove SK et al (1999) Salinity and hyperosmotic stress induce rapid increases in phosphatidylinositol 4,5-bisphosphate, diacylglycerol pyrophosphate, and phosphatidylcholine in Arabidopsis thaliana cells. J Biol Chem 274:38232–38240

Pokotylo I, Kolesnikov Y, Kravets V et al (2014) Plant phosphoinositide-dependent phospholipases C: variations around a canonical theme. Biochimie 96:144–157

Preuss ML, Schmitz AJ, Thole JM et al (2006) A role for the RabA4b effector protein PI-4Kβ1 in polarized expansion of root hair cells in Arabidopsis thaliana. J Cell Biol 172:991–998

Qin C, Wang X (2002) The Arabidopsis phospholipase D family. Characterization of a calcium-independent and phosphatidylcholine-selective PLD zeta 1 with distinct regulatory domains. Plant Physiol 128:1057–1068

Qin W, Pappan K, Wang X (1997) Molecular heterogeneity of phospholipase D (PLD). Biochemist 272:28267–28273

Serrazina S, Dias FV, Malhó R (2014) Characterization of FAB1 phosphatidylinositol kinases in Arabidopsis pollen tube growth and fertilization. New Phytol 203:784–793

Shisheva A (2008) PIKfyve: Partners, significance, debates and paradoxes. Cell Biol Int 32:591–604

Silva PA, Ul-Rehman R, Rato C et al (2010) Asymmetric localization of Arabidopsis SYP124 syntaxin at the pollen tube apical and sub-apical zones is involved in tip growth. BMC Plant Biol 10:179

Simon MLA, Platre MP, Assil S et al (2014) A multi-colour/multi-affinity marker set to visualize phosphoinositide dynamics in Arabidopsis. Plant J 77:322–337

Simon MLA, Platre MP, Marqués-Bueno MM et al (2016) A PI4P-driven electrostatic field controls cell membrane identity and signaling in plants. Nat Plants 2:16089

Sousa E, Kost B, Malho R (2008) Arabidopsis phosphatidylinositol-4-monophosphate 5-kinase 4 regulates pollen tube growth and polarity by modulating membrane recycling. Plant Cell 20:3050–3064

Stanislas T, Hüser A, Barbosa ICR et al (2015) Arabidopsis D6PK is a lipid domain-dependent mediator of root epidermal planar polarity. Nat Plants 1:15162

Stenzel I, Ischebeck T, König S et al (2008) The type B phosphatidylinositol-4-phosphate 5-kinase 3 is essential for root hair formation in Arabidopsis thaliana. Plant Cell 20:124–141

Stevenson JM, Perera IY, Boss WF (1998) A phosphatidylinositol 4-kinase pleckstrin homology domain that binds phosphatidylinositol 4-monophosphate. J Biol Chem 273:22761–22767

Stevenson-Paulik J, Love J, Boss WF et al (2003) Differential regulation of two Arabidopsis type III phosphatidylinositol 4-kinase isoforms. A regulatory role for the pleckstrin homology domain. Plant Physiol 132:1053–1064

Suzuki T, Matsushima C, Nishimura S et al (2016) Identification of phosphoinositide-binding protein PATELLIN2 as a substrate of Arabidopsis MPK4 MAP kinase during septum formation in cytokinesis. Plant Cell Physiol 57:1744–1755

Tasma IM, Brendel V, Whitham SA et al (2008) Expression and evolution of the phosphoinositide-specific phospholipase C gene family in *Arabidopsis thaliana*. Plant Physiol Biochem 46:627–637

Tejos R, Sauer M, Vanneste S et al (2014) Bipolar plasma membrane distribution of phosphoinositides and their requirement for auxin-mediated cell polarity and patterning in Arabidopsis. Plant Cell 26:2114–2128

Thole JM, Vermeer JEM, Zhang Y et al (2008) ROOT HAIR DEFECTIVE4 encodes a phosphatidylinositol-4-phosphate phosphatase required for proper root hair development in *Arabidopsis thaliana*. Plant Cell 20:381–395

Ugalde J-M, Rodriguez-Furlán C, Rycke R et al (2016) Phosphatidylinositol 4-phosphate 5-kinases 1 and 2 are involved in the regulation of vacuole morphology during Arabidopsis thaliana pollen development. Plant Sci 250:10–19

Van Leeuwen W, Vermeer JEM, Gadella TWJ et al (2007) Visualization of phosphatidylinositol 4,5-bisphosphate in the plasma membrane of suspension-cultured tobacco BY-2 cells and whole Arabidopsis seedlings. Plant J 52:1014–1026

Vermeer JEM, van Leeuwen W, Tobeña-Santamaria R et al (2006) Visualization of PtdIns3P dynamics in living plant cells. Plant J 47:687–700

Vernoud V, Horton AC, Yang Z et al (2003) Analysis of the small GTPase gene superfamily of Arabidopsis. Plant Physiol 131:1191–1208

Vorwerk S, Schiff C, Santamaria M et al (2007) EDR2 negatively regulates salicylic acid-based defenses and cell death during powdery mildew infections of *Arabidopsis thaliana*. BMC Plant Biol 7:35

Wada Y, Kusano H, Tsuge T et al (2015) Phosphatidylinositol phosphate 5-kinase genes respond to phosphate deficiency for root hair elongation in *Arabidopsis thaliana*. Plant J 81:426–437

Wang P, Hussey PJ (2015) Interactions between plant endomembrane systems and the actin cytoskeleton. Front Plant Sci 6:422

Wang WY, Zhang L, Xing S (2012) Arabidopsis AtVPS15 plays essential roles in pollen germination possibly by interacting with AtVPS34. J Genet Genomics 39:81–92

Whitley P, Hinz S, Doughty J (2009) Arabidopsis FAB1/PIKfyve proteins are essential for development of viable pollen. Plant Physiol 151:1812–1822

Williams ME (2005) Mutations in the Arabidopsis phosphoinositide phosphatase gene SAC9 lead to overaccumulation of PtdIns(4,5)P_2 and constitutive expression of the stress-response pathway. Plant Physiol 138:686–700

Xia K, Wang B, Zhang J et al (2017) *Arabidopsis* phosphoinositide-specific phospholipase C 4 negatively regulates seedling salt tolerance. Plant Cell Environ 40:1317–1331

Yamaguchi-Shinozaki K, Shinozaki K (2006) Transcriptional regulatory networks in cellular responses and tolerance to dehydration and cold stresses. Annu Rev Plant Biol 57:781–803

Yoo CM, Quan L, Cannon AE et al (2012) AGD1, a class 1 ARF-GAP, acts in common signaling pathways with phosphoinositide metabolism and the actin cytoskeleton in controlling Arabidopsis root hair polarity. Plant J 69:1064–1076

Zegzouti H, Li W, Lorenz TC et al (2006) Structural and functional insights into the regulation of Arabidopsis AGC VIIIa kinases. J Biol Chem 281:35520–35530

Zhao Y, Yan A, Feijó JA et al (2010) Phosphoinositides regulate clathrin-dependent endocytosis at the tip of pollen tubes in Arabidopsis and Tobacco. Plant Cell 22:4031–4044

Zheng SZ, Liu YL, Li B et al (2012) Phosphoinositide-specific phospholipase C9 is involved in the thermotolerance of Arabidopsis. Plant J 69:689–700

Zheng J, Han SW, Rodriguez-Welsh MF et al (2014) Homotypic vacuole fusion requires VTI11 and is regulated by phosphoinositides. Mol Plant 7:1026–1040

Zhong R, Ye Z (2003) The SAC domain-containing protein gene family in Arabidopsis. Plant Physiol 132:544–555

Zhong R, Burk DH, Morrison WH et al (2004) FRAGILE FIBER3, an Arabidopsis gene encoding a type II inositol polyphosphate 5-phosphatase, is required for secondary wall synthesis and actin organization in fiber cells. Plant Cell 16:3242–3259

Zhong R, Burk DH, Nairn CJ et al (2005) Mutation of SAC1, an Arabidopsis SAC domain phosphoinositide phosphatase, causes alterations in cell morphogenesis, cell wall synthesis, and actin organization. Plant Cell 17:1449–1466

Zhuang X, Wang H, Lam SK et al (2013) A BAR-domain protein SH3P2, which binds to phosphatidylinositol 3-phosphate and ATG8, regulates autophagosome formation in Arabidopsis. Plant Cell 25:4596–4615

Zhuang X, Chung KP, Cui Y et al (2017) ATG9 regulates autophagosome progression from the endoplasmic reticulum in *Arabidopsis*. Proc Natl Acad Sci U S A 114:E426–E435

Adv Exp Med Biol - Protein Reviews (2019) 20: 159–188
https://doi.org/10.1007/5584_2018_241
© Springer Nature Singapore Pte Ltd. 2018
Published online: 27 July 2018

Molecular Mechanisms of Vaspin Action – From Adipose Tissue to Skin and Bone, from Blood Vessels to the Brain

Juliane Weiner, Konstanze Zieger, Jan Pippel, and John T. Heiker

Abstract

Visceral adipose tissue-derived serine protease inhibitor (vaspin) or SERPINA12 according to the serpin nomenclature was identified together with other genes and gene products that were specifically expressed or overexpressed in the intra-abdominal or visceral adipose tissue (AT) of the Otsuka Long-Evans Tokushima fatty rat. These rats spontaneously develop visceral obesity, insulin resistance, hyperinsulinemia and -glycemia, as well as hypertension and thus represent a well suited animal model of obesity and related metabolic disorders such as type 2 diabetes.

The follow-up study reporting the cloning, expression and functional characterization of vaspin suggested the great and promising potential of this molecule to counteract obesity induced insulin resistance and inflammation and has since initiated over 300 publications, clinical and experimental, that have contributed to uncover the multifaceted functions and molecular mechanisms of vaspin action not only in the adipose, but in many different cells, tissues and organs. This review will give an update on mechanistic and structural aspects of vaspin with a focus on its serpin function, the physiology and regulation of vaspin expression, and will summarize the latest on vaspin function in various tissues such as the different adipose tissue depots as well as the vasculature, skin, bone and the brain.

Author contributed equally with all other contributors.
Juliane Weiner and Konstanze Zieger

J. Weiner and J. T. Heiker (✉)
Institute of Biochemistry, Faculty of Life Sciences,
University of Leipzig, Leipzig, Germany

Department of Medicine, University of Leipzig, Leipzig,
Germany
e-mail: jheiker@uni-leipzig.de

K. Zieger
Institute of Biochemistry, Faculty of Life Sciences,
University of Leipzig, Leipzig, Germany

J. Pippel
Structure and Function of Proteins, Helmholtz Centre for
Infection Research, Braunschweig, Germany

Keywords

Adiposity · Atherosclerosis · Crystal structure ·
Exosite · Inflammation · Insulin resistance ·
Metabolic syndrome · Serine proteases ·
Serpin

Abbreviations

AT	adipose tissue
ATM	adipose tissue macrophage

CART   Cocaine and amphetamine-regulated
        transcript
CRH   Corticotropin-releasing Hormone
DHEA-   dehydroepiandrosterone sulfate
S
HFD   high-fat diet
KLK7   kallikrein 7
MMP   matrix metalloproteinase
NMDA   N-Methyl-D-aspartat
T2D   type 2 diabetes

1 Vaspin Structure and Inhibition Mechanism

1.1 Serpin Mechanism

Based on a sequence homology of ~40% with anti-trypsin, vaspin was proposed to belong to the serpin family (Hida et al. 2000). The name serpin refers to the main function of this family as specific serine protease inhibitors (reviewed in (Gettins 2002)). The fold of inhibitory serpins is highly conserved and consists of three β-sheets (A-C), nine α-helices and, at the top of the molecule, a flexible, cleavable reactive center loop (RCL) which contains the protease recognition sequence. As a specific feature of the native state of inhibitory serpins, the central β-sheet A is composed of five β-strands (Silverman et al. 2001).

To initiate the inhibition reaction, the target protease recognizes its specific sequence in the RCL to form the non-covalent Michaelis complex. This orients the RCL into the active site of the target protease with the scissile bond (between P1 and P1') located in close proximity to the catalytic triad. Additional to interactions of the RCL with the target protease, the complex can also be stabilized by interactions outside the active site. Such interactions are referred to as exosites which are exploited by serpins to refine their specificity as well as to enhance their inhibition rate towards respective target proteases and can even compensate for possible detrimental effects resulting from an unfavorable RCL sequence (reviewed in (Gettins and Olson 2009)).

Hydrolysis of the peptide bond is initiated by the acylation reaction during which a covalent acyl-enzyme intermediate is formed between the catalytically active serine of the target protease and residue P1 which constitutes the N-terminal fragment of the now cleaved RCL. Before the reaction cycle is completed via the subsequent deacylation reaction, the RCL performs a fast movement and incorporates into the central β-sheet A as the sixth strand (Engh et al. 1989; Lawrence et al. 2000). Simultaneously, the RCL-bound protease is delocalized from the top of the serpin to the bottom by approximately 70 Å (Stratikos and Gettins 1999). As a consequence of the inhibition mechanism, the reactive center is severely distorted which prevents dissociation of the covalent complex (Huntington et al. 2000; Dementiev et al. 2006). Additionally, more distant parts of the target protease might become disordered (Huntington et al. 2000), making the complex susceptible to proteolytic digestion (Kaslik et al. 1995; Stavridi et al. 1996). In vivo, serpin-enzyme-complex (SEC) receptors located on the cell surface can bind serpin-protease complexes and mediate clearance from the circulation (Perlmutter et al. 1990; Joslin et al. 1992).

As the movement of the RCL is not reversible, the serpin inhibition mechanism represents a suicide-substrate mechanism (Huntington 2006). Thus, the protease may escape inhibition, if the peptide bond is hydrolyzed before the RCL is completely incorporated. For very efficient inhibition reactions the stoichiometry of inhibition is ~1, while for other more substrate-like reactions the SI is >1, and the SI represents an additional regulatory aspect to fine-tune protease activity (Gettins 2002). Serpins might also behave as substrate as a result of mutations which impede efficient and fast RCL incorporation (Gils et al. 1996; Carrell and Stein 1996). For vaspin, mutants T365R and A369P represent non-inhibitory variants (Heiker et al. 2013) and our data indeed revealed that RCL insertion is significantly slowed down but still takes place (Pippel et al. 2016).

1.2 Crystal Structures of Native and Cleaved Vaspin

To further elucidate the molecular mechanism underlying vaspin protease specificity and regulation of inhibitory action, four different crystal structures of human vaspin have been determined. This includes structures of the wt (Heiker et al. 2013) and two mutants (E379S and D305C/V383C) (Ulbricht et al. 2015) in the native uncleaved state as well as of the wt in the cleaved state (Pippel et al. 2016).

Notably, rat and murine vaspin share high sequence identities with the human protein (61.5% or 62.6% amino acid identity, respectively) and thus conclusions drawn from structures of human vaspin are likely relevant and true for the rodent vaspin proteins as well. The same holds true for known important residues and protein domains that will be discussed in the following paragraphs (e.g. RCL and cleavage site, exosites and heparin binding site) and which are also conserved in mouse and rat vaspin.

The structures of vaspin in the native state display the highly conserved structural elements of inhibitory serpins with a largely unresolved flexible RCL (Silverman et al. 2001) (Fig. 1a). However, compared to most other crystal structures of inhibitory serpins, the C-terminal part of the RCL (T375 – E379; residues P4 – P1' including the protease cleavage site) is defined by electron density and appears more rigid. The vaspin RCL is cleaved between M378 (P1) and E379 (P1') as described in more detail below (Heiker et al. 2013). In the native structure, residues E379 (P1') and T380 (P2') of the RCL interact with R302 of β-sheet C via a salt bridge and a water-mediated hydrogen bond, respectively. By manipulation of the RCL sequence via introduction of a serine at P1' (E379S) and an artificial disulfide bridge in proximity to the scissile bond (D305C/V383C), the flexibility of the C-terminal part of the RCL was enhanced as shown in the crystal structures.

An inherent feature of serpins in the native inhibitory state is their "poised ready" metastable conformation (Harrop et al. 1999) which provides the basis for the tremendous conformational changes during transition into the thermodynamically favored RCL-cleaved state (Huber and Carrell 1989; Ryu et al. 1996; Lee et al. 2000). The fold of RCL-cleaved serpins is also highly conserved and resembles that of the protein in the covalent complex (Huntington et al. 2000; Dementiev et al. 2006). The crystal structure of cleaved vaspin was determined in 2016 (Pippel et al. 2016).

Common to other structures of cleaved serpins, the N-terminal part of the RCL (hinge region residues P15 – P9) join the central β-sheet A as an additional strand (Fig. 1b, lower box) whereas the C-terminal cleavage site residues remain flexible and are thus partially unresolved in the crystal structure (Fig. 1b, upper box). The hinge region in serpins comprises a highly conserved GTEGAAx$_1$T sequence (P15 – P8; G$_{364}$TEGAAGT$_{371}$ for vaspin) which is fundamental for the observed conformational mobility (Irving et al. 2000). Additionally, residues of the breach region, which is located at top of β-sheet A and is the point of initial strand insertion, form a number of partially conserved interactions with hinge region residues described further below in more detail. Mutations in both regions mostly have adverse effects on the inhibitory activity of serpins (Hopkins et al. 1993; Stein and Carrell 1995). In contrast, residues 372–378 (P7 – P1) are not conserved among inhibitory serpins and accordingly, the environment of these amino acids is less constrained in vaspin with M378 (P1) as the last residue of the inserted strand being solvent exposed (Fig. 1b, lower box) (Pippel et al. 2016).

The metastable nature of inhibitory serpins allows these proteins to undergo large conformational changes but on the other hand makes them prone to inactivation by polymerization or latency transition without prior cleavage of the RCL. In these states, the RCL is not accessible for the

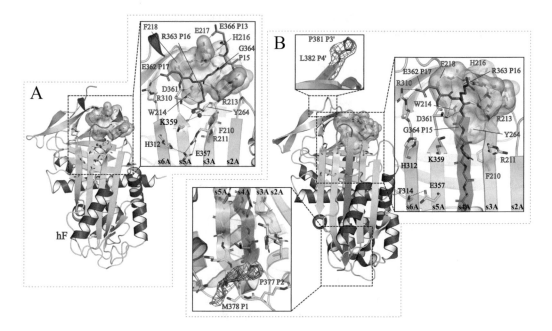

Fig. 1 Structural overview and details for native and cleaved vaspin. X-ray structures of native (**a**, pdb code 4IF8, (Heiker et al. 2013)) and cleaved vaspin (**b**, pdb code 5EI0, (Pippel et al. 2016)) with close-up presentations for interactions of the breach with the hinge region (middle box) and for selected residues of the inserted RCL (upper and lower box). Residues of the breach region are shown in yellow. Residues of the hinge region together with the RCL are depicted in orange. The groove is shown as grey surface. 2Fo-Fc electron density (contoured to 1 σ) for selected RCL residues is shown as red mesh. Of note, in the closed-up presentations, helix F (hF) was hidden for clarity and viewing angles were slightly changed compared to the presentation of the complete molecules. Figures were generated using PyMOL (http://www.pymol.org)

target proteases, resulting in non-inhibitory serpins (Gettins and Olson 2016). In contrast to several other human inhibitory serpins, native vaspin represents a remarkably thermostable serpin with only little tendency towards these inactivation mechanisms (Pippel et al. 2016; Ulbricht et al. 2015). In this regard, key features in the breach and hinge region were identified which may support this behavior and the most noticeable will be described in the following. As mentioned above, the breach region is located at the upper part of s3A and s5A and in vaspin, recalls the picture of a 'partial opened zipper'. In this region, a conserved FKGx$_1$Wx$_2$x$_3$x$_4$F sequence (F$_{210}$RARWKHEF$_{218}$ for vaspin) forms a prominent surface groove in which residues R363 – E366 (P16 – P13) of the RCL hinge region are rigidly embedded (Fig. 1a, middle box). This may prevent latency transition or polymerization of native vaspin. In both, the native and the RCL-cleaved state, R310 forms a

salt bridge with E362 (P17) and in consequence E362 determines the end of s5A and provides a hinge point for insertion of the new s4A (Fig. 1a, middle box, Fig. 1b, middle box). E362 is strictly conserved among inhibitory serpins and mutation of this residue (E342K, Z-variant) is responsible for 95% of all clinically recognized α$_1$-antitrypsin (α$_1$AT) deficiency cases (Stoller and Aboussouan 2012). The acute-phase protein α$_1$AT is produced mainly in the liver and functions as a major inhibitor of various proinflammatory proteases such as neutrophil elastase or cathepsin G. The Z-variant of α$_1$AT is prone for misfolding, latency and polymerization and is associated with cirrhosis or hepatocellular carcinoma (Lomas et al. 1992). Furthermore, R363 (P16) interacts with D361 and this salt bridge may prevent pre-insertion of the RCL into the open void between s3A and s5A (Fig. 1a, middle box) in the native state of vaspin (Pippel et al. 2016). However, until now, we can only speculate whether these interactions play major roles in the remarkable thermostability of vaspin and respective follow-up studies are necessary to verify our hypothesis.

1.3   Tested and Targeted Proteases

The finding of ~40% sequence identity of vaspin with the bona-fide serpin anti-trypsin suggested affiliation with the large serpin family of proteinase inhibitors and initiated the search for target proteases, as these (in addition to the serpin itself) represent promising, potentially new and direct drug targets which may lead to novel therapeutic strategies for the treatment of obesity and its related metabolic and cardiovascular diseases (Wada 2008; Athyros et al. 2010).

Vaspin did not inhibit well-known serine proteases, such as trypsin, elastase, urokinase, factor Xa, collagenase and dipeptidyl peptidase (Hida et al. 2005). We later screened selected members of the kallikrein family (KLKs) and we found KLK7 to be inhibited by vaspin via the classical serpin mechanism. The closely related kallikreins KLK4 and 5 were not inhibited (Heiker et al. 2013). KLK7 is a chymotryptic serine protease that was initially identified in skin where it is involved in the skin desquamation process (Lundstrom and Egelrud 1991) and aberrant activity of KLK7 is related to the pathogenesis of inflammatory skin diseases such as psoriasis (Ekholm and Egelrud 1999) and acne rosacea (Yamasaki et al. 2007). Furthermore, the trypsin- and chymotrypsin-like protease KLK14 was recently identified as the second member of the kallikrein family that is targeted by vaspin (Ulbricht et al. 2018). KLK14, together with the main players in KLK7 and KLK5, is also involved in skin desquamation (de Veer et al. 2017).

1.4   Determinants of Specificity – RCL Sequence and Exosites for KLK7 Inhibition

The scissile bond within the vaspin RCL is located between a methionine and a glutamate residue and this P1′ glutamate stands out when investigating potential protease targets of vaspin, as it is a highly unusual residue at this position and likely determining protease specificity. Measuring the inhibition kinetics for KLK7 by vaspin revealed a rather slow inhibition reaction with a second order rate (k_i) of 7.3×10^3 M$^{-1} \times$ s^{-1}, when compared with k_i values in the range of 10^6 M$^{-1} \times$ s^{-1} for other serpin-protease pairs (Heiker et al. 2013).

Using a synthetic combinatorial peptide library, Debela et al. investigated the S4-S1 specificity of KLK7 and demonstrated chymotryptic-like specificity with a preference for the residues Tyr, Ala and Met at the P1 position (in that order) (Debela et al. 2006). The KLK7 specificity profile was expanded by Oliviera et al. to the S1'-S3' sites using soluble FRET peptide libraries. This study confirmed Tyr residues to be preferred at P1 and the preference for hydrophilic amino acids, particularly Arg at P1' (Oliveira et al. 2015). These data implicate the vaspin RCL sequence of TPM-ETP (from P3 to P3') as a rather unfavorable cleavage site. Indeed, a peptide comprising the RCL sequence is not cleaved by KLK7 (Ulbricht et al. 2015). In line with these findings, the E379S mutant with a more favored serine residue in P1 has a significantly increased k_i value of 4×10^5 M$^{-1} \times$ s^{-1} for KLK7 (Ulbricht et al. 2015). These data clearly demonstrate that E379 is the major negative regulatory element purposely limiting the rate of inhibition while also increasing specificity. This seems of great importance with respect to the low serum levels and distinct expression pattern of vaspin.

As discussed above, in the crystal structure of native vaspin we were intrigued by a potential ionic interaction between the P1' residue E379 and R302, which is located in close proximity of the RCL. This interaction resulted in a rather constraint C-terminal part or the RCL including the cleavage site. With the intention to increase flexibility of the RCL, we mutated R302 but found these vaspin mutants, R302A and R302E, to be fully inactive. In contrast to the inactive substrate mutants vaspin A369P or T365R (Heiker et al. 2013), these mutants were not cleaved by KLK7, indicating R302 as an essential exosite, enabling KLK7 inhibition and further ensuring specificity (Ulbricht et al. 2015). In addition to the positive exosite R302, the negatively charged D305 in close proximity to the RCL is suppressing KLK7 inhibition, as the mutant D305A exhibits an 8-fold increase in inhibition rate (Ulbricht et al. 2018). The second protease KLK14 has the same cleavage site within the vaspin RCL (after M378) and is inhibited with a ki value of 1×10^3 $M^{-1} \times s^{-1}$. Interestingly, inhibition of KLK14 is not dependent on the exosites that enable or regulate KLK7 inhibition (R302 and D305, respectively) (Ulbricht et al. 2018).

Very likely, further exosites are contributing to the specificity of vaspin towards its protease targets and future work on the structural elucidation of vaspin in the Michaelis complex with these proteases is needed to fully understand the regulatory elements.

1.5 Vaspin Cofactors and Non-Protease Interacting Proteins

Many serpins (and their respective target proteases) are able to bind to glycosaminoglycans (GAG) such as heparin, heparan sulfate or dermatan sulfate. Binding of these cofactors enables formation of ternary complexes, with the GAG molecule bridging serpin and protease (for a model see Fig. 2), and induces Michaelis complex formation and thus accelerates protease inhibition (Gettins and Olson 2009).

Vaspin and KLK7 are both able to bind heparin (Ulbricht et al. 2015) and KLK7 is furthermore moderately activated by heparin (Oliveira et al. 2015). Vaspin has a high affinity for heparin (K_D of 21 nM) but it does not bind other GAGs such as dermatan sulfate or chondroitin sulfate. Inhibition of KLK7 by vaspin was accelerated 5-fold upon addition of heparin and a bell-shaped dose strongly suggested the template/bridging mechanism as underlying mechanism (Ulbricht et al. 2017).

Most heparin-binding serpins interact with the GAG molecule via basic residues of the D-helix or H-helix (Fig. 2). While vaspin is a highly basic protein with a pI of 9.3, there is only one large basic patch that is located on top of the central β-sheet A (Fig. 2a, lower box). Mutations of basic residues K359 and R211 within this patch decreased heparin affinity (K_D ~1 μM) and reduced heparin-mediated acceleration of the inhibition reaction by ~40% (Ulbricht et al. 2017). While the heparin-boost in KLK7 inhibition was moderate, cofactor binding at this site did not impede the serpin inhibition mechanism, i.e. insertion of the RCL into sheet A, and the SI in the presence of heparin remained unaltered (Ulbricht et al. 2017). Furthermore, the presence of heparin did not accelerate inhibition of KLK14 but impeded complex formation with vaspin (Ulbricht et al. 2018).

To investigate whether heparin binding of vaspin has physiological relevance in addition to regulating serpin-protease interaction, vaspin expressing HaCaT cells (Saalbach et al. 2016) were used to analyze potential localization of vaspin in the extracellular matrix (ECM). Indeed, a high-salt wash or heparinase treatment of these cells increased vaspin levels measured in cell culture supernatants indicating that a significant amount of vaspin is localized in the ECM of cells (Ulbricht et al. 2017). High concentrations of vaspin were also found in the membrane fractions of liver tissue lysates of transgenic vaspin mice where it is co-localized with α-2-macroglobulin, an important inhibitor of ECM degrading matrix metalloproteases (Nakatsuka et al. 2012).

Furthermore, with the 78 kDa glucose-regulated protein (GRP78) a first non-protease interaction partner of vaspin was identified using tandem-affinity tag purification in HepG2 liver cells (Nakatsuka et al. 2012). Interaction studies using chimeric vaspin-antitrypsin proteins (vaspin[21–249]/AT254–417 or AT[26–253]/vaspin [250–413]) indicate that GRP78 is likely bound via the helical domains of the vaspin N-terminus and not via the RCL region (Nakatsuka et al. 2012). The binding site of vaspin in the GRP78 molecule

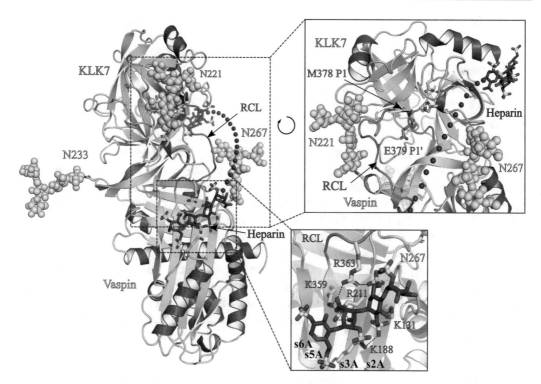

Fig. 2 **Model of the non-covalent Michaelis-complex between KLK7 and N-glycosylated vaspin bridged by heparin.** Interaction of KLK7 (pdb code 2QXI, (Debela et al. 2007)) and vaspin (pdb code 4IF8, (Heiker et al. 2013)), as well as heparin binding modes for both proteins were modeled using the ClusPro web server (Mottarella et al. 2014) and both tetrasaccharides were connected by a dashed line indicating binding of a long-chained heparin bridging both proteins. KLK7 is colored in grey and red whereas vaspin is colored in blue and red. N-glycosylation and heparin tetrasaccharides are presented as yellow spheres and purple sticks, respectively. Close-up presentations show models of vaspin N-glycosylation near the RCL (N221 and N267; yellow) (upper box) as

well as the binding mode of a heparin tetrasaccharide to basic residues of β-sheet A in vaspin (shown as blue sticks, lower box). N-glycosylation was generated using the NetNGlyc 1.0 web server (Blom et al. 2004) and the structure was produced with GlyProt (Bohne-Lang and von der Lieth 2005). The RCL of vaspin was constructed by homology modelling based on a template RCL structure (pdb code 1UHG (Yamasaki et al. 2003)) and is shown in orange with residues P1 and P1' shown as sticks. Residues of the catalytic triad in KLK7 are shown as grey sticks. Of note, in the close-up presentations, viewing angles were changed compared to the presentation of the complete molecules. Figures were generated using PyMOL (http://www.pymol.org)

remains unclear. Binding was impaired by antibodies against N- and C-terminal domains of GRP78 as well as high concentrations of α-2-macroglobulin (Nakatsuka et al. 2012), which interacts with N-terminal residues of GRP78 (Gonzalez-Gronow et al. 2006).

GRP78 is primarily known as an endoplasmic reticulum (ER) chaperone protein and important player in the unfolded protein response (Haas 1994). In recent years, GRP78 was also found on the cell surface of various tissues interacting with a

variety of protein and peptide ligands (reviewed in (Gonzalez-Gronow et al. 2009)). Together with GRP78, likely functioning as a tissue-specific signaling hub, vaspin was found to be co-localized with the DnaJ-like protein MTJ1 in liver tissue (Nakatsuka et al. 2012) and with the voltage-dependent anion channel (VDAC) in human aortic endothelial cells (Nakatsuka et al. 2013).

It has been shown that the interaction of GRP78 with MTJ1 is necessary for translocation of GRP78 from the ER to the plasma membrane

(Misra et al. 2005) where this complex also is a receptor for activated α-2-macroglobulin (Misra et al. 2004) and mediates intracellular signaling via PI3-kinase, AKT and NFkB (Gonzalez-Gronow et al. 2009). In endothelial cells, GRP78 together with VDAC serves as a receptor for kringel 5 (K5) (Gonzalez-Gronow et al. 2003; Davidson et al. 2005) to increase intracellular Ca^{2+} levels and apoptosis (Lu et al. 1999). Binding of vaspin to GRP78 prevented apoptotic K5 effects while inducing anti-apoptotic intracellular signaling cascades via AKT kinase (Nakatsuka et al. 2013). Vice versa, GRP78 binding was impaired and beneficial vaspin effects were decreased by K5. While the changes of intracellular Ca^{2+} levels are mediated by binding of K5 to VDAC (Gonzalez-Gronow et al. 2007), K5 is able to bind both, GRP78 and VDAC, and thus the binding site of GRP78 for vaspin remains unclear.

Investigation of liver tissue plasma membrane binding using ^{125}I-labelled vaspin revealed a high-affinity K_D of 22 nM, which is in the same range for heparin. Binding affinity to plasma membranes was not decreased by siRNA mediated GRP78 knock-down, though the maximum binding capacity was slightly reduced (Nakatsuka et al. 2012).

Together, these experiments indicate high affinity binding of vaspin to plasma membranes and the various intracellular signaling cascades activated by recombinant vaspin in in vitro cell models may in parts be mediated by binding of ECM GAGs or direct interaction with GRP78 and other molecules (Fig. 3). Along these lines, GAG binding may direct and regulate vaspin activity as a serpin or as a ligand for cell surface molecules and non-heparin binding vaspin mutants as well as a better understanding of the interaction

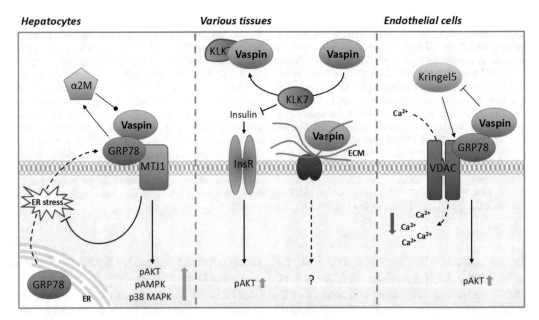

Fig. 3 Molecular interactions of vaspin and tissue specific signaling. Direct interactions of vaspin with the GRP78/MTJ1 complex provoke intracellular signaling cascades such as AKT, AMPK and MAPK in hepatocytes. This interaction also ameliorates ER stress response by inhibiting the translocation of GRP78 from the ER to the plasma membrane. Inhibition of the insulin-degrading enzyme KLK7 via the serpin mechanism is linked to increased insulin signaling. Localization of vaspin in the ECM may direct vaspin interactions with other cell surface proteins in various tissues. In endothelial cells, vaspin exerts anti-apoptotic effects by preventing Kringel5 binding to the GRP78/VDAC complex, impeding the increase of intracellular Ca^{2+} levels. Vaspin-mediated activation or inhibition of pathways or molecules is indicated by green or red arrows, respectively. *α2M* alpha-2 macroglobulin

between vaspin and GRP78 will help to delineate these mechanisms.

1.6   Glycosylation

Human vaspin has three sites for N-linked glycosylation with the consensus motif Asn-X-Ser/Thr (N221, N233 and N267) (Fig. 2) and all of them are utilized when vaspin is expressed in eukaryotic expression systems (Oertwig et al. 2017). For various serpins, these glycan modifications have been shown to be of great importance with respect to serpin secretion or plasma stability, serpin activity, inactivation via polymerization/aggregation and the interaction with GAGs. In contrast to other serpin glycoproteins, two of these glycosylation sites are located in close proximity to the RCL and near domains important for the extensive conformational changes underlying the serpin inhibition mechanism (Fig. 2, upper box). Interestingly, rat and murine vaspin lack the two glycosylation sites near the RCL, but have an additional predicted site near the bottom of helix C (N92) on the back side of the serpin molecule. The N267 glycosylation site is conserved.

For human vaspin, glycosylation does not alter anti-protease or more specifically anti-KLK7 activity and while the affinity for heparin is slightly reduced, heparin still accelerates the protease-inhibition reaction (Oertwig et al. 2017). Glycosylation seems to be important for proper vaspin secretion. While intracellular expression of the glycan-null mutant in HEK cells was similar to wild type vaspin, secretion into cell supernatant was very low (Oertwig et al. 2017). An important function of protein glycosylation is related to protein stability in the circulation. And while vaspin serum stability has not been thoroughly investigated, *in vitro* serum stability tests using fluorescently labelled vaspin derived from *E. coli* indicated high serum stability also for unglycosylated vaspin (Oertwig et al. 2017).

Importantly, consistent *in vitro* and *in vivo* findings have been reported in studies using unglycosylated (expressed in *E. coli* cells) or glycosylated (expressed in HEK293 cells) recombinant vaspin. It remains to be investigated whether physiologic functions of the glycoprotein vaspin are dependent or specifically affected by glycosylation.

2   Physiology of Vaspin

2.1   Tissue Expression of Vaspin

As mentioned above, Hida *et al.* identified vaspin mRNA as specifically expressed in visceral AT of the Otsuka Long-Evans Tokushima fatty (OLETF) rat and later reported the circulating form as a putative member of the serpin family (Hida et al. 2000; Hida et al. 2005). In the OLETF rat, vaspin expression in visceral AT and vaspin serum concentration were highest when insulin resistance and obesity reached their peak and decreased with the exacerbation of diabetes and concomitant body weight loss (Hida et al. 2005). In Wistar rats, an animal model for diet-induced obesity (DIO), vaspin expression was observed in visceral as well as subcutaneous AT, but vaspin mRNA levels and protein expression were higher in the visceral depot and positively correlated with body weight (Shaker and Sadik 2013). Furthermore, increased vaspin expression was accompanied by elevated serum leptin levels (Shaker and Sadik 2013).

In the first study investigating vaspin expression in human AT, vaspin mRNA expression was analyzed in 196 subjects, that comprised a wide range of BMI and insulin sensitivity, and included paired samples of visceral and subcutaneous AT (Kloting et al. 2006). Vaspin mRNA expression was only detectable in 23% of the visceral AT samples and in 15% of the subcutaneous AT samples, and an absence of vaspin expression was most frequent in healthy and lean subjects (BMI < 25). Multivariate regression analysis revealed increased body fat mass as the strongest predictor for visceral vaspin gene expression, while in subcutaneous AT, insulin sensitivity was the most powerful predictor for vaspin expression (Kloting et al. 2006). These data give further indications, that vaspin expression is induced by increased body weight, impaired glycemic control and metabolic syndrome and as such may represent a compensatory molecule in obesity, and related disorders such as inflammation and insulin resistance (Wada 2008).

In a cohort of normal weight Korean women (mean BMI of 24), vaspin gene expression was found to be significantly greater in the adipocyte fraction than in the stromal vascular fraction of visceral (omental) adipose tissue (Lee et al. 2011). Also, in the 40 subjects analyzed by computed tomography scan, vaspin expression was higher in the subcutaneous AT compared to the visceral. In contrast to the study of Lee *et al.* (Lee et al. 2011), a comparison of mRNA distribution of different proteins isolated from human omental AT of morbidly obese women (mean BMI of 46) undergoing bariatric surgery, demonstrated that vaspin is, in addition to preadipocytes and differentiated adipocytes, predominantly expressed in nonfat cells of the stromal vascular fraction (Fain et al. 2008). In the OLETF rat, Western and Northern Blot analysis of vaspin expression in subcutaneous and intra-abdominal AT depots clearly showed that vaspin mRNA as well as protein are expressed primarily in the adipocyte fraction and marginally in the SVF (Hida et al. 2005). Vaspin is furthermore significantly expressed in brown adipose tissue (BAT) and brown adipocytes (Weiner et al. 2017).

Importantly, and in addition to ATs, multiple studies have demonstrated expression of vaspin in various other tissues, such as human skin (Saalbach et al. 2012; Toulza et al. 2007), liver, pancreas (Korner et al. 2011), placenta (Caminos et al. 2009), stomach, the cerebrospinal fluid and the hypothalamus of *ob/ob* and C57BL/6 N mice (Kloting et al. 2011). Using different methods including qPCR, Western and Northern blots, ELISA and immunohistochemistry. To obtain comparable expression data we have recently analyzed tissue-specific expression in mice and found highest expression levels of vaspin in skin, liver and brain compared to modest expression in ATs, spleen and low or non-detectable

expression in bone marrow, muscle and kidney (Weiner et al. 2017).

2.2 Serum Levels of Vaspin

Average vaspin serum levels in healthy subjects range from 0.18 to 1.55 ng/ml (meta-analysis in (Feng et al. 2014)) and a subpopulation of ~7% of the Japanese and ~1% of a European population exhibit vaspin levels of >10 ng/ml and up to >30 ng/ml (Teshigawara et al. 2012). Elevated vaspin serum levels in humans were shown to be associated with body-mass-index and insulin resistance (Teshigawara et al. 2012; Youn et al. 2008) and low vaspin serum concentrations represent a risk factor for the progression of type 2 diabetes (T2D) (Jian et al. 2014). A number of studies found higher vaspin serum levels in obese (Cho et al. 2010; Derosa et al. 2013) and T2D patients (Teshigawara et al. 2012; Zhang et al. 2011) but others could not confirm these results (Youn et al. 2008; Seeger et al. 2008; Jeong et al. 2010; Gulcelik et al. 2009). Due to the inconsistency of data, Feng *et al.* performed a large scale meta-analysis to evaluate the relationship between vaspin serum levels, obesity and T2D (Feng et al. 2014). The meta-analysis addressing the relationship between vaspin serum levels and obesity enclosed six studies comprising 1833 participants and revealed a significantly increase of vaspin levels in obese patients with a mean difference of 0.51 ng/ml compared to non-obese controls. With respect to T2D, 1500 participants from 11 studies were included and again vaspin serum levels were significantly higher in T2D patients with a mean difference of 0.36 ng/ml vaspin between groups. In conclusion, these analyses confirmed the presence of higher vaspin serum levels in obese and T2D patients (Feng et al. 2014).

With 24 h serum profiles, *Jeong et al.* addressed the influence of circadian rhythms on vaspin serum levels in healthy human subjects (Jeong et al. 2010). They observed a diurnal rhythm comprising a preprandial rise and postprandial fall of circulating vaspin levels with a nadir at midafternoon and a nocturnal peak reaching 250% of the minimum levels. This rhythm was reciprocal to insulin levels and following the post-prandial rise of insulin levels vaspin serum levels decreased to a minimum. To exclude meal-related from circadian influences, 24 h profiles were analyzed in fasting patients as well as patients fed meals at unexpected times. In fasting patients, vaspin levels appeared to reach a maximum in the mid-afternoon and also unexpected meal ingestion caused suppression of vaspin serum concentrations subsequent to the postprandial insulin peak. Yet, the decrease of vaspin levels in the morning was still present in fasted patients, although delayed and with a lower amplitude. These data may indicate that vaspin levels are in part driven by a circadian clock (Jeong et al. 2010).

A number of studies have investigated effects of weight loss, e.g. after bariatric surgery or life-style intervention, on vaspin serum levels. Golpaie et *al.* found decreased vaspin levels short-term (6 weeks) after laparoscopic restrictive bariatric surgery in morbidly obese patients though serum vaspin levels did not correlate with metabolic parameters (Golpaie et al. 2011). Handisurya et *al.* also observed decreased vaspin levels in morbidly obese patients 12 months after Roux-en-Y gastric bypass, but here changes in vaspin serum levels correlated positively with metabolic parameters such as insulin, C-peptide and HbA1c levels and HOMA-IR (Handisurya et al. 2010). The correlation with HOMA-IR remained significant even after adjustment for surgery-induced BMI change. In contrast, Lu et *al.* observed stable vaspin levels in a subgroup of high-vaspin level patients (>2.5 ng/ml at baseline) and a gradual increase in vaspin levels in a subgroup of low-vaspin level patients (<2.5 ng/ml at baseline) over 12 months after biliopancreatic-diversion/duodenal-switch bariatric surgery (Lu et al. 2014). Overall, patients with high vaspin levels had the better metabolic profile throughout the study.

Results on vaspin regulation by weight-loss due to life-style intervention are more consistent and vaspin levels were decreased with weight-loss after short-term dietary intervention (Vink et al. 2017) as well as after long-term weight loss intervention (independent from type of diet) (Bluher et al. 2012). Also, after 12 weeks of caloric restriction in combination with physical activity and orlistat administration vaspin levels were decreased, though only in the responder group with >2% weight loss (Chang et al. 2010). Obese patients diagnosed with polycystic ovary syndrome (PCOS) are at higher risk for impaired glucose tolerance, insulin resistance, dyslipidemia and T2D. Serum vaspin levels were significantly higher in PCOS patients compared to healthy subjects (2.02 ng/ml versus 0.28 ng/ml; p = 0.048) (Koiou et al. 2011a). Interestingly, 6 months of dietary intervention in combination with metformin or orlistat treatment did not affect vaspin serum levels in both lean and obese patients with PCOS (Koiou et al. 2011a).

Studies investigating genetic variation identified a number of SNPs within and around the *SERPINA12* gene that affect vaspin serum levels (Teshigawara et al. 2012; Breitfeld et al. 2013a; Breitfeld et al. 2013b). Thus, the rare functional vaspin variant rs61757459 results in a stop codon causing lower circulating vaspin concentrations (Breitfeld et al. 2013b) while the minor allele sequence A of rs77060950 accounts for higher serum levels in 7% of the Japanese population (Teshigawara et al. 2012). With respect to genetic risk variants, a significant association of the SNP rs2236242 with T2D was identified in the German KORA study that was independent of obesity (Kempf et al. 2010).

Clearance of vaspin and vaspin-protease complexes from the circulation remains unknown. Vaspin levels have been measured in patients on chronic hemodialysis and were similar to control patients indicating that vaspin is likely not cleared in the kidneys via renal excretion (Seeger et al. 2008). Many serpin-enzyme complexes of serpins such as α1AT, ATIII or HCII are rapidly cleared from the circulation via the low density lipoprotein receptor-related protein 1 (LRP1) expressed on hepatocytes in the liver (Maekawa and Tollefsen 1996; Mast et al. 1991). It is unknown if vaspin-protease complexes can bind and be cleared via this receptor. In addition, vaspin interaction with GRP78 or other cell surface receptors in the liver and/or other tissues, as well as vaspin redistribution into the ECM may also contribute to vaspin clearance from the circulation. Furthermore, cell-surface or ECM localization of vaspin may contribute to the inconsistencies observed in many studies measuring circulating vaspin levels and also may disguise or complicate the finding of associations of circulating vaspin levels with aspects of the metabolic syndrome (Ebert et al. 2018), although the many beneficial vaspin effects investigated on the cellular or tissue level seem to be indicative of those observations.

2.3   Gene Regulation by Nutrients

Body fat content and metabolism are mainly determined by total energy expenditure and energy intake, but the dietary macronutrient composition plays a crucial role in promoting related diseases such as the metabolic syndrome (Lustig 2017). Dietary regulation of various adipokines has been investigated in many studies, demonstrating nutrition effects on adipokine expression and circulating levels (reviewed in (Chopra et al. 2014)).

In order to address nutrient-specific effects on vaspin, we analyzed vaspin expression in mice fed matched low glycemic, high-fat or high-sugar diets (Weiner et al. 2017). Vaspin mRNA expression is unaffected in subcutaneous and visceral white AT depots, but was significantly enhanced in BAT and liver after both obesogenic diets. Interestingly, mice on the high-sugar diet gained significantly less body weight and body fat compared to the HFD, indicating an effect on vaspin expression independent of body weight. Counterintuitively, vaspin plasma levels were reduced after both HFD and HSD and thus, circulating vaspin does not seem to be a suitable predictor for vaspin tissue expression (Weiner et al. 2017). In line with these findings, vaspin serum levels were also decreased in a rat model of obesity-related T2D after 6 weeks of HFD (Castro et al. 2017).

In addition to the macronutrient itself, vaspin expression is regulated by the nutritional status.

In rats, food restriction during gestation caused an increase in vaspin placenta expression (Caminos et al. 2009) and fasting periods of 24 h and 48 h provoke a significant decrease in visceral AT vaspin mRNA levels (Gonzalez et al. 2009). Serum vaspin concentrations in humans also correlated positively with restraint, hunger and disinhibition in human eating behavior (Breitfeld et al. 2013c). In accordance, undernourishment in girls with anorexia nervosa is accompanied with higher serum levels (Ostrowska et al. 2016). In contrast, lower vaspin levels were observed in a cohort of 44 underweight children compared to the healthy control group (Vehapoglu et al. 2015). With respect to fasting effects on vaspin, also differing results were reported in which circulating vaspin serum levels were neither affected by short-term nor by chronic energy deprivation (Kang et al. 2011).

2.4   Gene Regulation by Hormones/ Small Molecules

2.4.1   Leptin

Above mentioned dissimilarities may result from potential fasting induced alterations in other hormone levels which in turn may influence vaspin expression. For instance, the adipokine leptin is well known to correlate with AT mass and fasting provokes a marked decrease in leptin plasma levels (Boden et al. 1996). Along these lines, the fasting-induced decline in vaspin mRNA levels was partially reversible by leptin in rats, indicating an interrelationship between both adipokines (Gonzalez et al. 2009). However, neither acute nor chronic leptin treatment had an effect on vaspin serum levels in a human study (Kang et al. 2011).

Investigations in spontaneous dwarf rats, revealed the influence of growth hormones (GH) on vaspin expression with blunted gonadal AT mRNA levels in GH-deficiency. Mentionable, these dwarf rats are characterized by lower body weight and insulin levels (Gonzalez et al.

2009). As discussed above, the model used in this analysis may itself influence data.

2.4.2 Insulin/Insulin Sensitizers

In the OLETF diabetic rat model, vaspin serum levels increased after treatment with insulin and the insulin sensitizing PPARγ agonist pioglitazone (Hida et al. 2005). Pioglitazone significantly induced vaspin protein expression and secretion in 3T3-L1 cells (Handisurya et al. 2010) and rosiglitazone in immortalized BAT cells (Weiner et al. 2017). The PPARα agonist fenofibrate increased vaspin mRNA expression in 3T3-L1 cells, visceral AT depots in rats, as well as serum levels in rats and humans (Chen et al. 2014). Also, chronic treatment with the insulin-sensitizer metformin significantly increased vaspin expression in gonadal AT of rats (Gonzalez et al. 2009). Intriguingly, data from human studies revealed controversial results. As such, metformin treatment significantly decreased vaspin serum levels in women with PCOS (Tan et al. 2008) and T2D patients (Gulcelik et al. 2009). Consistent results were reported in T2D patients receiving a rosiglitazone (Zhang et al. 2011) or combined rosiglitazone and metformin therapy (Kadoglou et al. 2011). In patients with newly diagnosed T2D, short-term continuous subcutaneous insulin infusion significantly lowered vaspin serum levels associated with improved insulin sensitivity and changes in HOMA-IR (Li et al. 2011). In accordance, an acute insulin bolus caused a significant drop of vaspin serum concentration in healthy individuals (Kovacs et al. 2013). This is particularly interesting in the context of the already mentioned meal-related diurnal variation of vaspin (Jeong et al. 2010) and supports an insulin-dependent decrease in vaspin serum levels independent of nutritional influences.

2.4.3 Thyroid Hormones

Regulation of vaspin by thyroid hormones has been investigated with controversial results. Vaspin mRNA expression in gonadal AT of hypothyroid rats was increased while hyperthyroidism had the opposite effect (Gonzalez et al. 2009). A human study investigating the relationship of vaspin serum levels following weight loss by Roux-en-Y gastric bypass (RYGB) surgery reported a positive correlation between decreased thyroid-stimulating hormone (TSH) and vaspin serum levels (Handisurya et al. 2010). In contrast, a study in overt and subclinical hypothyroid patients found no significant association between vaspin levels and thyroid hormone status. Vaspin levels were neither changed by hypothyroidism nor after treatment with the thyroxine analog levothyroxine (Cinar et al. 2011). Together, there is no clear evidence for a regulation of vaspin by TSH (or vice versa) or whether indirect mechanisms are involved.

2.4.4 Sex Steroids

Analyzing gender-specific differences the majority of studies showed higher vaspin levels in females (Youn et al. 2008; Seeger et al. 2008; Gonzalez et al. 2009; Li et al. 2011; von Loeffelholz et al. 2010) while others did not (Saalbach et al. 2012). Interestingly, vaspin serum levels were significantly enhanced in women using oral contraceptives (von Loeffelholz et al. 2010) and also combined therapies using metformin and oestrogen/progestogen enhanced vaspin serum levels in non-obese women with hyperinsulinaemic androgen excess (Ibanez et al. 2009). Elevated serum levels and omental AT expression was also observed in obese women with PCOS and dehydroepiandrosterone sulfate (DHEA-S) significantly induced vaspin protein expression and secretion in omental adipose tissue explants (Tan et al. 2008). In children, circulating vaspin levels were higher in girls and increased with age and pubertal stage but did not correlate with sex-steroids (Korner et al. 2011). In accordance, no differences in serum levels were found between pre-and postmenopausal women (Handisurya et al. 2010) and mRNA expression in gonadal white AT was neither changed in pregnant nor in gonadoectomized rats (Gonzalez et al. 2009). Hence, more research is necessary to clarify the underlying mechanisms leading to the observed sexual dimorphism in vaspin levels.

Recently, multiple studies investigated the relationship between vaspin and PCOS, a common endocrine disorder affecting women in the reproductive age. Also here, some studies reported enhanced vaspin serum levels in PCOS patients (Koiou et al. 2011a; Tan et al. 2008; Cakal et al. 2011) while others found no association (Guvenc et al. 2016; Akbarzadeh et al. 2012). It is important to note, that PCOS is often associated with insulin resistance, obesity or T2D (Rojas et al. 2014), i.e. parameters with direct effects on vaspin levels and the variability in the data may arise from inhomogeneous study cohorts and also applied diagnostic criteria (Koiou et al. 2011b).

3 Multifaceted and Tissue-Specific Functions of Vaspin – *In Vivo* and *In Vitro* Data

3.1 Vaspin and Obesity Related Insulin-Resistance

In their pioneering study following the identification of the vaspin gene, Hida *et al.* demonstrated the beneficial effect of vaspin on insulin sensitivity and glycemic control (Hida et al. 2005). Administration of recombinant vaspin improved glucose tolerance and insulin sensitivity in diet-induced obese (DIO) and insulin resistant mice. In addition, vaspin application ameliorated expression of marker genes, such as *leptin, resistin, glucose transporter-4* and *Tnf-α*, reflecting amendment of diet-induced insulin resistance in AT. These findings lead to the hypothesis of vaspin acting as an insulin sensitizer also with anti-inflammatory effects counteracting obesity-related insulin resistance in AT. To further support this hypothesis, Nakatsuka *et al.* generated vaspin transgenic (tg) and knock-out (KO) mice to dissect the physiologic roles of vaspin in gain and loss of function animals (Nakatsuka et al. 2012). Vaspin expression was specifically induced in AT using the aP2 promotor (vaspin_tg), while the KO was a general knock-out (vaspin_KO). In line with the previous findings, transgenic overexpression of vaspin in adipose tissue of mice protected animals from HFD-induced AT inflammation and insulin resistance while KO animals displayed deteriorated

metabolic functions accompanied by adipocyte hypertrophy and AT inflammation under HFD. Unexpectedly, the most striking effects of increased vaspin expression after HFD were observed in the liver. Vaspin_tg mice were protected from hepatic steatosis, with less hepatic lipid incorporation and triglyceride content and showed improved insulin signaling and reduced mRNA expression of gluconeogenic and lipogenic genes. These effects were at least in part mediated by the interaction of vaspin with cell surface protein GRP78 (see above and Fig. 3) and reduction of GRP78 expression reduced glucose tolerance in HFD-fed vaspin_tg mice (Nakatsuka et al. 2012).

With the identification of KLK7 as a target protease of vaspin, we investigated whether known beneficial vaspin effects on glucose metabolism in obesity may be related to inhibition of KLK7 or other proteases (Heiker et al. 2013). Importantly, inactive vaspin mutants (A369P, see above) failed to improve glucose tolerance in HFD insulin resistant mice, suggesting that this effect is at least in part based on protease inhibition by vaspin. And while hyperinsulinemic-euglycemic clamps did not show increased insulin sensitivity in vaspin treated mice, we observed increased insulin levels upon a glucose challenge in vaspin treated mice, as well as increased insulin levels in glucose stimulated and vaspin treated isolated pancreatic islets, both without increased C-peptide levels. These data indicated that improved glucose metabolism was also independent of insulin secretion. Interestingly, vaspin and KLK7 are co-expressed in pancreatic islets, and insulin is a substrate of KLK7. Thus, the compensatory action of vaspin in obesity-related insulin resistance is likely mediated by inhibition of KLK7 and potentially further proteases, thereby regulating insulin degradation and increasing insulin half-life (Fig. 3) (Heiker et al. 2013).

Following up, we generated an AT-specific Klk7 KO mice to investigate the role of KLK7 in AT, obesity and insulin resistance. AT-Klk7-KO mice gained less weight under HFD with expansion of predominantly subcutaneous fat and improved insulin sensitivity (Zieger et al. 2017a). In agreement with improved insulin sensitivity, HFD-induced local (in AT) and systemic inflammation was significantly reduced. As a consequence of reduced expression of proinflammatory cytokines, such as interleukine (IL)-1β, IL-6 and monocyte chemoattractant protein-1 (Mcp-1), an increased number of AT macrophages of the anti-inflammatory M2 type was present in AT of AT-Klk7-KO mice. Furthermore, indirect calorimetry revealed that KLK7 deficient HFD mice exhibit increased energy expenditure and food intake, which may be due to lower circulating leptin serum levels (Zieger et al. 2017a). Yet, further research is necessary to identify the physiological substrates of KLK7 in AT which contribute to the improvement of inflammation and insulin sensitivity. Many of the inflammatory cytokine candidates have already been identified as substrates of KLK7, such as IL-1β, chemerin, midkine or MMP-9 and may be relevant in the process. Together, these data demonstrate a previously unknown role of KLK7 in obesity-related inflammation of AT and insulin resistance (Fig. 4) and suggest specific small compound KLK7 inhibitors as promising therapeutic molecules to combat obesity-related metabolic diseases.

3.2 Vaspin in Adipose Tissue

In the AT of HFD-induced obese and insulin resistant mice, administration of intraperitoneal

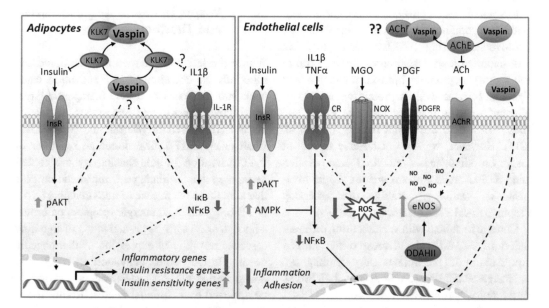

Fig. 4 Vaspin induced intracellular signaling in AT and vascular cells. In adipose tissue, vaspin improves insulin sensitivity and decreases inflammatory gene expression in white AT by enhancing AKT phosphorylation and it blunts cytokine-induced NF-κB signaling by reducing phosphorylation of the NF-κB-inhibitor IκB (and upstream kinase IKKα/β). Inhibition of KLK7 activity may underlie and contribute to both effects. Insulin signaling is also enhanced by vaspin in endothelial cells at the level of AKT phosphorylation. In parallel, ROS formation through inflammatory cytokines, MGO and PDGF, is reduced and NF-κB signaling is repressed. This leads to a decrease in gene expression of inflammatory and adhesion molecules. Furthermore, intracellular NO bioavailability is increased by vaspin through enhanced expression of DDAHII and increased AKT signaling. It is speculated, that vaspin may inhibit AChE via the serpin mechanism culminating in enhanced ACh-induced eNOS activation. Vaspin-mediated activation or inhibition of pathways or molecules is indicated by green or red arrows, respectively. *AMPK* AMP-activated protein kinase, *CR* cytokine receptor, *NOX* NADPH oxidase (NOX), *MGO* methylglyoxal, *PDGF* platelet-derived growth factor (PDGF), *PDGFR* PDGF receptor, *ROS* reactive oxygen species, *ACh* acetylcholine, *AChE* acetylcholine esterase, *AChR* ACh-receptor, *eNOS* endothelial NOS, *DDAHII* dimethylaminohydrolase II

(i.p.) vaspin resulted in an improved expression profile of genes associated with AT inflammation and insulin resistance such as *leptin*, *resistin*, *Tnf-α* and *glucose transporter-4* (Hida et al. 2005). Transgenic overexpression of vaspin in AT of mice resulted in ameliorated local AT inflammatory gene expression and also in reduced systemic IL-6 levels both under chow and HFD diet (Nakatsuka et al. 2012). Also, HFD-induced adipocyte hypertrophy was lower in vaspin tg mice, while vaspin KO increased adipocyte diameters also under chow diet (Nakatsuka et al. 2012). This lead to the hypothesis that vaspin may act as a compensatory molecule against AT inflammation and insulin resistance in obesity. A few studies have since addressed auto- or paracrine functions of vaspin on adipocytes.

Results of vaspin effects on adipogenesis and lipid incorporation into mature adipocytes are controversial. Using 3T3-L1 adipocytes, Liu *et al.* reported high vaspin concentrations (up to 200 ng/ml) to promote lipid accumulation and higher mRNA as well as protein expression of adipocyte-specific marker genes like *Pparγ* and *Fabp-4* in a dose dependent manner (Liu et al. 2015). However, we did not observe effects of vaspin on adipogenesis and lipid incorporation using 3T3-L1 cells with exogenous vaspin treatment as well as stably overexpressing Vaspin_3T3-L1 cells (Zieger et al. 2017b).

Chronic treatment with vaspin during differentiation of 3T3-L1 cells decreased the mRNA expression of the pro-inflammatory cytokine *Il-6* and increased *Glut-4* expression (Liu et al. 2015). Furthermore, vaspin overexpression in 3T3-L1 cells significantly reduced inflammatory cytokine action in these cells (Zieger et al. 2017b). IL-1-β-induced mRNA expression of *Mcp-1* and *Il-6* and secretion of IL-6 and TNF-α was significantly diminished in the vaspin-expressing adipocytes. These effects were likely mediated by blunted intracellular NFκB-signaling due to decreased phosphorylation of the upstream kinase IKKα/β and the NFκB-inhibitor IκB following IL-1β treatment in vaspin expressing adipocytes (Zieger et al. 2017b).

In addition, we found significantly increased AKT phosphorylation after insulin-stimulation of vaspin overexpressing differentiated 3T3-L1 adipocytes (Zieger et al. 2017b). AKT phosphorylation was not different under basal conditions in these cells, though Liu *et al.* observed a slight increase in basal AKT phosphorylation in premature 3T3-L1 cells after short term vaspin treatment (Liu et al. 2015).

Taken together, vaspin seems to have direct anti-inflammatory effects on adipocytes which may contribute to improved insulin sensitivity and glucose uptake (Fig. 4). However, the mechanism underlying the induction of intracellular signaling cascades, whether protease mediated or via direct interaction with cell surface receptors or both, remain unclear.

3.3 Vaspin in Brown Adipose Tissue and Thermogenesis

Within the last years, brown adipose tissue and especially its recruitment and activation came back into the focus as a new therapeutic target to combat obesity and ameliorate metabolic diseases (reviewed in (Harms and Seale 2013; Bhatt et al. 2017)). The dominant regulator of BAT activation is cold-sensing by the central nervous system, initiating a sympathetic outflow that stimulates the release of noradrenaline (NA). NA than agonizes adrenergic receptors on brown adipocytes and triggers the activation of signaling cascades resulting in expression of thermogenic genes. Besides cold, activation of BAT and non-shivering thermogenesis was also shown to be achieved by nutritional factors (Rothwell and Stock 1979) and a plethora of physiological and pharmacological agents (reviewed in (Kajimura and Saito 2014; Tamucci et al. 2017)).

In this context, a microarray study comparing intrinsic differences in cold induced gene expression of murine white and brown adipose depots disclosed vaspin under the top upregulated genes after cold exposure in BAT (Rosell et al. 2014). In our studies, we could confirm that vaspin mRNA and protein expression is specifically enhanced in activated BAT of mice subjected to either cold or diabetogenic diets (Weiner et al. 2017). In accordance, vaspin mRNA expression was significantly higher in immortalized brown compared to white adipocyte cell lines and further increased during adipocyte differentiation. Interestingly, β-adrenergic stimulation by NA and by the β(3)-adrenergic agonist CL316,243 failed to induce vaspin expression *in vitro*. Methylation analyses within the vaspin promoter further identified acute epigenetic changes in BAT upon cold exposure that may contribute to elevated vaspin mRNA expression. Notably, vaspin plasma levels where decreased upon BAT activation (Weiner et al. 2017). In light of BAT thermogenesis and AT browning as current targets to tackle obesity and diabetes, the dynamic and BAT-specific regulation of vaspin is a promising starting point to further investigate functional relevance of vaspin in BAT as well as in BAT function.

3.4   Vaspin in Vascular Cells

Multiple lines of evidence over the last years have established the role of vaspin in improving the function of endothelial and smooth muscle cells (SMC) under hyperlipidemic, hyperglycemic and proinflammatory conditions associated with obesity and insulin resistance by counteracting vascular inflammation and oxidative stress.

Chronic inflammation in obesity results in AT dysfunction and the secretion of proinflammatory adipokines contributes to endothelial dysfunction. Although a first study obtained negative results, various subsequent studies revealed potent anti-inflammatory action of vaspin in endothelial cells. Initially, vaspin was shown to not inhibit TNF-α triggered inflammatory signaling pathways in human umbilical vein endothelial cells (HUVECs) (Fu et al. 2009). Both, acute TNF-α-induced activation of intracellular c-Jun N-terminal kinase (JNK), mitogen-activated protein kinase (p38), and NF-κB, and long-term TNF-α mediated changes in mRNA expression of cellular adhesion molecules as well as MCP-1 were unaffected by vaspin (Fu et al. 2009). In other vascular cells though, it was clearly

demonstrated that vaspin inhibited TNF-α induced IκBα degradation and subsequent NF-κB activation in human aortic endothelial cells (HAECs) as well as in vascular SMC (Phalitakul et al. 2011; Jung et al. 2014). Luciferase reporter gene assays in human endothelial EA.hy926 cells showed that vaspin decreased TNF-α and IL-1β induced NF-κB transcriptional activity, resulting in diminished expression of pro-inflammatory cytokines such as IL-1 and IL-6 (Liu et al. 2014). Inflammatory cytokines TNFα and IL-1β induce expression of cellular adhesion molecules ICAM and VCAM, which in turn promote monocyte and platelet activation, adhesion and migration. Multiple studies demonstrated that vaspin prevented TNF-α-induced expression of intercellular adhesion molecule (ICAM), vascular cell adhesion molecule (VCAM) and E-selectin as well as MCP-1 expression which resulted in reduced monocyte adhesion (Phalitakul et al. 2011; Jung et al. 2014; Liu et al. 2014). siRNA-mediated knock down of AMP-activated protein kinase (AMPK) expression attenuated the effect of vaspin on cytokine-induced NF-κB activation and indicated the activation of AMPK as an signaling pathway (Jung et al. 2014). Also in pulmonary artery SMC vaspin inhibited IL-1β-induced activation of matrix metalloproteinase (MMP-2) and generation of reactive oxygen species (ROS) (Sakamoto et al. 2017).

Dyslipidemia is a major contributor to insulin resistance and endothelial dysfunction. Vaspin ameliorated the apoptotic effect of increased free fatty acids (FFA) in insulin-stimulated HAECS (Jung et al. 2011). FFA such as linoleic acid (LA) decreased insulin-sensitivity and induced apoptosis and caspase 3 activity in vascular endothelial cells. Both effects were prevented by vaspin treatment. Furthermore, pretreatment with the PI3-inhibitor Wortmannin indicated that vaspin prevents FFA-induced cell death via upregulation of the PI3-Akt signaling cascade (Jung et al. 2011).

Under hyperglycemic conditions vaspin was shown to reduced hyperglycemia (HG)-induced vascular smooth muscle cell proliferation and chemokinesis (Li et al. 2013). These effects were mediated by inhibition of HG-induced activation and phosphorylation of intracellular signaling pathways, such as p38-MAPK, the insulin receptor signaling axis and NF-κB signaling, leading to decreased production of ROS (Li et al. 2013). The effect of vaspin on insulin signaling seems to be different in smooth muscle cells (SMC) compared to endothelial cells, though these alterations may also be due to the different glucose stimulation conditions (chronic vs. acute).

In T2D patients, pathological elevated serum levels of methylglyoxal (MGO), a highly reactive dicarbonyl metabolite, contribute to the formation of advanced glycation end products (AEG). Consequently, AEGs are partly responsible for diabetes associated microvascular events, impaired insulin signaling and hinder host self-defense mechanism against vascular inflammation (Cai et al. 2012; Vulesevic et al. 2016). Treatment of cultured HUVECs with MGO resulted in caspase-dependent apoptosis, while pretreatment with vaspin depleted MGO-induced cell death by preventing the generation of MGO-induced ROS, activation of NADPH oxidase (NOX) and subsequent activation of caspase-3 (Phalitakul et al. 2013). Also, endothelial proliferation after vascular injury was ameliorated by application of vaspin or vaspin_adenovirus in streptozotocin-induced diabetic Wistar-Kyoto rats as well as in arteries of vaspin transgenic mice (Nakatsuka et al. 2013). Vaspin treatment induced proliferation and reduced apoptosis in isolated HAEC, where the interaction with GRP78 mediated vaspin-induced activation of AKT signaling (Nakatsuka et al. 2013). Also, sh-RNAs targeting GRP78 and VDAC inhibited the vaspin-induced AKT activation and downstream anti-apoptotic signaling. Thus, the authors suggested the vaspin-GRP78-VDAC cell surface complex as a possible key signaling hub mediating the vaspin signal into the cell with subsequent anti-apoptotic, anti-inflammatory and thus cardioprotective effects (Nakatsuka et al. 2013).

Nitric oxide (NO), generated by endothelial NO synthase (eNOS), plays the central role in vasodilatation and reduced NO-bioavailability and excessive ROS generation mainly contribute to endothelial dysfunction in obesity (reviewed in (Deanfield et al. 2007; Van Gaal et al. 2006). As mentioned above, vaspin has been shown to reduce ROS generation induced by hyperglycemia and increased MGO/AGE, conditions typical for T2D. Also, vaspin attenuated platelet-derived growth factor-BB (PDGF-BB)-induced migration of VSMC and this effect was mediated by inhibition of intracellular ROS generation, which blunted phosphorylation of p38-MAPK and HSP27 signaling (Phalitakul et al. 2012). In addition, vaspin has been shown to enhance bioavailability of NO in endothelial cells. Vaspin increased both eNOS expression and activation via STAT3 signaling and, in parallel, induction of dimethylarginine dimethylaminohydrolase (DDAH) II expression, the enzyme responsible for the degradation of asymmetric dimethyl-L-arginine (ADMA), the endogenous inhibitor of eNOS (Jung et al. 2012). Vaspin counteracted hyperglycemia induced eNOS inhibition in endothelial progenitor cells by inducing eNOS expression and activity via activation of the PI3-AKT pathway (Sun et al. 2015). Vasodilation by NO is induced by increased intracellular Ca^{2+} levels or by acetylcholine receptor agonists like acetylcholine (ACh). Vaspin increased ACh-induced vasorelaxation and ACh-induced eNOS phosphorylation in rat mesenteric arteries, while it had no effect on relaxation induced by other NO-inducing vasodilators such as histamine or carbachol and did not affect vasocontraction (Kameshima et al. 2016a). Interestingly, these effects were induced by vaspin mediated inhibition of ACh esterase (AChE) activity by ~25%. With respect to the catalytic mechanism of AChE to convert ACh to choline and acetic acid, which has similarities to serine proteinases, it was speculated that vaspin may directly inhibit AChE via the serpin mechanism (Kameshima et al. 2016a).

These *in vitro* findings were confirmed in spontaneous hypertensive rats (SHR) *in vivo*. Rats receiving long-term intraperitoneal vaspin application over 4 weeks maintained significantly lower systolic blood pressure (SBP). In isolated artery from SHR rats, vaspin did not alter vessel reactivity, but reduced arterial wall hypertrophy and ameliorated inflammatory TNF-a expression and ROS generation (Kameshima et al. 2016b). In accordance, in an animal model of monocrotaline (MCT)-induced pulmonary arterial hypertension (PAH), vaspin ameliorated PA pressure. In isolated PA, vaspin averted MCT-induced fibrosis but not hypertrophy of the vessel wall and reduced MCT-induced ROS generation and MMP-2 activation in lung tissues (Sakamoto et al. 2017).

In apolipoprotein E deficient mice, a mouse model spontaneously developing atherosclerotic lesions, vaspin_lentivirus application inhibited the progression of atherosclerotic plaque development (Lin et al. 2016). Vaspin_lentivirus application did not cause major alterations in the blood lipid profile, but reduced ER-stress in macrophages of ApoE-deficient mice (Lin et al. 2016).

Dysfunction of the endothelial barrier of pulmonary endothelial cells entails excessive leakage of fluid in the acute respiratory distress syndrome (ARDS). Using a lipopolysaccharide (LPS)-induced mouse model of ARDS, it was shown that vaspin dampened pulmonary inflammation and improved endothelial barrier function accompanied by activation of the AKT/GSK3 pathway (Qi et al. 2017). In isolated human pulmonary endothelial cells, vaspin treatment blunted LPS-induced generation of ROS, ameliorated inflammation and reduced apoptosis with contributions via the AKT signaling (Qi et al. 2017).

Together, these data clearly established beneficial effects of vaspin counteracting vascular inflammation, excessive ROS production and NO bioavailability (Fig. 4) and thus demonstrate its potential to preserve vascular function, lower blood pressure and the endothelial barrier.

3.5 Vaspin in the Brain – Central Action of Vaspin

In addition to their direct effects on peripheral tissues, many prominent hormones and adipokines such as insulin, leptin and adiponectin regulate energy expenditure and glucose homeostasis through central actions in the brain (reviewed in (Ahima and Lazar 2008; Varela

and Horvath 2012)). Based on the detection of vaspin in cerebrospinal fluid, Klöting et al. first reported central effects of vaspin (Kloting et al. 2011). Intracerebroventricular (icv) administration of vaspin acutely decreased food intake and lowered blood glucose levels in healthy C57BL/6 N and obese *db/db* mice. Interestingly, in obese and insulin resistant *db/db* mice glucose levels remained lower for 6 days after a single icv injection of vaspin (Kloting et al. 2011). Also, intrahypothalamic injection of vaspin in healthy and lean Wistar rats reduced food intake 24 h after injection (Brunetti et al. 2011). In the hypothalamus, vaspin injection reduced expression of orexigenic neuropeptide Y accompanied by a parallel increase in anorexigenic proopiomelanocortin expression 24 h after injection (Brunetti et al. 2011). Expression of AgRP, orexin-A, CART and CRH remained unchanged.

Luo et al. focused in detail on the central vaspin effects on glucose homeostasis after icv application in rats fed a chow or HFD (Luo et al. 2016). Again and under both diet regimes, food intake was acutely decreased for 24 h after a single vaspin bolus. In the HFD-fed rats, icv vaspin also increased the metabolic rate and elevated hypothalamic Fos expression. Neuronal Fos is expressed following voltage gated calcium influx and rapidly and transiently induced by neuronal excitation (Morgan et al. 1987). Continuous icv infusion of vaspin during hyperinsulinemic-euglycemic clamps increased glucose infusion rates necessary to maintain euglycemia, but only in the HFD animals. The increased insulin sensitivity was based on reduced hepatic glucose production, with the reduction of gluconeogenesis and glycogenolysis contributing 60% and 40%, respectively. Together, the suppression of hepatic glucose flux indicated improved hepatic insulin sensitivity and was accompanied by increased phosphorylation and activation of the insulin signaling cascade in HFD fed rats after icv vaspin (Luo et al. 2016). Signaling via the hepatic branch of the vagus nerve was investigated as an underlying mechanism of these regulatory effects. Indeed, hepatic vagotomy prevented central vaspin effects in clamp experiments and demonstrated that hepatic innervation is required for the glucose lowering effects of central vaspin (Luo et al. 2016). In the brain, pharmacologic inhibition of NMDA receptors in dorsal vagal complex (DVC) also prevented icv vaspin effects on glucose infusion rates and hepatic glucose production, indicating that vaspin signaling is transmitted by NMDA receptor expressing neurons in the DVC to nucleus of the solitary tract and subsequently to the liver via the hepatic branch of the vagus nerve (Luo et al. 2016). The finding of a neural signaling axis from the brain to the liver mediating central vaspin effects on energy balance, glucose homeostasis and insulin signaling (Fig. 5) raises many new questions. It is unknown whether these effects are mediated by vaspin expressed in peripheral tissues or by specific neurons in the brain. Additionally, it remains unclear, whether these effects are direct effects of vaspin acting on specific neurons by binding to cell surface receptors such as GPR78 (Fig. 3). Also indirect

mechanisms via the regulation of protease activities and their substrates (e.g. KLK7 and insulin) could contribute as well, as central insulin is known to decrease food intake and to increase energy expenditure via POMC neurons and the DVC (Filippi et al. 2014; Chen et al. 2017).

3.6   Vaspin in Bone

There is a wealth of experimental evidence for the modulation of metabolic bone function by adipokines (reviewed in (Gomez-Ambrosi et al. 2008)). In patients with rheumatoid arthritis (RA), a chronic inflammatory joint disease, vaspin serum levels were increased (Ozgen et al. 2010) and vaspin levels were also elevated in synovial fluid of RA, but not in osteoarthritis (OA) patients (Senolt et al. 2010). In line, a follow-up study investigating the risk of developing RA in auto-

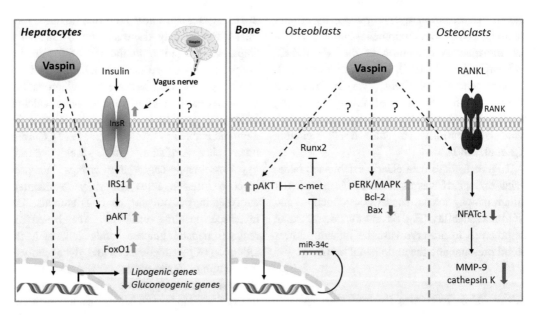

Fig. 5 Vaspin induced intracellular signaling in liver and bone. Vaspins central effects are relayed to the liver via the hepatic branch of the vagus nerve to decrease hepatic glucose flux, i.e. by reducing lipogenic and gluconeogenic gene expression and also activation of the insulin signaling cascade in hepatocytes. In bone, vaspin regulates bone remodeling by affecting bone-forming osteoblasts as well as bone-resorbing osteoclasts. RANKL-induced osteoclastogenesis is diminished via

decreased expression of the key mediator NFATc1 and accompanied by a decreased expression of the proteases cathepsine K and MMP-9. In osteoblasts, vaspin enhances pERK/MAPK signaling and prevents apoptosis by induction of anti-apoptotic Bcl-2 and lowering of pro-apoptotic Bax protein expression. Vaspin also attenuates osteogenic differentiation via a PI3K-AKT/miR-34c signaling loop. Vaspin-mediated activation or inhibition of pathways or molecules is indicated by green or red arrows, respectively antibody positive patients found a significant association between vaspin serum levels and the pathogenesis of RA, even after adjustment for overweight (Maijer et al. 2015). Studies evaluating the relationship between vaspin and bone mineral density (BMD) obtained diverging results. In inflammatory disease states, no correlation was found between vaspin and BMD in patients with multiple sclerosis (Assadi et al. 2011) or inflammatory bowel disease (Terzoudis et al. 2016). In contrast, a recent study in postmenopausal women (22% of the cohort had osteoporosis) revealed a significant positive association between vaspin serum levels and BMD in femoral neck (FN) and total hip after adjustment for age, body composition and various other parameters (Tanna et al. 2017).

The role of vaspin in bone metabolism has been investigated in *in vitro* studies demonstrating bilateral effects on both, bone-forming osteoblasts and bone-resorbing osteoclasts. *In vitro*, vaspin protected human osteoblasts from apoptosis in a dose-dependent manner (Zhu et al. 2013). On the cellular level, vaspin treatment resulted in enhanced protein expression of anti-apoptotic Bcl-2 and attenuated pro-apoptotic Bax expression via activation of the MAPK/ERK signaling pathway. Interestingly, vaspin treatment had no effects on PI3K-AKT, p38 or JNK pathways in human osteoblasts (Zhu et al. 2013). On the other hand, recent studies in the murine pre-osteoblast MC3T3-E1 cell line revealed an inhibitory or at least a modulating effect of vaspin on osteogenic differentiation (Liu et al. 2016). In particular, vaspin dose-dependently attenuated osteogenic differentiation by activation of the PI3K-signaling pathway leading to elevated miR-34c levels which reversely modulate the activation of PI3K-AKT signaling by targeting the tyrosine kinase receptor c-met. Thus, vaspin seems to participate in a regulatory PI3K-AKT/miR-34c loop involved in osteogenic differentiation (Liu et al. 2016). Along these lines, RANKL-induced osteoclastogenesis of RAW264.7 and bone marrow-derived cells was prevented by vaspin treatment through inhibition of the master regulator of osteoclastogenesis, nuclear factor of activated T cells c1 (NFATc1)

(Kamio et al. 2013). In RAW264.7 cells, vaspin inhibited RANKL-induced expression of bone resorption-related proteases cathepsin K and MMP-9 (Kamio et al. 2013). While the molecular mechanisms remain to be fully understood, these data indicate a dynamic regulation of bone metabolism by vaspin with induction of bone-formation and suppression of bone erosion (Fig. 5). This dual action profile suggests therapeutic value for the treatment of osteoporosis or rheumatoid arthritis.

3.7 Vaspin in Skin

Psoriasis is a chronic inflammatory skin disease and commonly associated with other pathological conditions such as obesity and T2D (Azfar and Gelfand 2008). In 2012, Saalbach *et al.* reported first evidence linking vaspin expression in skin with psoriatic skin disease (Saalbach et al. 2012). They identified keratinocytes as the major source of skin-derived vaspin and compared psoriatic skin biopsies of lesional and non-lesional areas. Interestingly, in non-lesional skin, vaspin expression was detectable across all epidermal layers, while it was decreased in lesional psoriatic skin (Saalbach et al. 2012). However, an association of vaspin serum levels and psoriasis is still discussed controversial. While Saalbach *et al.* did not observe differences in vaspin levels in healthy and psoriatic patients (Saalbach et al. 2012), other studies observed vaspin serum levels significantly decreased in patients with psoriasis (Ataseven and Kesli 2016) or systemic sclerosis with digital ulcers (Miura et al. 2015). After the identification of KLK7 as a protease target of vaspin, subsequent studies revealed distinct co-expression of vaspin with its target protease KLK7 in healthy skin as well as non-lesional skin of psoriasis patients while co-localization and expression was decreased in lesional psoriatic skin sections (Schultz et al. 2013). KLK7 is a chymotryptic serine protease that was initially identified in skin, where it is involved in the skin desquamation process (Lundstrom and Egelrud 1991) and overexpression and aberrant activity of KLK7 is related to pathogenesis of inflammatory skin diseases such as psoriasis (Ekholm and Egelrud 1999) and acne rosacea (Yamasaki et al. 2007). This lead to the hypothesis that vaspin may be involved in the fine-tuned balance of proteases involved in the regulation of inflammatory processes in the skin.

Recently, Saalbach *et al.* also investigated the role of vaspin in dysregulated immune cell and keratinocyte interaction leading to chronic inflammatory skin diseases (Saalbach et al. 2016). Vaspin expression was highly dynamic during the differentiation stages of the epidermis

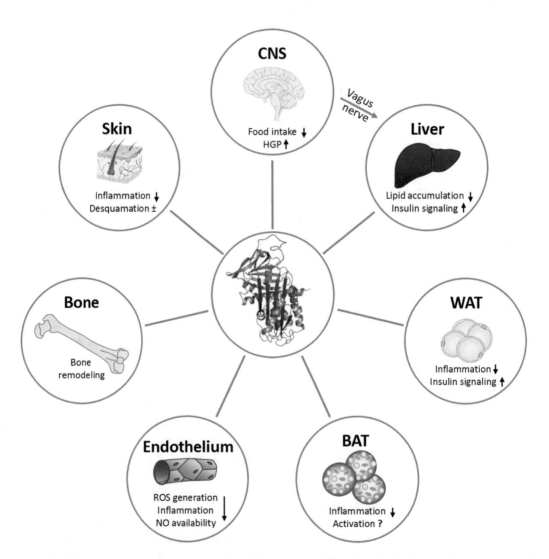

Fig. 6 Beneficial effects of vaspin in different tissues and organs that counteract obesity-induced inflammation, insulin resistance and other associated diseases. Central vaspin (application) attenuates food intake and reduces hepatic glucose production (HGP) via the hepatic branch of the vagus nerve accompanied by reduced hepatic lipid accumulation and increased insulin signaling in the liver. In white adipose tissue (WAT), vaspin reduces inflammation and enhances insulin signaling and vaspin expression in brown adipose tissue (BAT) is specifically upregulated upon BAT-activating stimuli. Vaspin counteracts endothelial dysfunction via the combination of its anti-inflammatory effects and by reducing ROS generation while enhancing NO availability. In bone, vaspin modulates bone formation via the regulation of osteoblasts and osteoclasts while in skin vaspin alleviates inflammatory processes and affects desquamation by regulating the activity of its target protease KLK7. *CNS* central nervous system

and down-regulation of vaspin expression in keratinocytes led to the activation of an altered gene cluster, characterized by reduced expression of differentiation-associated genes and an increased expression of proinflammatory cytokines. Moreover, vaspin expression in keratinocytes modulated the crosstalk between keratinocytes and immune cells in co-culture, thereby affecting the cytokine secretion profile of the inflammatory cells. These data suggest a protective role of vaspin in inflammatory skin diseases which was supported by *in vivo* experiments, in which vaspin application prevented myeloid cell infiltration in a psoriasis-like mouse model (Saalbach et al. 2016).

4 Concluding Remarks

In conclusion, the recent years have established vaspin as a multifaceted molecule with beneficial functions in various tissues, whether in auto- or endocrine ways or signaling from the brain to target tissues such as the liver (Fig. 6). We have discussed the diverse effects of vaspin *in vivo* and *in vitro*, and the data suggests that the main principle of vaspin function is to protect cells and tissues from inflammation under proinflammatory conditions found in obesity. And this protective, compensating or counteracting function, whether in adipocytes, vascular, skin or bone cells, seems to be executed via both, the regulation of protease activity and the interaction with cell surface receptors such as GRP78. Yet, the exact underlying mechanisms of signal transduction for many of the reported effects of vaspin remain to be elucidated and more research is needed to dissect the contributing pathways. Very likely, further receptors and target proteases are to be identified. A better understanding of these divers signaling pathways utilized by vaspin will help to employ this molecule, agonists of its receptors or inhibitors of its target protease(s) for new pharmacologic strategies to treat obesity associated inflammation, insulin resistance and atherosclerosis.

Acknowledgements Dr. Rene Meier and Sabina Kanton are kindly acknowledged for the homology modelling of the vaspin RCL. Our work is supported by the Deutsche Forschungsgemeinschaft SFB1052 "Obesity Mechanisms" (C07).

Author Contributions JW, KZ and JP wrote the manuscript. JW and JP created figures. JTH critically edited the manuscript.

Conflicts of Interest No conflicts of interest are declared by the authors.

Ethical Approval This article does not contain any studies with human participants or animals performed by any of the authors.

References

Ahima RS, Lazar MA (2008) Adipokines and the peripheral and neural control of energy balance. Mol Endocrinol 22(5):1023–1031

Akbarzadeh S, Nabipour I, Jafari SM, Movahed A, Motamed N, Assadi M, Hajian N (2012) Serum visfatin and vaspin levels in normoglycemic first-degree relatives of Iranian patients with type 2 diabetes mellitus. Diabetes Res Clin Pract 95(1):132–138

Assadi M, Salimipour H, Akbarzadeh S, Nemati R, Jafari SM, Bargahi A, Samani Z, Seyedabadi M, Sanjdideh Z, Nabipour I (2011) Correlation of circulating omentin-1 with bone mineral density in multiple sclerosis: the crosstalk between bone and adipose tissue. PLoS One 6(9):e24240

Ataseven A, Kesli R (2016) Novel inflammatory markers in psoriasis vulgaris: vaspin, vascular adhesion protein-1 (VAP-1), and YKL-40. G Ital Dermatol Venereol 151(3):244–250

Athyros VG, Tziomalos K, Karagiannis A, Anagnostis P, Mikhailidis DP (2010) Should adipokines be considered in the choice of the treatment of obesity-related health problems? Curr Drug Targets 11(1):122–135

Azfar RS, Gelfand JM (2008) Psoriasis and metabolic disease: epidemiology and pathophysiology. Curr Opin Rheumatol 20(4):416–422

Bhatt PS, Dhillo WS, Salem V (2017) Human brown adipose tissue-function and therapeutic potential in metabolic disease. Curr Opin Pharmacol 37:1–9

Blom N, Sicheritz-Ponten T, Gupta R, Gammeltoft S, Brunak S (2004) Prediction of post-translational glycosylation and phosphorylation of proteins from the amino acid sequence. Proteomics 4(6):1633–1649

Bluher M, Rudich A, Kloting N, Golan R, Henkin Y, Rubin E, Schwarzfuchs D, Gepner Y, Stampfer MJ, Fiedler M, Thiery J, Stumvoll M, Shai I (2012) Two patterns of adipokine and other biomarker dynamics in a long-term weight loss intervention. Diabetes Care 35 (2):342–349

Boden G, Chen X, Mozzoli M, Ryan I (1996) Effect of fasting on serum leptin in normal human subjects. J Clin Endocrinol Metab 81(9):3419–3423

Bohne-Lang A, von der Lieth CW (2005) GlyProt: in silico glycosylation of proteins. Nucleic Acids Res 33 (Web Server issue):W214–W219

Breitfeld J, Tonjes A, Bottcher Y, Schleinitz D, Wiele N, Marzi C, Brockhaus C, Rathmann W, Huth C, Grallert H, Illig T, Bluher M, Kovacs P, Stumvoll M (2013a) Genetic variation in the vaspin gene affects circulating serum vaspin concentrations. Int J Obes 37 (6):861–866

Breitfeld J, Heiker JT, Bottcher Y, Schleinitz D, Tonjes A, Weidle K, Krause K, Kuettner EB, Scholz M, Kiess W, Strater N, Beck-Sickinger AG, Stumvoll M, Korner A, Bluher M, Kovacs P (2013b) Analysis of a rare functional truncating mutation rs61757459 in vaspin (SERPINA12) on circulating vaspin levels. J Mol Med (Berl) 91(11):1285–1292

Breitfeld J, Tonjes A, Gast MT, Schleinitz D, Bluher M, Stumvoll M, Kovacs P, Bottcher Y (2013c) Role of vaspin in human eating behaviour. PLoS One 8(1): e54140

Brunetti L, Di Nisio C, Recinella L, Chiavaroli A, Leone S, Ferrante C, Orlando G, Vacca M (2011) Effects of vaspin, chemerin and omentin-1 on feeding behavior and hypothalamic peptide gene expression in the rat. Peptides 32(9):1866–1871

Cai W, Ramdas M, Zhu L, Chen X, Striker GE, Vlassara H (2012) Oral advanced glycation endproducts (AGEs) promote insulin resistance and diabetes by depleting the antioxidant defenses AGE receptor-1 and sirtuin 1. Proc Natl Acad Sci U S A 109(39):15888–15893

Cakal E, Ustun Y, Engin-Ustun Y, Ozkaya M, Kilinc M (2011) Serum vaspin and C-reactive protein levels in women with polycystic ovaries and polycystic ovary syndrome. Gynecol Endocrinol 27(7):491–495

Caminos JE, Bravo SB, Garces MF, Gonzalez CR, Cepeda LA, Gonzalez AC, Nogueiras R, Gallego R, Garcia-Caballero T, Cordido F, Lopez M, Dieguez C (2009) Vaspin and amylin are expressed in human and rat placenta and regulated by nutritional status. Histol Histopathol 24(8):979–990

Carrell RW, Stein PE (1996) The biostructural pathology of the serpins: critical function of sheet opening mechanism. Biol Chem Hoppe Seyler 377(1):1–17

Castro CA, da Silva KA, Buffo MM, Pinto KNZ, Duarte FO, Nonaka KO, Anibal FF, Duarte A (2017) Experimental type 2 diabetes induction reduces serum vaspin, but not serum omentin, in Wistar rats. Int J Exp Pathol 98(1):26–33

Chang HM, Lee HJ, Park HS, Kang JH, Kim KS, Song YS, Jang YJ (2010) Effects of weight reduction on serum vaspin concentrations in obese subjects:

modification by insulin resistance. Obesity (Silver Spring) 18(11):2105–2110

Chen M, Deng D, Fang Z, Xu M, Hu H, Luo L, Wang Y (2014) Fenofibrate increases serum vaspin by upregulating its expression in adipose tissue. Endocrine 45(3):409–421

Chen W, Balland E, Cowley MA (2017) Hypothalamic insulin resistance in obesity: effects on glucose homeostasis. Neuroendocrinology 104(4):364–381

Cho JK, Han TK, Kang HS (2010) Combined effects of body mass index and cardio/respiratory fitness on serum vaspin concentrations in Korean young men. Eur J Appl Physiol 108(2):347–353

Chopra M, Siddhu A, Tandon N (2014) Effect of nutritional regulation on adipokines in obesity: a review. Am J Food Nutr 2(4):66–70

Cinar N, Gulcelik NE, Aydin K, Akin S, Usman A, Gurlek A (2011) Serum vaspin levels in hypothyroid patients. Eur J Endocrinol 165(4):563–569

Davidson DJ, Haskell C, Majest S, Kherzai A, Egan DA, Walter KA, Schneider A, Gubbins EF, Solomon L, Chen Z, Lesniewski R, Henkin J (2005) Kringle 5 of human plasminogen induces apoptosis of endothelial and tumor cells through surface-expressed glucose-regulated protein 78. Cancer Res 65(11):4663–4672

de Veer SJ, Furio L, Swedberg JE, Munro CA, Brattsand M, Clements JA, Hovnanian A, Harris JM (2017) Selective substrates and inhibitors for Kallikrein-related peptidase 7 (KLK7) shed light on KLK proteolytic activity in the stratum corneum. J Invest Dermatol 137(2):430–439

Deanfield JE, Halcox JP, Rabelink TJ (2007) Endothelial function and dysfunction: testing and clinical relevance. Circulation 115(10):1285–1295

Debela M, Magdolen V, Schechter N, Valachova M, Lottspeich F, Craik CS, Choe Y, Bode W, Goettig P (2006) Specificity profiling of seven human tissue kallikreins reveals individual subsite preferences. J Biol Chem 281(35):25678–25688

Debela M, Hess P, Magdolen V, Schechter NM, Steiner T, Huber R, Bode W, Goettig P (2007) Chymotryptic specificity determinants in the 1.0 A structure of the zinc-inhibited human tissue kallikrein 7. Proc Natl Acad Sci U S A 104(41):16086–16091

Dementiev A, Dobo J, Gettins PG (2006) Active site distortion is sufficient for proteinase inhibition by serpins: structure of the covalent complex of alpha1-proteinase inhibitor with porcine pancreatic elastase. J Biol Chem 281(6):3452–3457

Derosa G, Fogari E, D'Angelo A, Bianchi L, Bonaventura A, Romano D, Maffioli P (2013) Adipocytokine levels in obese and non-obese subjects: an observational study. Inflammation 36(4):914–920

Ebert T, Gebhardt C, Scholz M, Wohland T, Schleinitz D, Fasshauer M, Bluher M, Stumvoll M, Kovacs P, Tonjes A (2018) Relationship between 12 adipocytokines and distinct components of the metabolic syndrome. J Clin Endocrinol Metab 103 (3):1015–1023

Ekholm E, Egelrud T (1999) Stratum corneum chymotryptic enzyme in psoriasis. Arch Dermatol Res 291 (4):195–200

Engh R, Lobermann H, Schneider M, Wiegand G, Huber R, Laurell CB (1989) The S variant of human alpha 1-antitrypsin, structure and implications for function and metabolism. Protein Eng 2(6):407–415

Fain JN, Buehrer B, Bahouth SW, Tichansky DS, Madan AK (2008) Comparison of messenger RNA distribution for 60 proteins in fat cells vs the nonfat cells of human omental adipose tissue. Metabolism 57 (7):1005–1015

Feng R, Li Y, Wang C, Luo C, Liu L, Chuo F, Li Q, Sun C (2014) Higher vaspin levels in subjects with obesity and type 2 diabetes mellitus: a meta-analysis. Diabetes Res Clin Pract 106(1):88–94

Filippi BM, Bassiri A, Abraham MA, Duca FA, Yue JT, Lam TK (2014) Insulin signals through the dorsal vagal complex to regulate energy balance. Diabetes 63(3):892–899

Fu BD, Yamawaki H, Okada M, Hara Y (2009) Vaspin can not inhibit TNF-alpha-induced inflammation of human umbilical vein endothelial cells. J Vet Med Sci 71(9):1201–1207

Gettins PG (2002) Serpin structure, mechanism, and function. Chem Rev 102(12):4751–4804

Gettins PG, Olson ST (2009) Exosite determinants of serpin specificity. J Biol Chem 284(31):20441–20445

Gettins PGW, Olson ST (2016) Inhibitory serpins. New insights into their folding, polymerization, regulation and clearance. Biochem J 473:2273–2293

Gils A, Knockaert I, Declerck PJ (1996) Substrate behavior of plasminogen activator inhibitor-1 is not associated with a lack of insertion of the reactive site loop. Biochemistry 35(23):7474–7481

Golpaie A, Tajik N, Masoudkabir F, Karbaschian Z, Talebpour M, Hoseini M, Hosseinzadeh-Attar MJ (2011) Short-term effect of weight loss through restrictive bariatric surgery on serum levels of vaspin in morbidly obese subjects. Eur Cytokine Netw 22 (4):181–186

Gomez-Ambrosi J, Rodriguez A, Catalan V, Fruhbeck G (2008) The bone-adipose axis in obesity and weight loss. Obes Surg 18(9):1134–1143

Gonzalez CR, Caminos JE, Vazquez MJ, Garces MF, Cepeda LA, Angel A, Gonzalez AC, Garcia-Rendueles ME, Sangiao-Alvarellos S, Lopez M, Bravo SB, Nogueiras R, Dieguez C (2009) Regulation of visceral adipose tissue-derived serine protease inhibitor by nutritional status, metformin, gender and pituitary factors in rat white adipose tissue. J Physiol 587 (Pt 14):3741–3750

Gonzalez-Gronow M, Kalfa T, Johnson CE, Gawdi G, Pizzo SV (2003) The voltage-dependent anion channel is a receptor for plasminogen kringle 5 on human endothelial cells. J Biol Chem 278(29):27312–27318

Gonzalez-Gronow M, Cuchacovich M, Llanos C, Urzua C, Gawdi G, Pizzo SV (2006) Prostate cancer cell proliferation in vitro is modulated by antibodies against glucose-regulated protein 78 isolated from patient serum. Cancer Res 66(23):11424–11431

Gonzalez-Gronow M, Kaczowka SJ, Payne S, Wang F, Gawdi G, Pizzo SV (2007) Plasminogen structural domains exhibit different functions when associated with cell surface GRP78 or the voltage-dependent anion channel. J Biol Chem 282(45):32811–32820

Gonzalez-Gronow M, Selim MA, Papalas J, Pizzo SV (2009) GRP78: a multifunctional receptor on the cell surface. Antioxid Redox Signal 11(9):2299–2306

Gulcelik NE, Karakaya J, Gedik A, Usman A, Gurlek A (2009) Serum vaspin levels in type 2 diabetic women in relation to microvascular complications. Eur J Endocrinol 160(1):65–70

Guvenc Y, Var A, Goker A, Kuscu NK (2016) Assessment of serum chemerin, vaspin and omentin-1 levels in patients with polycystic ovary syndrome. J Int Med Res 44(4):796–805

Haas IG (1994) BiP (GRP78), an essential hsp70 resident protein in the endoplasmic reticulum. Experientia 50 (11–12):1012–1020

Handisurya A, Riedl M, Vila G, Maier C, Clodi M, Prikoszovich T, Ludvik B, Prager G, Luger A, Kautzky-Willer A (2010) Serum vaspin concentrations in relation to insulin sensitivity following RYGB-induced weight loss. Obes Surg 20(2):198–203

Harms M, Seale P (2013) Brown and beige fat: development, function and therapeutic potential. Nat Med 19 (10):1252–1263

Harrop SJ, Jankova L, Coles M, Jardine D, Whittaker JS, Gould AR, Meister A, King GC, Mabbutt BC, Curmi PM (1999) The crystal structure of plasminogen activator inhibitor 2 at 2.0 A resolution: implications for serpin function. Structure 7(1):43–54

Heiker JT, Kloting N, Kovacs P, Kuettner EB, Strater N, Schultz S, Kern M, Stumvoll M, Bluher M, Beck-Sickinger AG (2013) Vaspin inhibits kallikrein 7 by serpin mechanism. Cell Mol Life Sci 70 (14):2569–2583

Hida K, Wada J, Zhang H, Hiragushi K, Tsuchiyama Y, Shikata K, Makino H (2000) Identification of genes specifically expressed in the accumulated visceral adipose tissue of OLETF rats. J Lipid Res 41 (10):1615–1622

Hida K, Wada J, Eguchi J, Zhang H, Baba M, Seida A, Hashimoto I, Okada T, Yasuhara A, Nakatsuka A, Shikata K, Hourai S, Futami J, Watanabe E, Matsuki Y, Hiramatsu R, Akagi S, Makino H, Kanwar YS (2005) Visceral adipose tissue-derived serine protease inhibitor: a unique insulin-sensitizing adipocytokine in obesity. Proc Natl Acad Sci U S A 102(30):10610–10615

Hopkins PC, Carrell RW, Stone SR (1993) Effects of mutations in the hinge region of serpins. Biochemistry 32(30):7650–7657

Huber R, Carrell RW (1989) Implications of the three-dimensional structure of alpha 1-antitrypsin for structure and function of serpins. Biochemistry 28 (23):8951–8966

Huntington JA (2006) Shape-shifting serpins--advantages of a mobile mechanism. Trends Biochem Sci 31 (8):427–435

Huntington JA, Read RJ, Carrell RW (2000) Structure of a serpin-protease complex shows inhibition by deformation. Nature 407(6806):923–926

Ibanez L, Lopez-Bermejo A, Diaz M, Enriquez G, del Rio L, de Zegher F (2009) Low-dose pioglitazone and low-dose flutamide added to metformin and oestro-progestagens for hyperinsulinaemic women with androgen excess: add-on benefits disclosed by a randomized double-placebo study over 24 months. Clin Endocrinol 71(3):351–357

Irving JA, Pike RN, Lesk AM, Whisstock JC (2000) Phylogeny of the serpin superfamily: implications of patterns of amino acid conservation for structure and function. Genome Res 10(12):1845–1864

Jeong E, Youn BS, Kim DW, Kim EH, Park JW, Namkoong C, Jeong JY, Yoon SY, Park JY, Lee KU, Kim MS (2010) Circadian rhythm of serum vaspin in healthy male volunteers: relation to meals. J Clin Endocrinol Metab 95(4):1869–1875

Jian W, Peng W, Xiao S, Li H, Jin J, Qin L, Dong Y, Su Q (2014) Role of serum vaspin in progression of type 2 diabetes: a 2-year cohort study. PLoS One 9(4): e94763

Joslin G, Griffin GL, August AM, Adams S, Fallon RJ, Senior RM, Perlmutter DH (1992) The serpin-enzyme complex (SEC) receptor mediates the neutrophil chemotactic effect of alpha-1 antitrypsin-elastase complexes and amyloid-beta peptide. J Clin Invest 90 (3):1150–1154

Jung CH, Lee WJ, Hwang JY, Seol SM, Kim YM, Lee YL, Park JY (2011) Vaspin protects vascular endothelial cells against free fatty acid-induced apoptosis through a phosphatidylinositol 3-kinase/Akt pathway. Biochem Biophys Res Commun 413(2):264–269

Jung CH, Lee WJ, Hwang JY, Lee MJ, Seol SM, Kim YM, Lee YL, Kim HS, Kim MS, Park JY (2012) Vaspin increases nitric oxide bioavailability through the reduction of asymmetric dimethylarginine in vascular endothelial cells. PLoS One 7(12):e52346

Jung CH, Lee MJ, Kang YM, Lee YL, Yoon HK, Kang SW, Lee WJ, Park JY (2014) Vaspin inhibits cytokine-induced nuclear factor-kappa B activation and adhesion molecule expression via AMP-activated protein kinase activation in vascular endothelial cells. Cardiovasc Diabetol 13:41

Kadoglou NP, Kapelouzou A, Tsanikidis H, Vitta I, Liapis CD, Sailer N (2011) Effects of rosiglitazone/metformin fixed-dose combination therapy and metformin monotherapy on serum vaspin, adiponectin and IL-6 levels in drug-naive patients with type 2 diabetes. Exp Clin Endocrinol Diabetes 119(2):63–68

Kajimura S, Saito M (2014) A new era in brown adipose tissue biology: molecular control of brown fat development and energy homeostasis. Annu Rev Physiol 76:225–249

Kameshima S, Yamada K, Morita T, Okada M, Yamawaki H (2016a) Visceral adipose tissue-derived serine protease inhibitor augments acetylcholine-induced relaxation via the inhibition of acetylcholine esterase activity in rat isolated mesenteric artery. Acta Physiol (Oxf) 216(2):203–210

Kameshima S, Sakamoto Y, Okada M, Yamawaki H (2016b) Vaspin prevents elevation of blood pressure through inhibition of peripheral vascular remodelling in spontaneously hypertensive rats. Acta Physiol (Oxf) 217(2):120–129

Kamio N, Kawato T, Tanabe N, Kitami S, Morita T, Ochiai K, Maeno M (2013) Vaspin attenuates RANKL-induced osteoclast formation in RAW264.7 cells. Connect Tissue Res 54(2):147–152

Kang ES, Magkos F, Sienkiewicz E, Mantzoros CS (2011) Circulating vaspin and visfatin are not affected by acute or chronic energy deficiency or leptin administration in humans. Eur J Endocrinol 164(6):911–917

Kaslik G, Patthy A, Balint M, Graf L (1995) Trypsin complexed with alpha 1-proteinase inhibitor has an increased structural flexibility. FEBS Lett 370 (3):179–183

Kempf K, Rose B, Illig T, Rathmann W, Strassburger K, Thorand B, Meisinger C, Wichmann HE, Herder C, Vollmert C (2010) Vaspin (SERPINA12) genotypes and risk of type 2 diabetes: results from the MONICA/KORA studies. Exp Clin Endocrinol Diabetes 118(3):184–189

Kloting N, Berndt J, Kralisch S, Kovacs P, Fasshauer M, Schon MR, Stumvoll M, Bluher M (2006) Vaspin gene expression in human adipose tissue: association with obesity and type 2 diabetes. Biochem Biophys Res Commun 339(1):430–436

Kloting N, Kovacs P, Kern M, Heiker JT, Fasshauer M, Schon MR, Stumvoll M, Beck-Sickinger AG, Bluher M (2011) Central vaspin administration acutely reduces food intake and has sustained blood glucose-lowering effects. Diabetologia 54(7):1819–1823

Koiou E, Tziomalos K, Dinas K, Katsikis I, Kalaitzakis E, Delkos D, Kandaraki EA, Panidis D (2011a) The effect of weight loss and treatment with metformin on serum vaspin levels in women with polycystic ovary syndrome. Endocr J 58(4):237–246

Koiou E, Dinas K, Tziomalos K, Toulis K, Kandaraki EA, Kalaitzakis E, Katsikis I, Panidis D (2011b) The phenotypes of polycystic ovary syndrome defined by the 1990 diagnostic criteria are associated with higher serum vaspin levels than the phenotypes introduced by the 2003 criteria. Obes Facts 4(2):145–150

Korner A, Neef M, Friebe D, Erbs S, Kratzsch J, Dittrich K, Bluher S, Kapellen TM, Kovacs P, Stumvoll M, Bluher M, Kiess W (2011) Vaspin is related to gender, puberty and deteriorating insulin sensitivity in children. Int J Obes 35(4):578–586

Kovacs P, Miehle K, Sandner B, Stumvoll M, Bluher M (2013) Insulin administration acutely decreases vaspin serum concentrations in humans. Obes Facts 6 (1):86–88

Lawrence DA, Olson ST, Muhammad S, Day DE, Kvassman JO, Ginsburg D, Shore JD (2000) Partitioning of serpin-proteinase reactions between stable inhibition and substrate cleavage is regulated by the rate of serpin reactive center loop insertion into beta-sheet A. J Biol Chem 275(8):5839–5844

Lee C, Park SH, Lee MY, Yu MH (2000) Regulation of protein function by native metastability. Proc Natl Acad Sci U S A 97(14):7727–7731

Lee JA, Park HS, Song YS, Jang YJ, Kim JH, Lee YJ, Heo YS (2011) Relationship between vaspin gene expression and abdominal fat distribution of Korean women. Endocr J 58(8):639–646

Li K, Li L, Yang M, Liu H, Liu D, Yang H, Boden G, Yang G (2011) Short-term continuous subcutaneous insulin infusion decreases the plasma vaspin levels in patients with type 2 diabetes mellitus concomitant with improvement in insulin sensitivity. Eur J Endocrinol 164(6):905–910

Li H, Peng W, Zhuang J, Lu Y, Jian W, Wei Y, Li W, Xu Y (2013) Vaspin attenuates high glucose-induced vascular smooth muscle cells proliferation and chemokinesis by inhibiting the MAPK, PI3K/Akt, and NF-kappaB signaling pathways. Atherosclerosis 228(1):61–68

Lin Y, Zhuang J, Li H, Zhu G, Zhou S, Li W, Peng W, Xu Y (2016) Vaspin attenuates the progression of atherosclerosis by inhibiting ER stress-induced macrophage apoptosis in apoE/mice. Mol Med Rep 13 (2):1509–1516

Liu S, Dong Y, Wang T, Zhao S, Yang K, Chen X, Zheng C (2014) Vaspin inhibited proinflammatory cytokine induced activation of nuclear factor-kappa B and its downstream molecules in human endothelial EA. hy926 cells. Diabetes Res Clin Pract 103(3):482–488

Liu P, Li G, Wu J, Zhou·X, Wang L, Han W, Lv Y, Sun C (2015) Vaspin promotes 3T3-L1 preadipocyte differentiation. Exp Biol Med (Maywood) 240 (11):1520–1527

Liu Y, Xu F, Pei HX, Zhu X, Lin X, Song CY, Liang QH, Liao EY, Yuan LQ (2016) Vaspin regulates the osteogenic differentiation of MC3T3-E1 through the PI3K-Akt/miR-34c loop. Sci Rep 6:25578

Lomas DA, Evans DL, Finch JT, Carrell RW (1992) The mechanism of Z alpha 1-antitrypsin accumulation in the liver. Nature 357(6379):605–607

Lu H, Dhanabal M, Volk R, Waterman MJ, Ramchandran R, Knebelmann B, Segal M, Sukhatme VP (1999) Kringle 5 causes cell cycle arrest and apoptosis of endothelial cells. Biochem Biophys Res Commun 258(3):668–673

Lu H, Fouejeu Wamba PC, Lapointe M, Poirier P, Martin J, Bastien M, Cianflone K (2014) Increased vaspin levels are associated with beneficial metabolic outcome pre- and post-bariatric surgery. PLoS One 9 (10):e111002

Lundstrom A, Egelrud T (1991) Stratum corneum chymotryptic enzyme: a proteinase which may be generally present in the stratum corneum and with a possible involvement in desquamation. Acta Derm Venereol 71(6):471–474

Luo X, Li K, Zhang C, Yang G, Yang M, Jia Y, Zhang L, Ma ZA, Boden G, Li L (2016) Central administration of vaspin inhibits glucose production and augments hepatic insulin signaling in high-fat-diet-fed rat. Int J Obes 40(6):947–954

Lustig RH (2017) Processed food-an experiment that failed. JAMA Pediatr 171(3):212–214

Maekawa H, Tollefsen DM (1996) Role of the proposed serpin-enzyme complex receptor recognition site in binding and internalization of thrombin-heparin cofactor II complexes by hepatocytes. J Biol Chem 271 (31):18604–18609

Maijer KI, Neumann E, Muller-Ladner U, Drop DA, Ramwadhdoebe TH, Choi IY, Gerlag DM, de Hair MJ, Tak PP (2015) Serum vaspin levels are associated with the development of clinically manifest arthritis in autoantibody-positive individuals. PLoS One 10(12): e0144932

Mast AE, Enghild JJ, Pizzo SV, Salvesen G (1991) Analysis of the plasma elimination kinetics and conformational stabilities of native, proteinase-complexed, and reactive site cleaved serpins: comparison of alpha 1-proteinase inhibitor, alpha 1-antichymotrypsin, antithrombin III, alpha 2-antiplasmin, angiotensinogen, and ovalbumin. Biochemistry 30(6):1723–1730

Misra UK, Gonzalez-Gronow M, Gawdi G, Wang F, Pizzo SV (2004) A novel receptor function for the heat shock protein Grp78: silencing of Grp78 gene expression attenuates alpha2M*-induced signalling. Cell Signal 16(8):929–938

Misra UK, Gonzalez-Gronow M, Gawdi G, Pizzo SV (2005) The role of MTJ-1 in cell surface translocation of GRP78, a receptor for alpha 2-macroglobulin-dependent signaling. J Immunol 174(4):2092–2097

Miura S, Asano Y, Saigusa R, Yamashita T, Taniguchi T, Takahashi T, Ichimura Y, Toyama T, Tamaki Z, Tada Y, Sugaya M, Sato S, Kadono T (2015) Serum vaspin levels: a possible correlation with digital ulcers in patients with systemic sclerosis. J Dermatol 42 (5):528–531

Morgan JI, Cohen DR, Hempstead JL, Curran T (1987) Mapping patterns of c-fos expression in the central nervous system after seizure. Science 237 (4811):192–197

Mottarella SE, Beglov D, Beglova N, Nugent MA, Kozakov D, Vajda S (2014) Docking server for the identification of heparin binding sites on proteins. J Chem Inf Model 54(7):2068–2078

Nakatsuka A, Wada J, Iseda I, Teshigawara S, Higashio K, Murakami K, Kanzaki M, Inoue K, Terami T, Katayama A, Hida K, Eguchi J, Horiguchi CS, Ogawa D, Matsuki Y, Hiramatsu R, Yagita H, Kakuta S, Iwakura Y, Makino H (2012) Vaspin is an adipokine ameliorating ER stress in obesity as a ligand for cell-surface GRP78/MTJ-1 complex. Diabetes 61 (11):2823–2832

Nakatsuka A, Wada J, Iseda I, Teshigawara S, Higashio K, Murakami K, Kanzaki M, Inoue K, Terami T, Katayama A, Hida K, Eguchi J, Ogawa D, Matsuki Y, Hiramatsu R, Yagita H, Kakuta S, Iwakura Y, Makino H (2013) Visceral adipose tissue-derived serine proteinase inhibitor inhibits apoptosis of endothelial cells as a ligand for the cell-surface GRP78/voltage-dependent anion channel complex. Circ Res 112(5):771–780

Oertwig K, Ulbricht D, Hanke S, Pippel J, Bellmann-Sickert K, Strater N, Heiker JT (2017) Glycosylation of human vaspin (SERPINA12) and its impact on serpin activity, heparin binding and thermal stability. Biochim Biophys Acta 1865(9):1188–1194

Oliveira JR, Bertolin TC, Andrade D, Oliveira LC, Kondo MY, Santos JA, Blaber M, Juliano L, Severino B, Caliendo G, Santagada V, Juliano MA (2015) Specificity studies on Kallikrein-related peptidase 7 (KLK7) and effects of osmolytes and glycosaminoglycans on its peptidase activity. Biochim Biophys Acta 1854 (1):73–83

Ostrowska Z, Ziora K, Oswiecimska J, Swietochowska E, Marek B, Kajdaniuk D, Strzelczyk J, Golabek K, Morawiecka-Pietrzak M, Wolkowska-Pokrywa K, Kos-Kudla B (2016) Vaspin and selected indices of bone status in girls with anorexia nervosa. Endokrynol Pol 67(6):599–606

Ozgen M, Koca SS, Dagli N, Balin M, Ustundag B, Isik A (2010) Serum adiponectin and vaspin levels in rheumatoid arthritis. Arch Med Res 41(6):457–463

Perlmutter DH, Glover GI, Rivetna M, Schasteen CS, Fallon RJ (1990) Identification of a serpin-enzyme complex receptor on human hepatoma cells and human monocytes. Proc Natl Acad Sci U S A 87 (10):3753–3757

Phalitakul S, Okada M, Hara Y, Yamawaki H (2011) Vaspin prevents TNF-alpha-induced intracellular adhesion molecule-1 via inhibiting reactive oxygen species-dependent NF-kappaB and PKCtheta activation in cultured rat vascular smooth muscle cells. Pharmacol Res 64(5):493–500

Phalitakul S, Okada M, Hara Y, Yamawaki H (2012) A novel adipocytokine, vaspin inhibits platelet-derived growth factor-BB-induced migration of vascular smooth muscle cells. Biochem Biophys Res Commun 423(4):844–849

Phalitakul S, Okada M, Hara Y, Yamawaki H (2013) Vaspin prevents methylglyoxal-induced apoptosis in human vascular endothelial cells by inhibiting reactive oxygen species generation. Acta Physiol (Oxf) 209 (3):212–219

Pippel J, Kuettner EB, Ulbricht D, Daberger J, Schultz S, Heiker JT, Strater N (2016) Crystal structure of cleaved vaspin (serpinA12). Biol Chem 397(2):111–123

Qi D, Wang D, Zhang C, Tang X, He J, Zhao Y, Deng W, Deng X (2017) Vaspin protects against LPS-induced ARDS by inhibiting inflammation, apoptosis and reactive oxygen species generation in pulmonary endothelial cells via the Akt/GSK3beta pathway. Int J Mol Med 40(6):1803–1817

Rojas J, Chavez M, Olivar L, Rojas M, Morillo J, Mejias J, Calvo M, Bermudez V (2014) Polycystic ovary syndrome, insulin resistance, and obesity: navigating the pathophysiologic labyrinth. Int J Reprod Med 2014:719050

Rosell M, Kaforou M, Frontini A, Okolo A, Chan YW, Nikolopoulou E, Millership S, Fenech ME, MacIntyre D, Turner JO, Moore JD, Blackburn E, Gullick WJ, Cinti S, Montana G, Parker MG, Christian M (2014) Brown and white adipose tissues: intrinsic differences in gene expression and response to cold exposure in mice. Am J Physiol Endocrinol Metab 306(8):E945–E964

Rothwell NJ, Stock MJ (1979) A role for brown adipose tissue in diet-induced thermogenesis. Nature 281 (5726):31–35

Ryu SE, Choi HJ, Kwon KS, Lee KN, Yu MH (1996) The native strains in the hydrophobic core and flexible reactive loop of a serine protease inhibitor: crystal structure of an uncleaved alpha1-antitrypsin at 2.7 A. Structure 4(10):1181–1192

Saalbach A, Vester K, Rall K, Tremel J, Anderegg U, Beck-Sickinger AG, Bluher M, Simon JC (2012) Vaspin--a link of obesity and psoriasis? Exp Dermatol 21(4):309–312

Saalbach A, Tremel J, Herbert D, Schwede K, Wandel E, Schirmer C, Anderegg U, Beck-Sickinger AG, Heiker JT, Schultz S, Magin T, Simon JC (2016) Anti-inflammatory action of keratinocyte-derived vaspin: relevance for the pathogenesis of psoriasis. Am J Pathol 186(3):639–651

Sakamoto Y, Kameshima S, Kakuda C, Okamura Y, Kodama T, Okada M, Yamawaki H (2017) Visceral adipose tissue-derived serine protease inhibitor prevents the development of monocrotaline-induced pulmonary arterial hypertension in rats. Pflugers Arch 469(11):1425–1432

Schultz S, Saalbach A, Heiker JT, Meier R, Zellmann T, Simon JC, Beck-Sickinger AG (2013) Proteolytic activation of prochemerin by kallikrein 7 breaks an ionic linkage and results in C-terminal rearrangement. Biochem J 452(2):271–280

Seeger J, Ziegelmeier M, Bachmann A, Lossner U, Kratzsch J, Bluher M, Stumvoll M, Fasshauer M (2008) Serum levels of the adipokine vaspin in relation to metabolic and renal parameters. J Clin Endocrinol Metab 93(1):247–251

Senolt L, Polanska M, Filkova M, Cerezo LA, Pavelka K, Gay S, Haluzik M, Vencovsky J (2010) Vaspin and omentin: new adipokines differentially regulated at the site of inflammation in rheumatoid arthritis. Ann Rheum Dis 69(7):1410–1411

Shaker OG, Sadik NA (2013) Vaspin gene in rat adipose tissue: relation to obesity-induced insulin resistance. Mol Cell Biochem 373(1–2):229–239

Silverman GA, Bird PI, Carrell RW, Church FC, Coughlin PB, Gettins PG, Irving JA, Lomas DA, Luke CJ, Moyer RW, Pemberton PA, Remold-O'Donnell E, Salvesen GS, Travis J, Whisstock JC (2001) The serpins are an expanding superfamily of structurally similar but functionally diverse proteins. Evolution, mechanism of inhibition, novel functions, and a revised nomenclature. J Biol Chem 276 (36):33293–33296

Stavridi ES, O'Malley K, Lukacs CM, Moore WT, Lambris JD, Christianson DW, Rubin H, Cooperman BS (1996) Structural change in alpha-chymotrypsin induced by complexation with alpha 1-antichymotrypsin as seen by enhanced sensitivity to proteolysis. Biochemistry 35(33):10608–10615

Stein PE, Carrell RW (1995) What do dysfunctional serpins tell us about molecular mobility and disease. Nat Struct Biol 2(2):96–113

Stoller JK, Aboussouan LS (2012) A review of alpha1-antitrypsin deficiency. Am J Respir Crit Care Med 185 (3):246–259

Stratikos E, Gettins PG (1999) Formation of the covalent serpin-proteinase complex involves translocation of the proteinase by more than 70 A and full insertion of the reactive center loop into beta-sheet A. Proc Natl Acad Sci U S A 96(9):4808–4813

Sun N, Wang H, Wang L (2015) Vaspin alleviates dysfunction of endothelial progenitor cells induced by high glucose via PI3K/Akt/eNOS pathway. Int J Clin Exp Pathol 8(1):482–489

Tamucci KA, Namwanje M, Fan L, Qiang L (2017) The dark side of browning. Protein Cell 9(2):152–163

Tan BK, Heutling D, Chen J, Farhatullah S, Adya R, Keay SD, Kennedy CR, Lehnert H, Randeva HS (2008) Metformin decreases the adipokine vaspin in overweight women with polycystic ovary syndrome concomitant with improvement in insulin sensitivity and a decrease in insulin resistance. Diabetes 57 (6):1501–1507

Tanna N, Patel K, Moore AE, Dulnoan D, Edwards S, Hampson G (2017) The relationship between circulating adiponectin, leptin and vaspin with bone mineral density (BMD), arterial calcification and stiffness: a cross-sectional study in post-menopausal women. J Endocrinol Invest 40(12):1345–1353

Terzoudis S, Malliaraki N, Damilakis J, Dimitriadou DA, Zavos C, Koutroubakis IE (2016) Chemerin, visfatin, and vaspin serum levels in relation to bone mineral density in patients with inflammatory bowel disease. Eur J Gastroenterol Hepatol 28(7):814–819

Teshigawara S, Wada J, Hida K, Nakatsuka A, Eguchi J, Murakami K, Kanzaki M, Inoue K, Terami T, Katayama A, Iseda I, Matsushita Y, Miyatake N, McDonald JF, Hotta K, Makino H (2012) Serum vaspin concentrations are closely related to insulin resistance, and rs77060950 at SERPINA12 genetically defines distinct group with higher serum levels in Japanese population. J Clin Endocrinol Metab 97(7): E1202–E1207

Toulza E, Mattiuzzo NR, Galliano MF, Jonca N, Dossat C, Jacob D, de Daruvar A, Wincker P, Serre G, Guerrin M (2007) Large-scale identification of human genes implicated in epidermal barrier function. Genome Biol 8(6):R107

Ulbricht D, Pippel J, Schultz S, Meier R, Strater N, Heiker JT (2015) A unique serpin P1' glutamate and a conserved beta-sheet C arginine are key residues for activity, protease recognition and stability of serpinA12 (vaspin). Biochem J 470(3):357–367

Ulbricht D, Oertwig K, Arnsburg K, Saalbach A, Pippel J, Strater N, Heiker JT (2017) Basic residues of beta-sheet a contribute to heparin binding and activation of vaspin (Serpin A12). J Biol Chem 292(3):994–1004

Ulbricht D, Tindall CA, Oertwig K, Hanke S, Strater N, Heiker JT (2018) Kallikrein-related peptidase 14 is the second KLK protease targeted by the serpin vaspin. Biol Chem. https://doi.org/10.1515/hsz-2018-0108

Van Gaal LF, Mertens IL, De Block CE (2006) Mechanisms linking obesity with cardiovascular disease. Nature 444(7121):875–880

Varela L, Horvath TL (2012) Leptin and insulin pathways in POMC and AgRP neurons that modulate energy balance and glucose homeostasis. EMBO Rep 13 (12):1079–1086

Vehapoglu A, Ustabas F, Ozgen TI, Terzioglu S, Cermik BB, Ozen OF (2015) Role of circulating adipocytokines vaspin, apelin, and visfatin in the loss of appetite in underweight children: a pilot trial. J Pediatr Endocrinol Metab 28(9–10):1065–1071

Vink RG, Roumans NJ, Mariman EC, van Baak MA (2017) Dietary weight loss-induced changes in RBP4, FFA, and ACE predict weight regain in people with overweight and obesity. Phys Rep 5(21):e13450

von Loeffelholz C, Mohlig M, Arafat AM, Isken F, Spranger J, Mai K, Randeva HS, Pfeiffer AFH, Weickert MO (2010) Circulating vaspin is unrelated to insulin sensitivity in a cohort of nondiabetic humans. Eur J Endocrinol 162(3):507–513

Vulesevic B, McNeill B, Giacco F, Maeda K, Blackburn NJ, Brownlee M, Milne RW, Suuronen EJ (2016) Methylglyoxal-induced endothelial cell loss and inflammation contribute to the development of diabetic cardiomyopathy. Diabetes 65(6):1699–1713

Wada J (2008) Vaspin: a novel serpin with insulin-sensitizing effects. Expert Opin Investig Drugs 17 (3):327–333

Weiner J, Rohde K, Krause K, Zieger K, Kloting N, Kralisch S, Kovacs P, Stumvoll M, Bluher M, Bottcher Y, Heiker JT (2017) Brown adipose tissue (BAT) specific vaspin expression is increased after obesogenic diets and cold exposure and linked to acute changes in DNA-methylation. Mol Metab 6 (6):482–493

Yamasaki M, Takahashi N, Hirose M (2003) Crystal structure of S-ovalbumin as a non-loop-inserted thermostabilized serpin form. J Biol Chem 278 (37):35524–35530

Yamasaki K, Di Nardo A, Bardan A, Murakami M, Ohtake T, Coda A, Dorschner RA, Bonnart C, Descargues P, Hovnanian A, Morhenn VB, Gallo RL (2007) Increased serine protease activity and cathelicidin promotes skin inflammation in rosacea. Nat Med 13(8):975–980

Youn BS, Kloting N, Kratzsch J, Lee N, Park JW, Song ES, Ruschke K, Oberbach A, Fasshauer M, Stumvoll M, Bluher M (2008) Serum vaspin concentrations in human obesity and type 2 diabetes. Diabetes 57(2):372–377

Zhang L, Li L, Yang M, Liu H, Yang G (2011) Elevated circulating vaspin levels were decreased by rosiglitazone therapy in T2DM patients with poor glycemic control on metformin alone. Cytokine 56 (2):399–402

Zhu X, Jiang Y, Shan PF, Shen J, Liang QH, Cui RR, Liu Y, Liu GY, Wu SS, Lu Q, Xie H, Liu YS, Yuan LQ, Liao EY (2013) Vaspin attenuates the apoptosis of human osteoblasts through ERK signaling pathway. Amino Acids 44(3):961–968

Zieger K, Weiner J, Kunath A, Gericke M, Krause K, Kern M, Stumvoll M, Kloting N, Bluher M, Heiker JT (2017a) Ablation of kallikrein 7 (KLK7) in adipose tissue ameliorates metabolic consequences of high fat diet-induced obesity by counteracting adipose tissue inflammation in vivo. Cell Mol Life Sci 75(4):727–774

Zieger K, Weiner J, Krause K, Schwarz M, Kohn M, Stumvoll M, Bluher M, Heiker JT (2017b) Vaspin suppresses cytokine-induced inflammation in 3T3-L1 adipocytes via inhibition of NFkappaB pathway. Mol Cell Endocrinol 460:181–188

Adv Exp Med Biol - Protein Reviews (2019) 20: 189–203
https://doi.org/10.1007/5584_2018_273
© Springer Nature Switzerland AG 2018
Published online: 29 September 2018

Exceptionally Selective Substrate Targeting by the Metalloprotease Anthrax Lethal Factor

Benjamin E. Turk

Abstract

The zinc-dependent metalloprotease anthrax lethal factor (LF) is the enzymatic component of a toxin thought to have a major role in *Bacillus anthracis* infections. Like many bacterial toxins, LF is a secreted protein that functions within host cells. LF is a highly selective protease that cleaves a limited number of substrates in a site-specific manner, thereby impacting host signal transduction pathways. The major substrates of LF are mitogen-activated protein kinase kinases (MKKs), which lie in the middle of three-component phosphorylation cascades mediating numerous functions in a variety of cells and tissues. How LF targets its limited substrate repertoire has been an active area of investigation. LF recognizes a specific sequence motif surrounding the scissile bonds of substrate proteins. X-ray crystallography of the protease in complex with peptide substrates has revealed the structural basis of selectivity for the LF cleavage site motif. In addition to having interactions proximal to the cleavage site, LF binds directly to a more distal region in its substrates through a so-called exosite interaction. This exosite has been mapped to a surface within a non-catalytic domain of LF with previously unknown function. A putative LF-binding site has likewise been identified on the catalytic domains of MKKs. Here we review our current state of understanding of LF-substrate interactions and discuss the implications for the design and discovery of inhibitors that may have utility as anthrax therapeutics.

Keywords

Anthrax · Enzyme specificity · Exosite · Metalloprotease · Protease · Mitogen-activated protein kinase kinase

Abbreviations

AC	Adenylyl cyclase
BoNT	Botulinum neurotoxin
cAMP	Cyclic AMP
EdTx	Edema toxin
EF	Edema factor
HTS	High-throughput screening
LeTx	Lethal toxin
LF	Anthrax lethal factor
MAPK	Mitogen-activated protein kinase
MKK	Mitogen-activated protein kinase kinase
NLRP	NACHT domain, leucine rich repeat and pyrin domain containing protein
NOS1	Constitutive nitric oxide synthase
PA	Protective antigen
PAMP	Pathogen-associated molecular pattern

B. E. Turk (✉)
Department of Pharmacology, Yale School of Medicine, New Haven, CT, USA
e-mail: ben.turk@yale.edu

1   Introduction

Proteolytic enzymes are found in all kingdoms of life. Fundamentally, living things require a means to break down proteins and peptides, whether produced by the organism itself or obtained from the environment as a nutrient, to provide amino acid building blocks for new protein synthesis. Such digestive processes are mediated by dedicated non-specific proteases and peptidases. By contrast, all organisms also possess proteases that act as processing enzymes, precisely targeting specific sites of cleavage on a more limited repertoire of proteins. Even these more specific proteases typically act upon a relatively large number of substrates, albeit in a highly site-specific manner. For example, eukaryotic proprotein convertases in the furin family are responsible for limited proteolysis and consequent maturation of perhaps hundreds of proteins within the secretory pathway (Thomas 2002); upon initiation of apoptosis, caspases cleave numerous proteins to mediate the orderly process of programmed cell death (Julien and Wells 2017). Less common are highly specific proteases that act on only a small number of proteins. In eukaryotes for example, the mitotic protease separase cleaves only four substrates to effect sister chromatid separation at the transition to anaphase (Sullivan et al. 2004).

One context in which highly specific proteases have an important role is to act as virulence factors for pathogenic bacteria. These proteases are generally secreted from the bacteria and act on either intracellular or extracellular host proteins. Substrates cleaved by these proteases play essential roles in specific physiological processes that typically limit bacterial dissemination. Because they typically are a direct cause of disease pathology, protease virulence factors are generally classified as toxins. Prominent examples are the botulinum neurotoxin (BoNT) metalloproteinases produced by *Clostridium botulinum* (Pirazzini et al. 2017). BoNT isozymes cleave components of SNARE complexes critical for exocytic neurotransmitter release in synaptic transmission, causing flaccid paralysis associated with botulism.

The various BoNT serotypes display extremely stringent, yet remarkably distinct, substrate specificities. Three serotypes (A, C and E) selectively cleave the protein SNAP-25, and four others (B, D, F and G) exclusively cleave synaptobrevin. Remarkably, even isozymes that cleave a common substrate do so at different sites with precision. Other examples of highly selective bacterial protease toxins include *Bacteroides fragilysin*, which cleaves host E-cadherin to cause intestinal pathology (Wu et al. 1998), and the subtilase cytotoxin produced by some pathogenic strains of *E. coli*, which cleaves the chaperone BiP in the host endoplasmic reticulum to potently induce cell death (Paton et al. 2006).

The metalloprotease anthrax lethal factor (LF) is one such highly specific bacterial protease virulence factor. Anthrax infections are caused by the encapsulated, spore-forming bacteria *Bacillus anthracis* (Mock and Fouet 2001). Anthrax most commonly presents as a localized cutaneous infection that responds well to antibiotics. However, ingestion of the bacteria or inhalation of aerosolized spores causes a systemic, deadly infection. The high mortality rate associated with systemic anthrax has been attributed to the production of circulating toxins by the bacteria. *B. anthracis* secretes three plasmid-encoded toxin proteins: LF, edema factor (EF), and protective antigen (PA). These proteins function in pairs, with lethal toxin (LeTx) being the combination of LF and PA, and edema toxin (EdTx) being the combination of EF and PA. A variety of evidence supports an important role for the two toxins in anthrax pathogenesis. Intravenous injection of either LeTx or EdTx is sufficient to kill a variety of experimental animals (Moayeri et al. 2004; Klein et al. 1962; Firoved et al. 2005; Newman et al. 2010). PA is the immunogen in effective anthrax vaccines and is the target of the humanized monoclonal antibody obiltoxaximab, which has been approved by the FDA as anthrax therapy (Mohamed et al. 2005; Cybulski Jr. et al. 2009). Finally, in mouse models of the disease, strains harboring deletions in any of the toxin genes are attenuated, with LF-and PA-deficient strains being completely avirulent (Pezard et al. 1991; Liu et al. 2010).

LeTx and EdTx are classical A-B toxins in which an enzymatic subunit (LF or EF) functions with a targeting subunit (PA) required for uptake into host cells. The pathway for toxin uptake has been described in detail elsewhere (Abrami et al. 2005). Briefly, PA binds to one of two related receptors, TEM8 and CMG2, on the surface of target cells. Following limited proteolysis on the cell surface, PA molecules oligomerize into heptameric and octameric rings. LF and EF share a homologous N-terminal domain that binds directly to PA oligomers. LF-PA and EF-PA complexes are internalized by clathrin-mediated endocytosis. Acidification of endocytic vesicles triggers the transition of the oligomeric PA ring into a membrane-inserted pore. LF and EF are translocated through the pore, either directly into the cytosol or indirectly through multivesicular bodies that fuse with the endosomal membrane. The small size of the PA pore indicates that LF and EF must partially unfold for translocation to occur. There is ample evidence for cytosolic activity of both enzymes, indicating that refolding occurs in the cytosol. Whether the refolding process is spontaneous or requires the assistance of host factors has not been determined.

Though they share a region required for entry into host cells, LF and EF have completely distinct catalytic activities (Turk 2007). EF is a Ca^{2+}/calmodulin-activated adenylyl cyclase (AC) that acts by flooding the cell with high levels of cyclic AMP (cAMP). The biochemistry and biology of EF has been covered in other reviews and will not be discussed further here (Moayeri and Leppla 2009; Tang and Guo 2009). LF is a Zn^{2+}-dependent metalloprotease whose primary substrates are protein kinases in the mitogen-activated protein kinase kinase (MKK) family (Duesbery et al. 1998; Vitale et al. 1998, 2000). MKKs are dual specificity threonine/tyrosine kinases that lie in the middle of three-component mitogen-activated protein kinase (MAPK) phosphorylation cascades, which have key roles in signal transduction in essentially all human cells (Johnson and Lapadat 2002). MAPK cascades consist of an upstream kinase (MKKK or MAP3K) that activates an MKK through direct phosphorylation. The MKK in turn phosphorylates and activates its downstream MAPK. Activated MAPKs phosphorylate hundreds of substrates that act as their effectors in mediating changes to cellular physiology and gene transcription. Humans have four distinct MAPK pathways (the ERK1/2, ERK5, JNK and p38 MAPK cascades) that differ in their activating stimuli and signaling output. LF cleavage functionally inactivates MKKs to completely abolish downstream signaling. LF acts on six of the seven human MKKs (all but MEK5), thus blocking the ERK1/2, JNK, and p38 MAPK pathways in target cells. Blockade of MAPK pathways is a general theme in microbial pathogenesis, being targeted by virulence factors from a number of bacteria. For example, the *Yersinia* YopJ and *Vibrio parahaemolyticus* VopA proteins are acetyltransferases that modify Ser residues on MKKs that are normally phosphorylated by MKKKs, thereby preventing activation (Krachler et al. 2011). The *Shigella* OspF proteins are phosphothreonine lyases that catalyze β-elimination of a phosphorylated residue on the ERK MAPKs, causing their irreversible inactivation (Li et al. 2007). The recurrent inhibition of MAPK pathways in promoting virulence reflects their essential roles in maintaining host defense against bacterial infections.

How LeTx contributes to the severity of anthrax infection has been an area of active research and some debate. In keeping with the ubiquitous roles for MAPK pathways in mammalian cells, LeTx can functionally impair a multitude of cell types *in vitro*, including cells of the innate and adaptive immune system, epithelial cells, and cells of the vascular system (Moayeri and Leppla 2009). Early studies suggested macrophages as an important target of LeTx (Friedlander 1986). LeTx potently and rapidly kills macrophages from some, though not all, inbred strains of mice and rats (Newman et al. 2010; Watters and Dietrich 2001). While initially assumed to reflect general toxicity due to inactivation of multiple MKKs, LeTx-induced macrophage death is now recognized to proceed independently of MAPK pathways, through a type of programmed cell death called pyroptosis

(Fink et al. 2008). Pyroptosis, a process specific to macrophages and other cells of the innate immune system, is associated with inflammatory signaling pathways culminating in activation of caspase-1. Insight into mechanisms underlying LeTx-induced pyroptosis emerged from genetic studies of mouse macrophage susceptibility to the toxin. Strains of mice with LeTx-susceptible macrophages were found to have specific alleles of the gene encoding NACHT domain, leucine rich repeat and pyrin domain containing protein 1B (NLRP1B) (Boyden and Dietrich 2006). NLRP1B belongs to a family of intracellular ATPases that recognize various conserved components of infectious microbes termed pathogen-associated molecular patterns (PAMPs) (Tschopp et al. 2003). NLRPs are components of inflammasomes, large multiprotein complexes that induce caspase-1 activation in response to the presence of PAMPs. LF directly cleaves NLRP1B to trigger inflammasome activation and consequent pyroptosis (Levinsohn et al. 2012; Hellmich et al. 2012; Chavarria-Smith and Vance 2013). Somewhat counterintuitively, macrophage sensitivity among mouse strains correlates inversely with susceptibility to challenge with live anthrax spores (Welkos et al. 1986). Accordingly, macrophage sensitivity to LeTx has been proposed to be an evolutionary adaptive strategy to limit infection. Because anthrax spores germinate in macrophages, their rapid lytic death may curtail the outgrowth of vegetative bacteria. Though well-established in rodents, LF does not cleave human NLRP1 or induce pyroptosis in human macrophages.

Though MAPK pathways are important for the proper functioning of many cell types, critical cellular targets of LeTx *in vivo* have only recently come to light. Much insight into the specific tissues targeted by LeTx (and EdTx) has come from the use of genetically engineered mice in which the gene encoding the major toxin receptor CMG2 has been conditionally knocked out (Liu et al. 2009, 2010, 2013). Ablation of CMG2 in the myeloid lineage (including macrophages and neutrophils) rendered mice completely resistant to infection with *B. anthracis* spores (Liu et al.

2010). Depletion of neutrophils re-sensitized these mice to anthrax spores, suggesting this cell type to be a major cellular target of the toxins *in vivo*. In keeping with these observations, prior *in vitro* studies indicated that LeTx inhibition of the ERK and p38 MAPK pathways impairs neutrophil superoxide production and actin-based motility, processes important for bacterial killing (Crawford et al. 2006; During et al. 2005). However, it should be noted that by elevating cAMP levels, EdTx also functionally impairs neutrophils and other myeloid cells, and is likely to cooperate with LeTx in establishing infection (Baldari et al. 2006). For example, EdTx alone inhibits neutrophil phagocytosis (O'Brien et al. 1985), and it acts synergistically with LeTx in blocking superoxide production (Wright and Mandell 1986). Furthermore, LeTx and EdTx inhibit the production of distinct cytokines by dendritic cells, suggesting that they cooperate to impair host defense (Tournier et al. 2005; Agrawal et al. 2003). Notably, myeloid-specific CMG2 deletion had no effect on the susceptibility of mice to intravenous challenge with purified LeTx. These observations suggest that LeTx and EdTx target neutrophils and possible other myeloid cells early in infection, impairing a critical host defense mechanism that would otherwise prevent establishment of a severe, systemic infection. Surprisingly, expression of CMG2 in vascular smooth muscle cells and cardiac myocytes in mice was both necessary and sufficient to mediate death from LeTx injection (Liu et al. 2013). These studies are consistent with prior observations of LeTx-induced cardiac toxicity in mice (Moayeri et al. 2009; Kandadi et al. 2010). While the specific LF substrates mediating cardiotoxicity have not been identified, inhibition of ERK signaling through MEK1/2 cleavage could decrease survival of cardiovascular muscle cells. Of possible relevance is a report showing that the constitutive nitric oxide synthase (NOS1) is a substrate of LF (Kim et al. 2008), particularly in light of observations that genetic deletion or pharmacological inhibition of NOS1 exacerbated LeTx-induced cardiotoxicity in mice (Moayeri et al. 2009). However, cleavage of NOS1 by LF has yet to be independently verified. Collectively, these studies suggest that despite widespread effects of LeTx on cultured cells, a limited number of cell types are likely to be relevant targets *in vivo*.

As described above, prior biochemical and biological studies indicate that LF functions by cleaving a small number of protein substrates, with only the six MKKs likely relevant to human infections. In general, highly specific proteases utilize multiple mechanisms to achieve selective substrate targeting. Invariably, these mechanisms involve either direct or indirect physical interactions between protease and substrate. Here, I review of our current understanding of the specific interactions by which LF targets a remarkably narrow repertoire of substrates.

2   LF Structure and Biochemistry

LF is a 94 kDa monomeric protein that includes a 26 kDa metalloprotease catalytic domain (termed domain IV) at its C-terminus (Fig. 1a, b) (Turk 2007). The LF protease domain shares a conserved active site HExxH sequence with many groups of structurally distinct metalloproteases (Cerda-Costa and Gomis-Ruth 2014). This motif is found within an α-helix within the catalytic cleft, with the two His residues (His686 and His690) serving to coordinate a Zn^{2+} ion (Fig. 1c) (Pannifer et al. 2001). A third Zn^{2+}-coordinating residue, Glu735, is located considerably downstream in the primary sequence of LF. The catalytic mechanism of LF is likely similar to that of other Zn^{2+}-dependent metalloproteases, in which a water molecule is activated through coordination of the Zn^{2+} ion to directly hydrolyze the peptide bond of the bound substrate (Cerda-Costa and Gomis-Ruth 2014). Glu687 within the HExxH motif acts as a general base during catalysis, removing a proton from the bound water. Distinct from many other HExxH motif metalloproteases, LF has an additional essential active site residue, Tyr728, which makes polar contacts through its hydroxyl group with both the catalytic Zn^{2+} ion and the carbonyl group of the bound substrate (Tonello et al. 2004). This residue is required for high affinity Zn^{2+} binding and is positioned to act as the "oxyanion hole", stabilizing the negatively charged tetrahedral intermediate formed during amide hydrolysis. Several other metalloproteases, including the BoNTs and members of the human meprin family, have a similarly positioned Tyr residue despite having catalytic domain folds substantially different from LF (Rawlings et al. 2018).

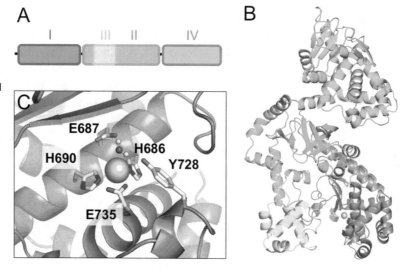

Fig. 1 LF structure. (**a**) Scheme showing the four domains of LF within its primary structure. (**b**) Three-dimensional structure of LF (PDB entry 1J7N). Domains are colored as in panel **a**. (**c**) LF catalytic center showing metal-coordinating residues, essential catalytic residues (Glu687 and Tyr728) and a water molecule bound to the active site zinc ion

LF was long thought to be a unique protease lacking homologs with structurally similar catalytic domains. Recently whole genome sequencing of bacterial isolates has revealed a large number of LF homologs among various strains of *B. cereus* and *B. thuringiensis*, the two species most closely related to *B. anthracis* (Grubbs et al. 2017). These putative proteases have 42% or lower sequence identity with LF, and most lack the N-terminal signal sequence and PA-binding domain. Presumably, these homologs have substrates distinct from those of LF, and potentially function within the bacterial cytoplasm. Recent structural studies have also identified two families of proteases with a three-dimensional fold in common with LF despite having little sequence similarity aside from the signature catalytic residues. Most closely related are a group of secreted Pro-Pro endopeptidases found in some *Bacillus* and *Clostridium* species, represented by Zmp1 of the pathogen *C. difficile* (Cafardi et al. 2013; Schacherl et al. 2015). In addition, the more distantly related MtfA family of transcriptional activators common in gram negative bacteria also share the LF fold and catalytic residues, though it has not yet been established whether they function as proteases (Xu et al. 2012).

LF has three additional domains located upstream of the catalytic domain (Fig. 1a, b). The N-terminal PA-binding domain (domain I) is necessary and sufficient for uptake of LF into cells (Arora and Leppla 1993). Interestingly, domain I has sequence and structural similarity to the metalloprotease domain (Pannifer et al. 2001). However, domain I is missing conserved residues required for catalysis and is predicted to lack enzymatic activity. The large central domain II resembles bacterial ADP-ribosyltransferase toxins, yet again lacks a key catalytic residue and is likely inactive. The function of domain II has remained elusive, but as discussed below may be important for substrate targeting. In the LF crystal structure, domains II and IV make intimate contact and appear to fold together as a single unit (Pannifer et al. 2001). Domain II may therefore be important for maintaining the structural integrity of the catalytic domain. Domain III is a small helical bundle inserted into domain II such that it is located near the catalytic cleft of domain IV. Domain III appears to be mobile, with its position dictated by ligands bound to the catalytic domain (Maize et al. 2014). In the apoenzyme structure, domain III covers a portion of the catalytic cleft, suggesting that it may play an autoinhibitory role by restricting substrate access.

3 LF Substrate Specificity: Catalytic Site Interactions

The catalytic clefts of proteases are generally broad, accommodating multiple amino acid residues on either side of the cleavage site. X-ray crystal structures of proteases in complex with peptide substrates have revealed that the peptide almost invariably binds in an extended conformation (Tyndall et al. 2005). This binding mode is facilitated by main chain hydrogen bonding interactions between the peptide and protease. LF and other HExxH motif-containing metalloproteases have an antiparallel β-sheet abutting the catalytic cleft (Cerda-Costa and Gomis-Ruth 2014) (Fig. 2a). The bound substrate effectively extends this β-sheet, providing some binding energy and enforcing the N-to-C orientation of the peptide that is likely essential for cleavage.

In addition to these universal features of protease-substrate interactions, proteases also recognize specific sequence motifs surrounding the cleavage site. By standard nomenclature, residues surrounding the cleavage site are designated ...P3-P2-P1-P1'-P2'-P3'..., where cleavage occurs between the P1 and P1' residues. The corresponding sites on the protease that interact with these residues are termed ...S3-S2-S1-S1'-S2'-S3'.... A consensus sequence recognized by LF is evident from alignment of the cleavage sites from its protein substrates (Duesbery et al. 1998; Vitale et al. 2000; Levinsohn et al. 2012; Hellmich et al. 2012; Chavarria-Smith and Vance 2013) (Table 1). This motif is characterized by a cluster of basic residues at multiple positions upstream of the cleavage site (P7 – P4), hydrophobic aliphatic residues at the P2 and P1' positions, and the

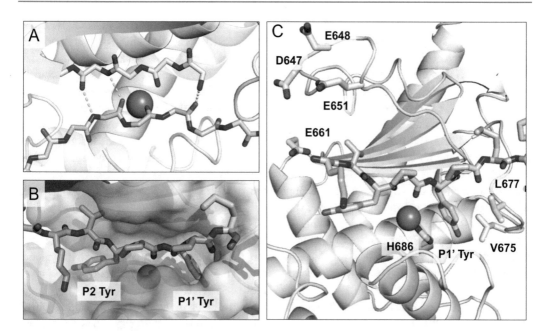

Fig. 2 LF-peptide interactions. (**a**) Main chain interactions between the β-strand abutting the LF catalytic cleft (pale blue) and a bound substrate peptide (beige) (PDB entry 1PWW). For clarity, amino acid side chains are not shown. (**b**) LF-peptide complex showing protrusion of the P1' Tyr residue into the deep S1' hydrophobic pocket. The P2 Tyr residue lies in a shallow indentation within the catalytic cleft. (**c**) LF-peptide complex showing substrate-interacting residues described in the main text

Table 1 LF cleavage sites. LF cleaves MKK4, MKK7, and mouse NLRP1B at two distinct sites. MKK sites correspond to human orthologs. Cleavage sites in NOS1 have not been precisely mapped. +, basic residue; φ, hydrophobic residue

Substrate	Cleavage site position										
	P7	P6	P5	P4	P3	P2	P1	P1'	P2'	P3'	P4'
MEK1	P	K	K	K	P	T	P	I	Q	L	N
MEK2	R	R	K	P	C	L	P	A	L	T	I
MKK3	K	R	K	K	D	L	R	I	S	C	M
MKK4	G	K	R	K	A	L	K	L	N	F	A
MKK4	P	F	K	S	T	A	R	F	T	L	N
MKK6	K	R	N	P	G	L	K	I	P	K	E
MKK7	R	P	R	P	T	L	Q	L	P	L	A
MKK7	R	P	R	H	M	L	G	L	P	S	T
NLRP1B (BALBc mouse)	E	L	K	H	R	P	K	L	E	R	H
NLRP1B (BALBc mouse)	K	L	E	R	H	L	K	L	G	M	I
NLRP1B (CDF rat)	S	K	P	R	P	R	P	L	P	R	V
Library optimized peptide	R	R	K	K	V	Y	P	Y	P	M	E
Consensus	+	+	+	+		φ	+/P	φ			

presence of either Pro or a basic residue at the P1 position. Notably, MEK5 lacks a recognizable LF consensus sequence in its N-terminus, perhaps explaining why among MKKs it is unique in being resistant to cleavage by LF. Synthetic peptide library and phage display experiments revealed a broadly similar sequence motif, with the main differences being that aromatic rather than aliphatic residues were preferred at the P2 and P1' positions (Turk et al. 2004; Li et al. 2011;

Zakharova et al. 2009). Measurement of enzyme kinetics using short peptide substrates confirmed that an optimized substrate with Tyr residues at these positions was cleaved 50-fold faster than a peptide derived from the MEK1 cleavage site (Turk et al. 2004). Longer peptides incorporating additional basic residues extending to the P7 position provided a >300-fold rate enhancement, indicating that the basic cluster contributes substantially to cleavage efficiency. Individual alanine substitutions at the P5, P2 and P1' positions within peptide substrates decreased the cleavage rate, with loss of the P5 basic residue having the most dramatic impact. In the context of full length MKK substrates, mutagenesis of the P1' hydrophobic residues to negatively charged residues completely abolished cleavage by LF, confirming this position to be key to catalytic site recognition by LF (Lee et al. 2011).

X-ray crystal structures of LF in complex with peptide substrates and peptide-derived inhibitors have provided insight into how the protease recognizes its cleavage motif (Fig. 2b). LF was first crystallized in complex with a peptide derived from the MEK2 cleavage site (Pannifer et al. 2001). However, in that structure the peptide was located proximal to, but not within, the catalytic cleft. Furthermore, the MEK2 peptide was bound in the reverse orientation compared to the canonical mode of substrate binding for HExxH motif metalloproteases. The MEK2 peptide was therefore not bound in a manner that would be compatible with substrate cleavage, and in all likelihood the structure reflected a crystallographic artifact. This artifact may have been a consequence of the low affinity of LF for peptides corresponding to its native substrates. Indeed, co-crystallography with an optimized peptide substrate selected from peptide library experiments was more successful, with the scissile amide bond positioned in close proximity to the active site Zn^{2+} ion (Turk et al. 2004). This structure revealed a hydrophobic S1' pocket defined by the Zn^{2+} ligand His686 and a loop comprising two aliphatic residues, Val675 and Leu677 (Fig. 2b). Notably, comparison with the structure of unliganded LF revealed that substrate engagement induced movement of this loop to expand the S1' pocket (Turk et al. 2004; Maize et al. 2015). The flexibility of this loop may explain why the LF S1' pocket can accommodate both smaller aliphatic and larger aromatic residues. The selectivity for hydrophobic amino acid residues at the P2 position appeared to be due to a shallow hydrophobic groove located near the catalytic center. The basis for the unique selectivity of LF for basic residues at multiple positions upstream of the cleavage site was less clear from the structure. While Lys residues at P4 and P5 could be modeled, density was not observed for Arg residues present at two positions further upstream. The flexibility provided by the long Arg and Lys sidechains may allow the peptide to assume multiple binding modes or to have a high degree of mobility. Notably, a patch of acidic residues in the catalytic domain (Asp647, Glu648, Glu651, Glu661, Fig. 2b) is present near the bound P5 Lys residue, suggesting a likely site of interaction for basic residues located further upstream.

In addition to contributing to overall cleavage efficiency, LF catalytic site specificity likely serves to direct cleavage to specific, functionally important sites within its substrate proteins (Fig. 3a). For example, LF cleaves MKKs near their N-termini, either within or just downstream of docking sites that directly interact with their respective MAPK substrates (Biondi and Nebreda 2003). LF cleavage thus disrupts the MKK-MAPK interaction, abolishing phosphorylation and activation of the MAPK (Bardwell et al. 2004). Interestingly, the sequence motif required for efficient binding to MAPKs is strikingly similar to the LF cleavage site consensus, consisting of a patch of basic residues upstream of a φ-x-φ sequence (where φ is a hydrophobic residue). This similarity is unlikely to be coincidental. Most mutations to MKKs that would abolish cleavage by LF would also be functionally deleterious, precluding a potential evolutionary strategy to protect against anthrax infections. As with the MKKs, LF cleavage sites in NLRP1B also fall close to its N-terminus (Fig. 3a). Cleavage within this region appears to be critical for inflammasome activation by LF. Indeed, N-terminal proteolysis has been proposed to be

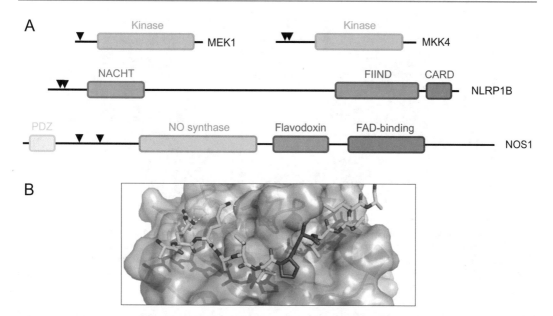

Fig. 3 LF substrates. (**a**) Schemes showing location of cleavage sites and folded domains in the primary structure of LF substrates. MKK substrates are represented by MEK1 and MKK7. Note that NOS1 has not been verified to be a true substrate of LF. (**b**) X-ray co-crystal structure of the complex between the MAPK ERK2 and the docking motif from MEK2 (PDB entry 4H3Q). Residues in magenta flank the LF cleavage site

a general activating stimulus for NLRP1 homologs, including human, potentially through the action of other microbial proteases (Chavarria-Smith et al. 2016). While cleavage near the N-terminus is of clear functional significance for LF targets, it does not appear that the protease specifically recognizes protein N-termini as MEK1, MKK6, and NLRP1B constructs produced as N-terminal fusion proteins are still efficient substrates (Bannwarth et al. 2012; Goldberg et al. 2017). Furthermore, LF cleaves NOS1 at sites downstream of its folded PDZ domain, though it is as yet unclear how its cleavage rate compares with the well-validated MKK substrates (Kim et al. 2008) (Fig. 3a).

4   LF–Substrate Exosite Interactions

The suboptimal nature of LF cleavage sites in MKKs suggests that additional interactions are important for efficient proteolysis. Furthermore, there are thousands of occurrences of the LF cleavage motif in the human proteome, indicating that catalytic site interactions alone cannot confer selectivity for specific protein substrates. Notably, MEK2 was initially discovered to be an LF substrate through a yeast two-hybrid screen, and the LF-interacting clone corresponded to a truncated kinase lacking the N-terminal cleavage sites (Vitale et al. 1998). Subsequent experiments demonstrated that LF cleaves MKK constructs lacking the catalytic domain with extremely low efficiency (Bannwarth et al. 2012; Chopra et al. 2003). Collectively, these observations indicate that MKKs have an LF-binding site, distal from the site of proteolysis, that strongly promotes cleavage. Correspondingly, LF itself must have an exosite (a region outside of the catalytic cleft) that serves to recruit substrates. In this section, I describe efforts to identify regions of LF and MKKs involved in this exosite interaction.

4.1   Mapping of the LF Exosite Region

The first report attempting to map the LF exosite noted several patches of hydrophobic residues on

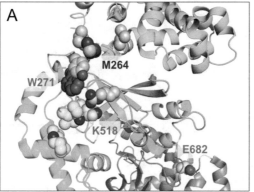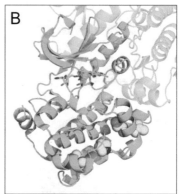

Fig. 4 LF exosite interactions. (**a**) Residues required for optimal cleavage of at least some LF substrates are shown in spacefill view. Residues in yellow were mapped by Liang et al. (2004), and those in beige by Goldberg et al. (2017). Labeled residues in magenta, blue and white have substrate-selective effects on cleavage as described in the main text. (**b**) Putative LF-interacting residues (yellow) mapped onto the X-ray crystal structure of MEK1 (green) in complex with BRAF (cyan) (PDB entry 4MNE)

the protein surface (Liang et al. 2004). Because hydrophobic residues are generally buried within globular proteins, these surface-exposed patches were surmised to be potential protein interaction sites. Mutation of some residues within and near a hydrophobic patch on domain II (Fig. 4a, residues in yellow) dramatically reduced toxicity to macrophages and cleavage of MEK2 in cells. Though not tested explicitly, the observation that these mutants could not kill macrophages suggests that they were also impaired for NLRP1B cleavage. Curiously, each of the mutants was able to cleave MEK1 *in vitro*, seemingly at odds with observations in cultured cells. It is possible that mutations within this region destabilize LF so that it fails to refold following translocation into cells. However, it would seem unlikely that mutating any of a number of surface residues of different properties (Asn, Leu, Lys, Arg) would produce such severe effects on protein stability. Perhaps the most likely explanation for the discrepancy is that the high concentrations of LF used for the *in vitro* cleavage assays masked authentic decreases in cleavage efficiency. Alternatively, the mutants could have selectively impaired cleavage of MEK2 and not MEK1. On balance, it is likely that this patch of residues contributes to the interaction between LF and at least MEK2 and NLRP1B.

Two subsequent studies identified additional residues on LF that were important for cleavage of only a subset of its substrates. The Mogridge group identified LF-K518E,E682G in a screen for mutants defective in causing macrophage pyroptosis (Ngai et al. 2010) (Fig. 4a, residues in white). This mutant was also impaired it its ability to downregulate MAPK pathways in cells, with the p38 MAPK pathway being most resistant to inhibition. Subsequent analysis of MKK cleavage in cells revealed that cleavage of MEK2 and MKK6 were most severely decreased, while MEK1, MKK3 and MKK4 were largely unaffected. This mutant provided the first evidence that LF may harbor exosite(s) that is utilized in a substrate-specific manner. Lys518 notably falls near the region of domain II required for macrophage cell death and MEK2 cleavage described by Liang et al. above, suggesting that it could be a part of a continuous substrate interaction surface. However, it should be noted that Glu682 lies within the S1' pocket of LF and could conceivably affect substrate proteolysis by affecting cleavage site recognition. Analysis of individual point mutants or comparative analysis of peptide substrates will be necessary to determine whether the substrate-selective impact of the double mutant is due to perturbation of exosite or catalytic site interactions.

Recently my group used algorithms designed to predict protein interaction interfaces to identify sites of LF-substrate interactions (Goldberg et al. 2017). These programs correctly identified the catalytic cleft in domain IV, as well as the PA-binding site in domain I, as predicted protein interaction sites. In addition to these known interaction interfaces, a surface spanning domains I and II was also predicted to be a site of protein-protein interaction. Indeed, we observed that deletion of domain I reduced MKK cleavage by LF approximately ten-fold, yet had no effect on cleavage of a peptide substrate. This observation was surprising given that domain I had been thought to function solely in toxin uptake into cells. Further point mutagenesis indicated that residues at the hinge joining domains I and II and the N-terminal helix of domain II (Fig. 4a, residues in beige, blue and magenta) were most important for LF cleavage of MEK1, MKK4 and MKK6 in vitro as well as MEK2 and MKK3 in LeTx-treated cells. As was observed in the study cited above, certain point mutations differentially affected cleavage of specific substrates. Most strikingly, Trp271 (Fig. 4a, magenta) was essential for cleavage of MKK3, MKK4 and MKK6 but was completely dispensable for cleavage of MEK1, MEK2 and NLRP1B. Similarly, mutation of Met264 (Fig. 4a, blue) abolished NLRP1B cleavage while having less dramatic effects on MKK cleavage. Taken as a whole, the three studies described above identified a continuous surface on LF that appears to harbor a critical exosite. While all substrates appear to interact with this surface, the precise mode of binding must differ among LF substrates, leading to selective requirements for specific residues within the region.

4.2 Sites on MKKs Implicated in Interactions with LF

Additional studies have attempted to identify the regions of MKKs that bind to the LF exosite(s). Early observations that N-terminally truncated forms of MEK1 interact directly with LF pointed to the MKK catalytic domain as a likely site of interaction (Vitale et al. 1998). My group found using MEK1/MKK6 chimeric proteins that the catalytic domain dictated susceptibility to cleavage by the LF-W271A mutant. This observation is consistent with an interaction between the MKK kinase domain and the exosite in domain II (Goldberg et al. 2017). To more precisely map the LF-interacting region, the Duesbery lab examined the ability of LF to cleave a series of deletion and truncation mutants of MEK1 (Chopra et al. 2003). They found that helix αG of the kinase domain and a loop immediately downstream were required for proteolysis by LF. Point mutation of four residues in this region in either MEK1 or MKK6 also reduced cleavage (Fig. 4b). Because three of these residues are buried within the hydrophobic core, their mutation may have affected cleavage by perturbing the local or overall structure of the catalytic domain. It is notable, however, that this region forms part of the binding interface with BRAF, the kinase that activates MEK1 (Haling et al. 2014). In addition, a protease inactive LF mutant can block BRAF phosphorylation of MEK1, consistent with the notion that LF and BRAF have an overlapping LF-binding interface (Chopra et al. 2003).

5 Exploiting Substrate Interactions for LF Inhibitor Design and Discovery

Antibiotics fail to cure most advanced cases of anthrax, and accordingly there has been interest in developing anti-toxin drugs to be used in combination with conventional antimicrobials. Given prior success in therapeutic targeting of proteases, substantial effort has been made toward the development of LF inhibitors (Goldberg and Turk 2016). Insight into LF substrate specificity has facilitated the discovery and design of such inhibitors. Almost all approved protease inhibitors are substrate competitive and mechanism based (Turk 2006). A common starting point in the development of protease inhibitors is to append a group that interacts covalently with the protease active site to either the N-terminal or C-terminal portion of a substrate peptide.

Metalloprotease inhibitors generally incorporate a metal chelating group that coordinates the active site Zn^{2+} ion. The most potent peptide-derived LF inhibitors incorporate a relatively long (~10-mer) N-terminal peptide sequence, to which a Zn^{2+}-binding hydroxamic acid group is appended to the C-terminus (Tonello et al. 2002). Incorporation of amino acid residues selected from substrate-based peptide libraries has led to increased potency of these inhibitors (Turk et al. 2004; Li et al. 2011). Lower molecular weight, more drug-like hydroxamic acids have been generated that primarily target the hydrophobic S1' pocket with an aromatic group, and these compounds have shown efficacy in animal models of anthrax (Shoop et al. 2005). Interestingly, subsequent elaboration of these compounds to append a group interacting with the S3 pocket produced a series of highly potent and selective LF inhibitors (Kim et al. 2011; Jiao et al. 2010, 2012). By accessing sites on both sides of the scissile bond, these compounds combined the high potency of long peptide-based hydroxamates with the favorable pharmacology of low molecular weight compounds, and would appear the most likely candidates for clinical advancement.

The presence of an exosite on LF provides an alternative to targeting the active site in inhibitor development. Targeting exosites offers a potential means to produce inhibitors more selective than canonical mechanism-based inhibitors. For example, metal chelating LF inhibitors incorporating a hydrophobic P1' group are prone to cross-react with host matrix metalloproteases and related enzymes (Turk et al. 2004). A variety of strategies have been used to produce exosite-targeting inhibitors of other proteases, including phage display selection of high affinity protease-binding peptides or antibodies (Bjorklund et al. 2004; Dennis et al. 2000; Atwal et al. 2011). My group recently used a high throughput screening (HTS) approach to identify exosite-binding LF inhibitors (Bannwarth et al. 2012). Exosite inhibitors are generally difficult to discover by HTS, in large part because short peptides are used as substrates, biasing strongly towards the discovery of substrate-competitive inhibitors. As an alternative, we developed an assay suitable for HTS that used full length MKK6 as a substrate.

From a screen of several thousand compounds, we identified the natural product stictic acid to inhibit LF cleavage of MKKs but not short peptide substrates (Bannwarth et al. 2012). This compound and related natural products also inhibited LeTx-induced macrophage cytolysis. While these compounds may lack drug-like properties, their discovery indicates that the exosite can be targeted by small molecule inhibitors.

6 Conclusions and Future Perspectives

Substrate recognition by LF involves a combination of catalytic site and exosite interactions, both of which appear to be critical for cleavage of its substrates. Interactions with the catalytic cleft are currently understood in great detail owing to a combination of biochemical and structural analysis, and these insights have been exploited in the development of LF inhibitors of potential therapeutic use. In contrast, though an exosite interaction has been mapped to specific regions of LF and MKKs, details regarding the precise mode of binding are currently lacking. In addition, the structural basis of the substrate-selective requirement for specific residues within the LF exosite is not understood. An X-ray crystal structure of an LF-MKK complex has been elusive, likely owing to the low affinity nature of the enzyme-substrate complex. Mass spectrometry-based techniques such as hydrogen-deuterium exchange or chemical crosslinking may be useful in the future for more finely mapping the LF-MKK interface. These approaches may reveal additional currently unknown sites of interaction between LF and its substrates. Additional insight into how LF targets its substrates will provide a more complete understanding of molecular mechanisms underlying anthrax pathogenesis, and may facilitate further development of LF inhibitors.

Acknowledgements This work was supported by National Institutes of Health grant R01 GM104047.

Conflicts of Interest The author declares no conflict of interests.

References

Abrami L, Reig N, van der Goot FG (2005) Anthrax toxin: the long and winding road that leads to the kill. Trends Microbiol 13(2):72–78

Agrawal A, Lingappa J, Leppla SH, Agrawal S, Jabbar A, Quinn C, Pulendran B (2003) Impairment of dendritic cells and adaptive immunity by anthrax lethal toxin. Nature 424(6946):329–334

Arora N, Leppla SH (1993) Residues 1-254 of anthrax toxin lethal factor are sufficient to cause cellular uptake of fused polypeptides. J Biol Chem 268(5):3334–3341

Atwal JK, Chen Y, Chiu C, Mortensen DL, Meilandt WJ, Liu Y, Heise CE, Hoyte K, Luk W, Lu Y, Peng K, Wu P, Rouge L, Zhang Y, Lazarus RA, Scearce-Levie K, Wang W, Wu Y, Tessier-Lavigne M, Watts RJ (2011) A therapeutic antibody targeting BACE1 inhibits amyloid-beta production in vivo. Sci Transl Med 3(84):84ra43

Baldari CT, Tonello F, Paccani SR, Montecucco C (2006) Anthrax toxins: a paradigm of bacterial immune suppression. Trends Immunol 27(9):434–440

Bannwarth L, Goldberg AB, Chen C, Turk BE (2012) Identification of exosite-targeting inhibitors of anthrax lethal factor by high-throughput screening. Chem Biol 19(7):875–882

Bardwell AJ, Abdollahi M, Bardwell L (2004) Anthrax lethal factor-cleavage products of MAPK (mitogen-activated protein kinase) kinases exhibit reduced binding to their cognate MAPKs. Biochem J 378 (Pt 2):569–577

Biondi RM, Nebreda AR (2003) Signalling specificity of Ser/Thr protein kinases through docking-site-mediated interactions. Biochem J 372 .(Pt 1:1–13

Bjorklund M, Heikkila P, Koivunen E (2004) Peptide inhibition of catalytic and noncatalytic activities of matrix metalloproteinase-9 blocks tumor cell migration and invasion. J Biol Chem 279(28):29589–29597

Boyden ED, Dietrich WF (2006) Nalp1b controls mouse macrophage susceptibility to anthrax lethal toxin. Nat Genet 38(2):240–244

Cafardi V, Biagini M, Martinelli M, Leuzzi R, Rubino JT, Cantini F, Norais N, Scarselli M, Serruto D, Unnikrishnan M (2013) Identification of a novel zinc metalloprotease through a global analysis of Clostridium difficile extracellular proteins. PLoS One 8(11): e81306

Cerda-Costa N, Gomis-Ruth FX (2014) Architecture and function of metallopeptidase catalytic domains. Protein Sci 23(2):123–144

Chavarria-Smith J, Vance RE (2013) Direct proteolytic cleavage of NLRP1B is necessary and sufficient for inflammasome activation by anthrax lethal factor. PLoS Pathog 9(6):e1003452

Chavarria-Smith J, Mitchell PS, Ho AM, Daugherty MD, Vance RE (2016) Functional and evolutionary analyses identify proteolysis as a general mechanism for NLRP1 inflammasome activation. PLoS Pathog 12 (12):e1006052

Chopra AP, Boone SA, Liang X, Duesbery NS (2003) Anthrax lethal factor proteolysis and inactivation of MAPK kinase. J Biol Chem 278(11):9402–9406

Crawford MA, Aylott CV, Bourdeau RW, Bokoch GM (2006) Bacillus anthracis toxins inhibit human neutrophil NADPH oxidase activity. J Immunol 176 (12):7557–7565

Cybulski RJ Jr, Sanz P, O'Brien AD (2009) Anthrax vaccination strategies. Mol Asp Med 30(6):490–502

Dennis MS, Eigenbrot C, Skelton NJ, Ultsch MH, Santell L, Dwyer MA, O'Connell MP, Lazarus RA (2000) Peptide exosite inhibitors of factor VIIa as anticoagulants. Nature 404(6777):465–470

Duesbery NS, Webb CP, Leppla SH, Gordon VM, Klimpel KR, Copeland TD, Ahn NG, Oskarsson MK, Fukasawa K, Paull KD, Vande Woude GF (1998) Proteolytic inactivation of MAP-kinase-kinase by anthrax lethal factor. Science 280(5364):734–737

During RL, Li W, Hao B, Koenig JM, Stephens DS, Quinn CP, Southwick FS (2005) Anthrax lethal toxin paralyzes neutrophil actin-based motility. J Infect Dis 192(5):837–845

Fink SL, Bergsbaken T, Cookson BT (2008) Anthrax lethal toxin and Salmonella elicit the common cell death pathway of caspase-1-dependent pyroptosis via distinct mechanisms. Proc Natl Acad Sci U S A 105 (11):4312–4317

Firoved AM, Miller GF, Moayeri M, Kakkar R, Shen Y, Wiggins JF, McNally EM, Tang WJ, Leppla SH (2005) Bacillus anthracis edema toxin causes extensive tissue lesions and rapid lethality in mice. Am J Pathol 167(5):1309–1320

Friedlander AM (1986) Macrophages are sensitive to anthrax lethal toxin through an acid-dependent process. J Biol Chem 261(16):7123–7126

Goldberg AB, Turk BE (2016) Inhibitors of the metalloproteinase anthrax lethal factor. Curr Top Med Chem 16(21):2350–2358

Goldberg AB, Cho E, Miller CJ, Lou HJ, Turk BE (2017) Identification of a substrate-selective exosite within the metalloproteinase anthrax lethal factor. J Biol Chem 292(3):814–825

Grubbs KJ, Bleich RM, Santa Maria KC, Allen SE, Farag S, AgBiome T, Shank EA, Bowers AA (2017) Large-scale bioinformatics analysis of Bacillus genomes uncovers conserved roles of natural products in bacterial physiology. mSystems 2(6):e00040

Haling JR, Sudhamsu J, Yen I, Sideris S, Sandoval W, Phung W, Bravo BJ, Giannetti AM, Peck A, Masselot A, Morales T, Smith D, Brandhuber BJ, Hymowitz SG, Malek S (2014) Structure of the BRAF-MEK complex reveals a kinase activity independent role for BRAF in MAPK signaling. Cancer Cell 26(3):402–413

Hellmich KA, Levinsohn JL, Fattah R, Newman ZL, Maier N, Sastalla I, Liu S, Leppla SH, Moayeri M (2012) Anthrax lethal factor cleaves mouse Nlrp1b in both toxin-sensitive and toxin-resistant macrophages. PLoS One 7(11):e49741

Jiao GS, Kim S, Moayeri M, Cregar-Hernandez L, McKasson L, Margosiak SA, Leppla SH, Johnson AT (2010) Antidotes to anthrax lethal factor intoxication. Part 1: discovery of potent lethal factor inhibitors with in vivo efficacy. Bioorg Med Chem Lett 20 (22):6850–6853

Jiao GS, Kim S, Moayeri M, Crown D, Thai A, Cregar-Hernandez L, McKasson L, Sankaran B, Lehrer A, Wong T, Johns L, Margosiak SA, Leppla SH, Johnson AT (2012) Antidotes to anthrax lethal factor intoxication. Part 3: evaluation of core structures and further modifications to the C2-side chain. Bioorg Med Chem Lett 22(6):2242–2246

Johnson GL, Lapadat R (2002) Mitogen-activated protein kinase pathways mediated by ERK, JNK, and p38 protein kinases. Science 298(5600):1911–1912

Julien O, Wells JA (2017) Caspases and their substrates. Cell Death Differ 24(8):1380–1389

Kandadi MR, Hua Y, Ma H, Li Q, Kuo SR, Frankel AE, Ren J (2010) Anthrax lethal toxin suppresses murine cardiomyocyte contractile function and intracellular Ca2+ handling via a NADPH oxidase-dependent mechanism. PLoS One 5(10):e13335

Kim J, Park H, Myung-Hyun J, Han SH, Chung H, Lee JS, Park JS, Yoon MY (2008) The effects of anthrax lethal factor on the macrophage proteome: potential activity on nitric oxide synthases. Arch Biochem Biophys 472 (1):58–64

Kim S, Jiao GS, Moayeri M, Crown D, Cregar-Hernandez-L, McKasson L, Margosiak SA, Leppla SH, Johnson AT (2011) Antidotes to anthrax lethal factor intoxication. Part 2: structural modifications leading to improved in vivo efficacy. Bioorg Med Chem Lett 21 (7):2030–2033

Klein F, Hodges DR, Mahlandt BG, Jones WI, Haines BW, Lincoln RE (1962) Anthrax toxin: causative agent in the death of rhesus monkeys. Science 138:1331–1333

Krachler AM, Woolery AR, Orth K (2011) Manipulation of kinase signaling by bacterial pathogens. J Cell Biol 195(7):1083–1092

Lee CS, Dykema KJ, Hawkins DM, Cherba DM, Webb CP, Furge KA, Duesbery NS (2011) MEK2 is sufficient but not necessary for proliferation and anchorage-independent growth of SK-MEL-28 melanoma cells. PLoS One 6(2):e17165

Levinsohn JL, Newman ZL, Hellmich KA, Fattah R, Getz MA, Liu S, Sastalla I, Leppla SH, Moayeri M (2012) Anthrax lethal factor cleavage of Nlrp1 is required for activation of the inflammasome. PLoS Pathog 8(3): e1002638

Li H, Xu H, Zhou Y, Zhang J, Long C, Li S, Chen S, Zhou JM, Shao F (2007) The phosphothreonine lyase activity of a bacterial type III effector family. Science 315 (5814):1000–1003

Li F, Terzyan S, Tang J (2011) Subsite specificity of anthrax lethal factor and its implications for inhibitor development. Biochem Biophys Res Commun 407 (2):400–405

Liang X, Young JJ, Boone SA, Waugh DS, Duesbery NS (2004) Involvement of domain II in toxicity of anthrax lethal factor. J Biol Chem 279(50):52473–52478

Liu S, Crown D, Miller-Randolph S, Moayeri M, Wang H, Hu H, Morley T, Leppla SH (2009) Capillary morphogenesis protein-2 is the major receptor mediating lethality of anthrax toxin in vivo. Proc Natl Acad Sci U S A 106(30):12424–12429

Liu S, Miller-Randolph S, Crown D, Moayeri M, Sastalla I, Okugawa S, Leppla SH (2010) Anthrax toxin targeting of myeloid cells through the CMG2 receptor is essential for establishment of Bacillus anthracis infections in mice. Cell Host Microbe 8 (5):455–462

Liu S, Zhang Y, Moayeri M, Liu J, Crown D, Fattah RJ, Wein AN, Yu ZX, Finkel T, Leppla SH (2013) Key tissue targets responsible for anthrax-toxin-induced lethality. Nature 501(7465):63–68

Maize KM, Kurbanov EK, De L, Mora-Rey T, Geders TW, Hwang DJ, Walters MA, Johnson RL, Amin EA, Finzel BC (2014) Anthrax toxin lethal factor domain 3 is highly mobile and responsive to ligand binding. Acta Crystallogr D Biol Crystallogr 70. (Pt 11:2813–2822

Maize KM, Kurbanov EK, Johnson RL, Amin EA, Finzel BC (2015) Ligand-induced expansion of the S1' site in the anthrax toxin lethal factor. FEBS Lett 589(24 Pt B):3836–3841

Moayeri M, Leppla SH (2009) Cellular and systemic effects of anthrax lethal toxin and edema toxin. Mol Asp Med 30(6):439–455

Moayeri M, Martinez NW, Wiggins J, Young HA, Leppla SH (2004) Mouse susceptibility to anthrax lethal toxin is influenced by genetic factors in addition to those controlling macrophage sensitivity. Infect Immun 72 (8):4439–4447

Moayeri M, Crown D, Dorward DW, Gardner D, Ward JM, Li Y, Cui X, Eichacker P, Leppla SH (2009) The heart is an early target of anthrax lethal toxin in mice: a protective role for neuronal nitric oxide synthase (nNOS). PLoS Pathog 5(5):e1000456

Mock M, Fouet A (2001) Anthrax. Annu Rev Microbiol 55:647–671

Mohamed N, Clagett M, Li J, Jones S, Pincus S, D'Alia G, Nardone L, Babin M, Spitalny G, Casey L (2005) A high-affinity monoclonal antibody to anthrax protective antigen passively protects rabbits before and after aerosolized Bacillus anthracis spore challenge. Infect Immun 73(2):795–802

Newman ZL, Printz MP, Liu S, Crown D, Breen L, Miller-Randolph S, Flodman P, Leppla SH, Moayeri M (2010) Susceptibility to anthrax lethal toxin-induced rat death is controlled by a single chromosome 10 locus that includes rNlrp1. PLoS Pathog 6(5): e1000906

Ngai S, Batty S, Liao KC, Mogridge J (2010) An anthrax lethal factor mutant that is defective at causing pyroptosis retains proapoptotic activity. FEBS J 277 (1):119–127

O'Brien J, Friedlander A, Dreier T, Ezzell J, Leppla S (1985) Effects of anthrax toxin components on human neutrophils. Infect Immun 47(1):306–310

Pannifer AD, Wong TY, Schwarzenbacher R, Renatus M, Petosa C, Bienkowska J, Lacy DB, Collier RJ, Park S, Leppla SH, Hanna P, Liddington RC (2001) Crystal structure of the anthrax lethal factor. Nature 414 (6860):229–233

Paton AW, Beddoe T, Thorpe CM, Whisstock JC, Wilce MC, Rossjohn J, Talbot UM, Paton JC (2006) AB5 subtilase cytotoxin inactivates the endoplasmic reticulum chaperone BiP. Nature 443(7111):548–552

Pezard C, Berche P, Mock M (1991) Contribution of individual toxin components to virulence of *Bacillus anthracis*. Infect Immun 59(10):3472–3477

Pirazzini M, Rossetto O, Eleopra R, Montecucco C (2017) Botulinum neurotoxins: biology, pharmacology, and toxicology. Pharmacol Rev 69(2):200–235

Rawlings ND, Barrett AJ, Thomas PD, Huang X, Bateman A, Finn RD (2018) The MEROPS database of proteolytic enzymes, their substrates and inhibitors in 2017 and a comparison with peptidases in the PANTHER database. Nucleic Acids Res 46(D1):D624–D632

Schacherl M, Pichlo C, Neundorf I, Baumann U (2015) Structural basis of proline-proline peptide bond specificity of the metalloprotease Zmp1 implicated in motility of *Clostridium difficile*. Structure 23(9):1632–1642

Shoop WL, Xiong Y, Wiltsie J, Woods A, Guo J, Pivnichny JV, Felcetto T, Michael BF, Bansal A, Cummings RT, Cunningham BR, Friedlander AM, Douglas CM, Patel SB, Wisniewski D, Scapin G, Salowe SP, Zaller DM, Chapman KT, Scolnick EM, Schmatz DM, Bartizal K, MacCoss M, Hermes JD (2005) Anthrax lethal factor inhibition. Proc Natl Acad Sci U S A 102(22):7958–7963

Sullivan M, Hornig NC, Porstmann T, Uhlmann F (2004) Studies on substrate recognition by the budding yeast separase. J Biol Chem 279(2):1191–1196

Tang WJ, Guo Q (2009) The adenylyl cyclase activity of anthrax edema factor. Mol Asp Med 30(6):423–430

Thomas G (2002) Furin at the cutting edge: from protein traffic to embryogenesis and disease. Nat Rev Mol Cell Biol 3(10):753–766

Tonello F, Seveso M, Marin O, Mock M, Montecucco C (2002) Screening inhibitors of anthrax lethal factor. Nature 418(6896):386

Tonello F, Naletto L, Romanello V, Dal Molin F, Montecucco C (2004) Tyrosine-728 and glutamic acid-735 are essential for the metalloproteolytic activity of the lethal factor of *Bacillus anthracis*. Biochem Biophys Res Commun 313(3):496–502

Tournier JN, Quesnel-Hellmann A, Mathieu J, Montecucco C, Tang WJ, Mock M, Vidal DR, Goossens PL (2005) Anthrax edema toxin cooperates with lethal toxin to impair cytokine secretion during infection of dendritic cells. J Immunol 174 (8):4934–4941

Tschopp J, Martinon F, Burns K (2003) NALPs: a novel protein family involved in inflammation. Nat Rev Mol Cell Biol 4(2):95–104

Turk B (2006) Targeting proteases: successes, failures and future prospects. Nat Rev Drug Discov 5(9):785–799

Turk BE (2007) Manipulation of host signalling pathways by anthrax toxins. Biochem J 402(3):405–417

Turk BE, Wong TY, Schwarzenbacher R, Jarrell ET, Leppla SH, Collier RJ, Liddington RC, Cantley LC (2004) The structural basis for substrate and inhibitor selectivity of the anthrax lethal factor. Nat Struct Mol Biol 11(1):60–66

Tyndall JDA, Nall T, Fairlie DP (2005) Proteases universally recognize beta strands in their active sites. Chem Rev 105(3):973–999

Vitale G, Pellizzari R, Recchi C, Napolitani G, Mock M, Montecucco C (1998) Anthrax lethal factor cleaves the N-terminus of MAPKKs and induces tyrosine/threonine phosphorylation of MAPKs in cultured macrophages. Biochem Biophys Res Commun 248 (3):706–711

Vitale G, Bernardi L, Napolitani G, Mock M, Montecucco C (2000) Susceptibility of mitogen-activated protein kinase kinase family members to proteolysis by anthrax lethal factor. Biochem J 352(Pt 3):739–745

Watters JW, Dietrich WF (2001) Genetic, physical, and transcript map of the *Ltxs1* region of mouse chromosome 11. Genomics 73(2):223–231

Welkos SL, Keener TJ, Gibbs PH (1986) Differences in susceptibility of inbred mice to *Bacillus anthracis*. Infect Immun 51(3):795–800

Wright GG, Mandell GL (1986) Anthrax toxin blocks priming of neutrophils by lipopolysaccharide and by muramyl dipeptide. J Exp Med 164(5):1700–1709

Wu S, Lim KC, Huang J, Saidi RF, Sears CL (1998) *Bacteroides fragilis* enterotoxin cleaves the zonula adherens protein, E-cadherin. Proc Natl Acad Sci U S A 95(25):14979–14984

Xu Q, Gohler AK, Kosfeld A, Carlton D, Chiu HJ, Klock HE, Knuth MW, Miller MD, Elsliger MA, Deacon AM, Godzik A, Lesley SA, Jahreis K, Wilson IA (2012) The structure of Mlc titration factor A (MtfA/YeeI) reveals a prototypical zinc metallopeptidase related to anthrax lethal factor. J Bacteriol 194 (11):2987–2999

Zakharova MY, Kuznetsov NA, Dubiley SA, Kozyr AV, Fedorova OS, Chudakov DM, Knorre DG, Shemyakin IG, Gabibov AG, Kolesnikov AV (2009) Substrate recognition of anthrax lethal factor examined by combinatorial and pre-steady-state kinetic approaches. J Biol Chem 284(27):17902–17913

Adv Exp Med Biol - Protein Reviews (2019) 20: 205–218
https://doi.org/10.1007/5584_2018_289
© Springer Nature Switzerland AG 2018
Published online: 10 November 2018

Salmonella, E. coli, and Citrobacter Type III Secretion System Effector Proteins that Alter Host Innate Immunity

Samir El Qaidi, Miaomiao Wu, Congrui Zhu,
and Philip R. Hardwidge

Abstract

Bacteria deliver virulence proteins termed 'effectors' to counteract host innate immunity. Protein-protein interactions within the host cell ultimately subvert the generation of an inflammatory response to the infecting pathogen. Here we briefly describe a subset of T3SS effectors produced by enterohemorrhagic *Escherichia coli* (EHEC), enteropathogenic *E. coli* (EPEC), *Citrobacter rodentium*, and *Salmonella enterica* that inhibit innate immune pathways. These effectors are interesting for structural and mechanistic reasons, as well as for their potential utility in being engineered to treat human autoimmune disorders associated with perturbations in NF-κB signaling.

Keywords

Bacterial pathogenesis · Effector · Innate immunity · Type III secretion · Virulence

1   Introduction

Bacterial pathogens must circumvent both the host immune system and competition from commensal bacteria to establish an infection. Innate

S. El Qaidi, M. Wu, C. Zhu, and P. R. Hardwidge (✉)
College of Veterinary Medicine, Kansas State University,
Manhattan, KS, USA
e-mail: hardwidg@vet.k-state.edu

immunity is a primary stage of defense against pathogens (Medzhitov and Janeway Jr 1997). In addition to physical and chemical barriers, the innate immune system also consists of surface proteins classified as pattern recognition receptors (PRRs) that sense components of pathogens known as pathogen-associated molecular patterns (PAMPs) (Suresh and Mosser 2013). These receptor-ligand associations lead to the induction of the NF-κB signaling pathway that ultimately culminates in the production of pro-inflammatory cytokines (Liu et al. 2017). In turn, bacterial pathogens have evolved strategies to target and inhibit NF-κB signaling pathway to promote their colonization (Le Negrate 2012).

Bacteria have evolved secretion systems to deliver arsenals of virulence proteins termed 'effectors' to counteract host innate immunity. The type three secretion system (T3SS) is a syringe-like secretion apparatus that injects effectors that interact with host components including proteins and small molecules such as inositol phosphate (Salomon et al. 2013). These interactions ultimately subvert the generation of an inflammatory response to the infecting pathogen (Daniell et al. 2001). However, the mere action of the T3SS can also lead to NF-κB activation (Litvak et al. 2017). Here we briefly describe a subset of T3SS effectors produced by enterohemorrhagic *Escherichia coli* (EHEC), enteropathogenic *E. coli* (EPEC), *Citrobacter rodentium*, and

Salmonella enterica that inhibit innate immune pathways.

EPEC is an attaching/effacing (A/E) pathogen that causes infantile diarrhea, primarily in developing countries (Clarke et al. 2002). EPEC adherence is mediated by an interaction between the bacterial Tir and intimin proteins to create intimate contact with host intestinal cells (Kenny et al. 1997). Polymerized actin accumulates under the A/E lesion, forming a pedestal-like structure (Campellone and Leong 2003). The EPEC chromosome harbors the locus of enterocyte effacement (LEE), a pathogenicity island that encodes a T3SS and multiple effectors (McDaniel et al. 1995). LEE encoded effectors are generally designated by the term 'Esp' to denote EPEC secreted proteins, while effectors encoded outside of the LEE are generally designated as 'Nles', to denote non-LEE encoded effectors. However, for many effectors the nomenclature remains ambiguous, as some effectors designated as 'Esps' are encoded outside of the LEE.

EHEC is a human pathogen that causes hemorrhagic colitis and hemolytic uremic syndrome (HUS) (Newell and La Ragione 2018). EHEC also contains the LEE and Nles, but also encodes Shiga-like toxins that are responsible for causing more serious disease.

C. rodentium is a natural pathogen of mice that was discovered during a collapse of mouse colonies due to transmissible murine colonic hyperplasia (TMCH) (Luperchio et al. 2000). The genomic and mechanistic similarities between *C. rodentium* and EPEC have made *C. rodentium* a model system for studying A/E pathogens *in vivo* (Silberger et al. 2017).

Salmonella enterica remains one of most frequently reported causes of foodborne illness (Yombi et al. 2015). Ingestion of contaminated foods, especially poultry, is the primary source of infection, although contaminated water and direct contact with animals are also risk factors. *Salmonella* encodes multiple T3SSs. T3SS1, encoded by the *Salmonella* pathogenicity island (SPI)-1, has many functions, including the invasion of epithelial cells and the induction of an inflammatory response at the intestinal surface (Marcus et al. 2000). T3SS2 is encoded by SPI-2 and is a

key virulence determinant for systemic disease in mice (Ochman et al. 1996). T3SS2 is induced intracellularly (Hensel 2000) and is required for *Salmonella*-containing vacuole (SCV) biogenesis (Steele-Mortimer 2008) (Fig. 1).

1.1 *E. coli* and *Citrobacter* Effectors

NleA inhibits Nod-Like Receptor 3 (NLRP3)-mediated inflammasome activation via a direct association with NLRP3. NleA binding to NLRP3 prevents its deubiquitination, which is required for inflammasome assembly and activation (Yen et al. 2015). As a consequence, interleukin 1 beta (IL-1β) secretion decreases due to caspase 1 inhibition (Yen et al. 2015). NleA also binds the Sec24 subunit of the mammalian coatamer protein II complex (COPII) to inhibit vesicle trafficking (Kim et al. 2007).

NleB is a glycosyltransferase that glycosylates host proteins with N-acetylglucosamine (GlcNAc) on arginine residues (El Qaidi et al. 2017; Li et al. 2013; Pearson et al. 2013). Deleting *nleB* significantly attenuates *C. rodentium* virulence (Kelly et al. 2006). While a single *nleB* gene is found in *C. rodentium*, two *nleB* genes, *nleB1* and *nleB2*, are found in most EPEC and EHEC strains. EPEC NleB1 glycosylates the tumor necrosis factor receptor (TNFR) type 1-associated DEATH domain protein (TRADD) and the FAS-associated death domain-containing protein (FADD) (Li et al. 2013; Pearson et al. 2013), with preferential affinity for FADD (Scott et al. 2017). Glycosylation of TRADD inhibits its dimerization, thus preventing the proper assembly of the TNFR complex and the activation of the NF-κB pathway (Li et al. 2013; Pearson et al. 2013). EHEC NleB1 and *C. rodentium* NleB also glycosylate glyceraldehyde-3-phosphate dehydrogenase (GAPDH) (El Qaidi et al. 2017; Gao et al. 2013), which reduces GAPDH binding to TRAF2, subsequently limiting the extent of TRAF2 ubiquitination, leading to reduced NF-κB activity (Gao et al. 2013).

NleC is a zinc metalloprotease that cleaves the NF-κB p65 subunit, leading to NF-κB pathway inhibition (Baruch et al. 2011; Giogha et al. 2015;

A

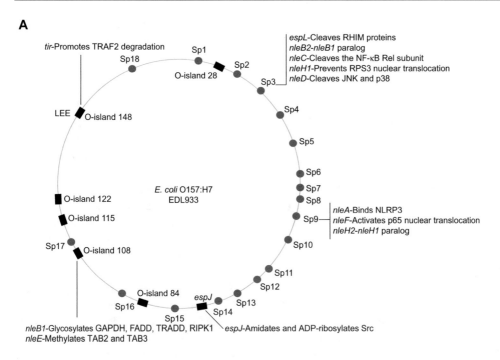

B

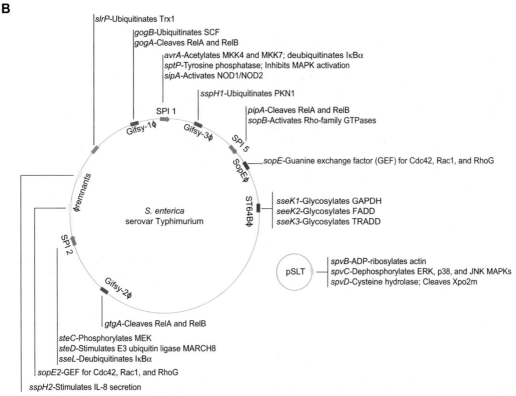

Fig. 1 EHEC and *Salmonella* genomes. (**a**) Schematic of the EHEC O157:H7 EDL933 genome. Black rectangles, O-islands; red dots, prophage elements. (**b**) Schematic of the *S.* Typhimurium genome. Red arrows, *Salmonella* pathogenicity islands (SPIs); Grey and yellow rectangles, prophage elements

Yen et al. 2010). NleC-mediated cleavage of p65 occurs between C38 and E39, generating a p65 $^{1-38}$ fragment that also competes with full-length p65 for binding to the ribosomal protein RPS3, an additional NF-κB subunit (Hodgson et al. 2015). This competition reduces the extent of RPS3 nuclear translocation, thus additionally inhibiting NF-κB pathway activation (Hodgson et al. 2015).

NleD is also a zinc metalloprotease that cleaves and inactivates the mitogen-activated protein kinase signaling proteins c-Jun N-terminal kinase (JNK) and p38 (Baruch et al. 2011). JNK cleavage by NleD occurs before Y185 in the TPY (T183-P184-Y185) motif within the kinase activation loop of JNK, whereas p38 cleavage occurs at residue M213 (Creuzburg et al. 2017). As p38/JNK activation influences pro-inflammatory cytokine gene expression, their cleavage by NleD contributes downregulated inflammatory responses. In a random mutagenesis study of NleD, R203 was found to be critical for protease activity against p38 but not JNK (Creuzburg et al. 2017). In another study, mutation of E143 within the metalloprotease motif ($H^{142}ELLH^{146}$) abrogated JNK cleavage (Baruch et al. 2011).

NleE was originally described to inhibit the nuclear translocation of the p65 NF-κB subunit by preventing IκBα phosphorylation (Newton et al. 2010). The C terminal IDSY(M/I)K motif was found to be critical for NleE-mediated inhibition of p65 translocation (Newton et al. 2010). It was later shown that NleE is an S-adenosylmethionine (SAM)-dependent methyltransferase that modifies a zinc-coordinating cysteine in the Npl4 zinc finger (NZF) domains in TAB2 and TAB3, which are two ubiquitin-chain sensory proteins involved in NF-κB signaling (Zhang et al. 2012). Upon NleE mediated methylation of C673 and C692 in TAB2 and TAB3 respectively, the two proteins lose their ability to bind zinc and ubiquitin chains (Zhang et al. 2012). The NleE crystal structure displays a SAM pocket that particularly involves two critical residues, R107 and Y212 (Yao et al. 2014). The same study also identified ZRANB3 as a target of NleE. NleE mediated methylation of the ZRANB3-NZF domain disrupted its ubiquitin chain binding without affecting ZRANB3 recruitment to sites of DNA damage (Yao et al. 2014).

NleF is important to EHEC colonization of germ-free piglets (Echtenkamp et al. 2008). NleF is a pro-inflammatory effector because it induces NF-κB activation during the early stages of the infection (Pallett et al. 2014; Pollock et al. 2017). The C-terminal region of NleF is essential for its interaction with caspase 4, 8, and 9 (Blasche et al. 2013). NleF counteracts the host inflammatory response by preventing the heteromerization of caspase-4-p19 and caspase-4-p10, which is essential for caspase-4 activation and the eventual expression of IL-1β (Blasche et al. 2013; Song et al. 2017). The transient pro-inflammatory activity of NleF is later overcome by the anti-inflammatory functions of other effectors such as NleE, NleH, and NleB.

NleH1 and **NleH2** are two paralog effectors (with 84% homology) that contain a C-terminal kinase domain (Grishin et al. 2014). Both proteins bind RPS3, but only NleH1 inhibits RPS3 translocation to the nucleus (Gao et al. 2009). NleH1 interaction with RPS3 prevents its phosphorylation on S209 by IKKβ (Wan et al. 2011). NleH1 phosphorylates the v-crk sarcoma virus CT10 oncogene-like protein (CRKL), which may play an important scaffolding role between NleH1 and IKKβ (Pham et al. 2013). K159 in NleH1 and K169 in NleH2 are critical residues for the kinase activity of these effectors (Gao et al. 2009).

EspJ was first described as a prophage-carried effector that influences the dynamics of pathogen clearance (Dahan et al. 2005) and inhibits opsono-phagocytosis (Marches et al. 2008). More recently, its mode of action was clarified by the demonstration that this effector functions as an adenosine diphosphate (ADP) ribosyltransferase that inhibits the host Src kinase by amidation and ADP ribosylation of a Src glutamic acid residue (Young et al. 2014). Additional host kinases in the Src, Tec, and Syk NRTK families have also been described as EspJ substrates *in vivo* (Pollard et al. 2018).

EspL is a cysteine protease that directly cleaves the receptor interacting protein (RIP) homotypic interaction motif (RHIM) domain-containing proteins RIPK1, RIPK3, TRIF, and ZBP1/DAI, leading to their rapid inactivation, with subsequent inhibition of necroptosis and inflammatory signaling (Pearson et al. 2017). A conserved cysteine protease motif with catalytic residues C47, H131, and D153 was defined (Pearson et al. 2017). The proteolytic activity of EspL is important to the duration of *C. rodentium* intestinal colonization of mice, with Δ*espL* being cleared faster than wild-type *C. rodentium*, despite similar initial colonization magnitudes (Pearson et al. 2017).

Tir. In addition to its well-studied role in bacterial adhesion, Tir also inhibits NF-κB pathway activation by interacting with TNFR-associated factor (TRAF) adaptor proteins leading to their proteasomal-independent degradation (Ruchaud-Sparagano et al. 2011) (Fig. 2).

1.2   *Salmonella* Effectors

GogB is the first ORF in the Gifsy-1 prophage in *Salmonella enterica* serovar Typhimurium (McClelland et al. 2001). GogB is a leucine-rich repeat protein that is secreted by both type three secretion systems encoded in Salmonella Pathogenicity Island-1 (SPI-1) and SPI-2 (Coombes et al. 2005). GogB is regulated by its transcriptional regulator SsrB through SPI-2 induction, and then localizes to the host cell cytoplasm (Xu and Hensel 2010). GogB alters the function of the host SKP, cullin, F-box containing complex (SCF) E3 type ubiquitin ligase by interacting with the human F-box only 22 (FBXO22) protein and S-phase kinase associated protein 1 (Skp1) (Pilar et al. 2012). GogB-deficient *Salmonella* fail to limit NF-κB activation in RAW264.7 cells, indicating that GogB is an anti-inflammatory factor (Pilar et al. 2012).

PipA, **GtgA**, and **GogA** comprise a family of proteases that cleave both the RelA (p65) and RelB NF-κB transcription factors, thus limiting host inflammatory responses (Sun et al. 2016). Salmonella mutants in the *pipA*, *gogA*, or *gtgA* genes show increased NF-κB pathway stimulation and virulence in C57/BL6 mice, as compared to wild-type *Salmonella* (Sun et al. 2016), indicating that this family of effectors inhibits host immune responses.

Salmonella encodes three members of the novel E3 Ligase (NEL) family, **SlrP**, **SspH1**, and **SspH2**. SlrP functions as an E3 ubiquitin ligase for mammalian thioredoxin (Bernal-Bayard and Ramos-Morales 2009), which may affect the redox state and activity of transcription factors. SlrP also binds to the chaperone ERdj3 in the endoplasmic reticulum (Bernal-Bayard et al. 2010), possible impacting the unfolded protein response and apoptosis. SspH1 and SspH2 have 39% and 38% identity to SlrP (Miao and Miller 2000). SspH1 inhibits NF-κB signaling (Haraga and Miller 2006). SspH1 interacts with and ubiquitinates human protein kinase N1 (PKN1), to activate its catalytic function (Keszei et al. 2014). Mutating SspH1 C492 abolishes its ability to ubiquitinate PKN1 (Keszei et al. 2014). However, the inhibition of NF-κB signaling is not due to the ubiquitination and proteasome-dependent degradation of PKN1 (Keszei et al. 2014). SspH2 increases Nod1-mediated IL-8 secretion by ubiquitinating Nod1 (Bhavsar et al. 2013). Mutating SspH2 C580 abolishes its ubiquitination activity (Quezada et al. 2009).

SpvB is encoded by the *spv* operon in the *Salmonella* virulence plasmid (Caldwell and Gulig 1991). SpvB ADP-ribosylates actin on R177 (Hochmann et al. 2006), causing both actin depolymerization and apoptosis (Guiney and Lesnick 2005). P-bodies are enzymes involved in mRNA turnover and post-transcriptional regulation. SpvB is also an important regulator of P-body disassembly (Eulalio et al. 2011).

SpvC is also encoded by the *spv* operon (Coynault et al. 1992) and is translocated by both T3SS-1 and T3SS-2 (Heiskanen et al. 1994). SpvC has a phosphothreonine lyase activity that targets ERK, p38, and JNK mitogen-activated protein kinases (MAPKs) (Mazurkiewicz et al. 2008). SpvC limits host inflammatory responses by reducing pro-inflammatory cytokine gene transcription (Haneda et al. 2012). Recently, kinetic studies have been performed on SpvC, revealing that this enzyme is extremely specific for MAPKs (Chambers et al. 2018).

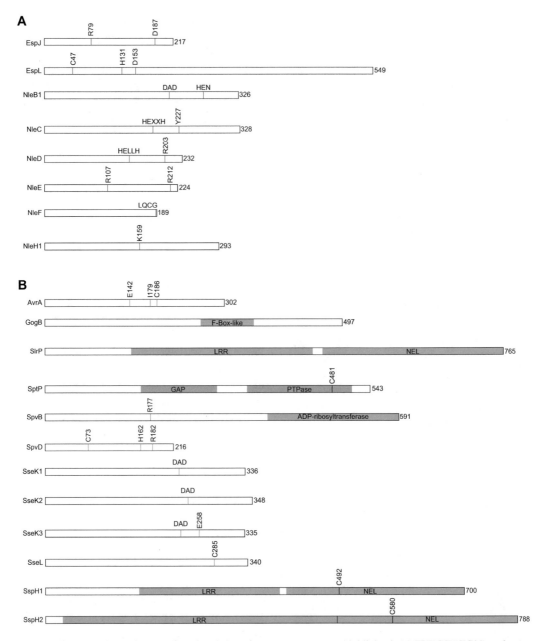

Fig. 2 Effector amino acids and functional domains.
(a) Schematic representation of T3SS effectors targeting host innate immunity. Critical amino acids and secondary structures are highlighted. (a) EHEC/EPEC/*C. rodentium* effectors. (b) *S.* Typhimurium effectors

SpvD functions as a cysteine hydrolase with a papain-like fold and a catalytic triad consisting of C73, H162, and D182 (Grabe et al. 2016). SpvD interacts with the exportin Xpo2m, which mediates nuclear-cytoplasmic protein recycling (Rolhion et al. 2016). The disruption of importin-alpha recycling from the nucleus results in p65 translocation deficiency and NF-κB inhibition (Rolhion et al. 2016).

SteC is encoded within SPI-2 and has amino acid similarity to human kinase Raf-1, a proto-oncogene serine/threonine-protein kinase (Poh et al. 2008). An *steC* mutant showed no decrease for bacterial replication and virulence, but was defective for SPI-2 dependent F-Actin meshwork formation (Poh et al. 2008). SteC phosphorylates MEK and activates a signaling pathway involving MEK and ERK, MLCK, and Myosin IIB (Odendall et al. 2012). An *steC* mutant shows greater replication in HeLa cells, as compared to wild-type Salmonella, indicating that SteC may limit bacterial replication by regulating F-Actin meshwork formation (Odendall et al. 2012).

SteD stimulates MARCH8, a member of the MARCH family E3 integral membrane ubiquitin ligases (Bayer-Santos et al. 2016). SteD causes ubiquitination and depletion of surface-localized mature MHC class II (mMHCII), thus helping to reduce T cell activation (Bayer-Santos et al. 2016).

Salmonella Secreted Effector L (**SseL**) is encoded within SPI-2 and is transported into the cytosol from the SCV via the T3SS-2. SseL functions as a deubiquitinase (DUB) to induce a delayed cytotoxic effect in *Salmonella*-infected macrophages (Rytkonen et al. 2007). NF-κB signaling in macrophages is significantly activated after infection with an *sseL* mutant, as compared to its activation by wild-type Salmonella (Le Negrate et al. 2008). Expression of SseL in mammalian cells impairs IκBα ubiquitination and degradation, and mutating C285 abolished this effect (Le Negrate et al. 2008).

AvrA is encoded on SPI-1 and is a member of the YopJ/Avr protein family (Hardt and Galan 1997). AvrA is a multifunctional effector that influences eukaryotic cell pathways by regulating the ubiquitination (Sun et al. 2004) and acetylation (Wu et al. 2010) of target proteins to modulate proliferation (Liu et al. 2010), inflammation (Du and Galan 2009), and apoptosis (Collier-Hyams et al. 2002). AvrA acetylates p53 (Wu et al. 2010) and also inhibits JNK signaling by acetylating MKK4 (Jones et al. 2008) and MKK7 (Du and Galan 2009). AvrA also has a deubiquitinase activity that targets β-catenin, resulting in stabilization of β-catenin and increasing intestinal epithelial cell proliferation (Wu et al. 2010). Mutating E142, I179, and C186 reduces AvrA acetyltransferase activity (Wu et al. 2010); C186 also appears to be critical for AvrA

deubiquitinase activity (Ye et al. 2007). It should be noted that the activity of AvrA has largely been reconciled to function as a acetyltransferase, similar to its homolog, the *Yersinia* outer protein J (YopJ) and many early observations regarding a potential AvrA protease activity may have been artifacts (Ma and Ma 2016) (Fig. 3).

Salmonella encodes up to three NleB orthologs named **SseK1**, **SseK2**, and **SseK3** (Brown et al. 2011). The NleB and SseK effectors contain a conserved DXD motif, which is required for enzymatic function of glycosyltransferases of the GT-A family (Gao et al. 2013). SseK1 and SseK2 inhibit TNF-induced NF-κB activation and cell death during infection in macrophages (Gunster et al. 2017). SseK3 binds, but does not glycosylate TRIM32, an E3 ubiquitin ligase involved in TNF and interferon signaling (Yang et al. 2015). SseK1 and SseK3 inhibit TNF-mediated NF-κB pathway activation. SseK1 glycosylates GAPDH, while SseK2 glycosylates FADD (El Qaidi et al. 2017). SseK3 glycosylates TRADD and has an overall fold that is similar to other glycosyltransferase type-A (GT-A) family members (Esposito et al. 2018). Mutating E258 impairs the ability of SseK3 to glycosylate TRADD (Esposito et al. 2018).

Salmonella protein tyrosine phosphatase (**SptP**) shares significant homology in its C-terminal domain to the catalytic domains of eukaryotic tyrosine phosphatases (PTPases) and the *Yersinia* protein tyrosine phosphatases YopH (Kaniga et al. 1996). SptP inhibits ERK activation by interfering with Raf1 activation, a MAPKKK (Murli et al. 2001) Inhibition is dependent on SptP-phosphatase activity, as a SptP C481S mutation no longer inhibits ERK activation (Lin et al. 2003). SptP phosphatase activity is also important for inhibiting TNF-α induction during *Salmonella* infection of macrophages (Lin et al. 2003).

Salmonella outer protein B (**SopB**, also known as **SigD**) has sequence homology to mammalian inositol polyphosphate 4-phosphatases and recombinant SopB functions as an inositol phosphate phosphatase (Norris et al. 1998). It is encoded within SPI-5 but delivered into cells via

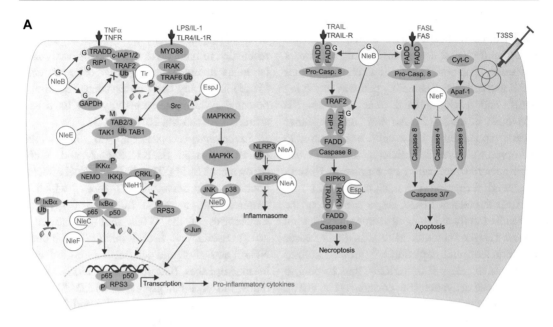

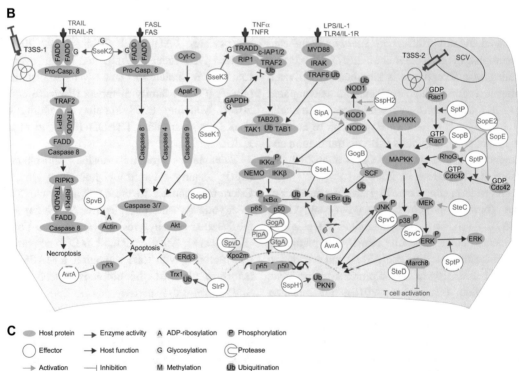

Fig. 3 Host inflammatory pathways and their targeting by T3SS effectors. Upon TNFα stimulation, TRADD forms a complex with TRAF2, leading to RIP1 and TRAF polyubiquitination followed by recruitment of TAK1. TAK1 activation results in activation of the IKK complex and phosphorylation of IκBα, which is then ubiquitinated and degraded, releasing NF-κB for nuclear translocation and pro-inflammatory gene transcription. Stimulation of TLR4 or IL-1R by LPS or IL-1β initiates a signaling cascade that leads to TAK1 activation. Stimulation with FASL results in FADD oligomerization and activation of caspase 8, leading to caspase 3/7 maturation and apoptosis. Stimulation with TRAIL initiates necroptosis through a pathway involving the formation of a complex among RIP1, RIP3, TRADD, FADD, and caspase 8. Upon extracellular stimulation, mitogen

T3SS SPI-1 (Knodler et al. 2002). **SopE** is 69% identical to **SopE2** (Bakshi et al. 2000). SopB, SopE, and SopE2, cooperate in a functionally redundant manner to activate Rho-family GTPases during *Salmonella* infection (Patel and Galan 2006), resulting in activation of MAPK and NF-κB signaling pathways (Bruno et al. 2009). SopB functions as a phosphoinositide phosphatase to activate the RhoG exchange factor, (Patel and Galan 2006), while SopE and SopE2 act as exchange factors for Cdc42, Rac1, and RhoG (Stender et al. 2000).

Salmonella invasion protein A (SipA) is encoded within SPI-1 (Kaniga et al. 1995) and is transported into mammalian cells via the T3SS-1 (Zhou et al. 1999a). SipA binds F-actin, contributing to *Salmonella* internalization and mediates the downstream activation of Rho GTPases (Zhou et al. 1999b). SipA induces NF-κB activation and intestinal inflammation by activating the NOD1/NOD2 signaling pathway (Keestra et al. 2011). SipA also induces IL-8 expression by inducing Jun and p38 MAPK phosphorylation (Figueiredo et al. 2009).

flora. Microbial genomes provide information about how co-evolution has shaped pathogen targeting of host immunity. Characterizing anti-inflammatory microbial proteins offers opportunities both to identify natural modulators of inflammation and to engineer these proteins for use as effective immunotherapeutics (Ruter and Hardwidge 2014). T3SS effectors represent a major pool of pathogen-derived immunomodulatory molecules that may have future utility in treating autoimmune disorders associated with perturbations in NF-κB signaling (Sun et al. 2013). Indeed, *Shigella flexneri* and *Yersinia pestis* effectors have already been exploited to selectively inhibit MAPK pathways and to tune T-cell responses (Wei et al. 2012).

Acknowledgements We apologize to authors whose work we could not cite due to journal-mandated space limitations. This work was supported by grants AI093913, AI127973, and AR072594 from the National Institutes of Health. The content is solely the responsibility of the authors and does not necessarily represent the official views of the National Institutes of Health. There are no ethical issues related to animal or human subjects research to disclose.

2   Summary

Here we have described many of the *E. coli* and *Salmonella* T3SS effectors that target host innate immunity. Many of these effectors apply their inhibitory effect in a cumulative or synergistic manner by blocking multiple steps within a signaling pathway. Effector translocation cell is not random but rather is temporally controlled by the pathogen (Lara-Tejero et al. 2011), allowing the pathogen to adjust the host inflammatory response not only to evade inflammation but also to alter the relative abundance of commensal

References

Bakshi CS, Singh VP, Wood MW, Jones PW, Wallis TS, Galyov EE (2000) Identification of SopE2, a Salmonella secreted protein which is highly homologous to SopE and involved in bacterial invasion of epithelial cells. J Bacteriol 182(8):2341–2344

Baruch K, Gur-Arie L, Nadler C, Koby S, Yerushalmi G, Ben-Neriah Y, Yogev O, Shaulian E, Guttman C, Zarivach R, Rosenshine I (2011) Metalloprotease type III effectors that specifically cleave JNK and NF-kappaB. EMBO J 30(1):221–231. https://doi.org/10.1038/emboj.2010.297

Bayer-Santos E, Durkin CH, Rigano LA, Kupz A, Alix E, Cerny O, Jennings E, Liu M, Ryan AS, Lapaque N,

---

Fig. 3 (continued) activated protein kinases (MAPKs) pathway activation begins with the binding of Ras/Rho family proteins to a mitogen activated protein kinase kinase kinase (MAPKKK), which in turn leads to the phosphorylation and activation of a mitogen activated protein kinase kinase (MAPKK), followed by the phosphorylation of the mitogen activated protein kinases (MAPKs) ERK, p38, and JNK. NOD1 and NOD2 recruit receptor-interacting serine/threonine-protein kinase 2 (RIPK2), which further activates MAPKs and the IKK complex. (**a**) Strategies used by EHEC/EPEC/*C. rodentium* to counteract inflammation. (**b**) Strategies used by *Salmonella enterica* Serovar Typhimurium to counteract inflammation

Kaufmann SHE, Holden DW (2016) The Salmonella effector SteD mediates MARCH8-dependent ubiquitination of MHC II molecules and inhibits T cell activation. Cell Host Microbe 20(5):584–595. https://doi.org/10.1016/j.chom.2016.10.007

Bernal-Bayard J, Ramos-Morales F (2009) Salmonella type III secretion effector SlrP is an E3 ubiquitin ligase for mammalian thioredoxin. J Biol Chem 284 (40):27587–27595. https://doi.org/10.1074/jbc.M109.010363

Bernal-Bayard J, Cardenal-Munoz E, Ramos-Morales F (2010) The Salmonella type III secretion effector, Salmonella leucine-rich repeat protein (SlrP), targets the human chaperone ERdj3. J Biol Chem 285 (21):16360–16368. https://doi.org/10.1074/jbc.M110.100669

Bhavsar AP, Brown NF, Stoepel J, Wiermer M, Martin DD, Hsu KJ, Imami K, Ross CJ, Hayden MR, Foster LJ, Li X, Hieter P, Finlay BB (2013) The Salmonella type III effector SspH2 specifically exploits the NLR co-chaperone activity of SGT1 to subvert immunity. PLoS Pathog 9(7):e1003518. https://doi.org/10.1371/journal.ppat.1003518

Blasche S, Mortl M, Steuber H, Siszler G, Nisa S, Schwarz F, Lavrik I, Gronewold TM, Maskos K, Donnenberg MS, Ullmann D, Uetz P, Kogl M (2013) The E. coli effector protein NleF is a caspase inhibitor. PLoS One 8(3):e58937. https://doi.org/10.1371/journal.pone.0058937

Brown NF, Coombes BK, Bishop JL, Wickham ME, Lowden MJ, Gal-Mor O, Goode DL, Boyle EC, Sanderson KL, Finlay BB (2011) Salmonella phage ST64B encodes a member of the SseK/NleB effector family. PLoS One 6(3):e17824. https://doi.org/10.1371/journal.pone.0017824

Bruno VM, Hannemann S, Lara-Tejero M, Flavell RA, Kleinstein SH, Galan JE (2009) Salmonella typhimurium type III secretion effectors stimulate innate immune responses in cultured epithelial cells. PLoS Pathog 5(8):e1000538. https://doi.org/10.1371/journal.ppat.1000538

Caldwell AL, Gulig PA (1991) The Salmonella typhimurium virulence plasmid encodes a positive regulator of a plasmid-encoded virulence gene. J Bacteriol 173(22):7176–7185

Campellone KG, Leong JM (2003) Tails of two Tirs: actin pedestal formation by enteropathogenic E. coli and enterohemorrhagic E. coli O157:H7. Curr Opin Microbiol 6(1):82–90

Chambers KA, Abularrage NS, Scheck RA (2018) Selectivity within a family of bacterial phosphothreonine lyases. Biochemistry 57(26):3790–3796. https://doi.org/10.1021/acs.biochem.8b00534

Clarke SC, Haigh RD, Freestone PP, Williams PH (2002) Enteropathogenic Escherichia coli infection: history and clinical aspects. Br J Biomed Sci 59(2):123–127

Collier-Hyams LS, Zeng H, Sun J, Tomlinson AD, Bao ZQ, Chen H, Madara JL, Orth K, Neish AS (2002) Cutting edge: salmonella AvrA effector inhibits the key proinflammatory, anti-apoptotic NF-kappa B pathway. J Immunol 169(6):2846–2850

Coombes BK, Wickham ME, Brown NF, Lemire S, Bossi L, Hsiao WW, Brinkman FS, Finlay BB (2005) Genetic and molecular analysis of GogB, a phage-encoded type III-secreted substrate in Salmonella enterica serovar typhimurium with autonomous expression from its associated phage. J Mol Biol 348 (4):817–830. https://doi.org/10.1016/j.jmb.2005.03.024

Coynault C, Robbe-Saule V, Popoff MY, Norel F (1992) Growth phase and SpvR regulation of transcription of Salmonella typhimurium spvABC virulence genes. Microb Pathog 13(2):133–143

Creuzburg K, Giogha C, Wong Fok Lung T, Scott NE, Muhlen S, Hartland EL, Pearson JS (2017) The type III effector NleD from Enteropathogenic Escherichia coli differentiates between host substrates p38 and JNK. Infect Immun 85(2):e00620–e00616. https://doi.org/10.1128/IAI.00620-16

Dahan S, Wiles S, La Ragione RM, Best A, Woodward MJ, Stevens MP, Shaw RK, Chong Y, Knutton S, Phillips A, Frankel G (2005) EspJ is a prophage-carried type III effector protein of attaching and effacing pathogens that modulates infection dynamics. Infect Immun 73(2):679–686. https://doi.org/10.1128/IAI.73.2.679-686.2005

Daniell SJ, Takahashi N, Wilson R, Friedberg D, Rosenshine I, Booy FP, Shaw RK, Knutton S, Frankel G, Aizawa S (2001) The filamentous type III secretion translocon of enteropathogenic Escherichia coli. Cell Microbiol 3(12):865–871

Du F, Galan JE (2009) Selective inhibition of type III secretion activated signaling by the Salmonella effector AvrA. PLoS Pathog 5(9):e1000595. https://doi.org/10.1371/journal.ppat.1000595

Echtenkamp F, Deng W, Wickham ME, Vazquez A, Puente JL, Thanabalasuriar A, Gruenheid S, Finlay BB, Hardwidge PR (2008) Characterization of the NleF effector protein from attaching and effacing bacterial pathogens. FEMS Microbiol Lett 281(1):98–107. https://doi.org/10.1111/j.1574-6968.2008.01088.x

El Qaidi S, Chen K, Halim A, Siukstaite L, Rueter C, Hurtado-Guerrero R, Clausen H, Hardwidge PR (2017) NleB/SseK effectors from Citrobacter rodentium, Escherichia coli, and Salmonella enterica display distinct differences in host substrate specificity. J Biol Chem 292(27):11423–11430. https://doi.org/10.1074/jbc.M117.790675

Esposito D, Gunster RA, Martino L, El Omari K, Wagner A, Thurston TLM, Rittinger K (2018) Structural basis for the glycosyltransferase activity of the Salmonella effector SseK3. J Biol Chem 293(14):5064–5078. https://doi.org/10.1074/jbc.RA118.001796

Eulalio A, Frohlich KS, Mano M, Giacca M, Vogel J (2011) A candidate approach implicates the secreted Salmonella effector protein SpvB in P-body disassembly. PLoS One 6(3):e17296. https://doi.org/10.1371/

journal.pone.0017296

Figueiredo JF, Lawhon SD, Gokulan K, Khare S, Raffatellu M, Tsolis RM, Baumler AJ, McCormick BA, Adams LG (2009) Salmonella enterica typhimurium SipA induces CXC-chemokine expression through p38MAPK and JUN pathways. Microbes Infect 11(2):302–310. https://doi.org/10.1016/j.micinf.2008.12.005

Gao X, Wan F, Mateo K, Callegari E, Wang D, Deng W, Puente J, Li F, Chaussee MS, Finlay BB, Lenardo MJ, Hardwidge PR (2009) Bacterial effector binding to ribosomal protein s3 subverts NF-kappaB function. PLoS Pathog 5(12):e1000708. https://doi.org/10.1371/journal.ppat.1000708

Gao X, Wang X, Pham TH, Feuerbacher LA, Lubos ML, Huang M, Olsen R, Mushegian A, Slawson C, Hardwidge PR (2013) NleB, a bacterial effector with glycosyltransferase activity, targets GAPDH function to inhibit NF-kappaB activation. Cell Host Microbe 13(1):87–99. https://doi.org/10.1016/j.chom.2012.11.010

Giogha C, Lung TW, Muhlen S, Pearson JS, Hartland EL (2015) Substrate recognition by the zinc metalloprotease effector NleC from enteropathogenic Escherichia coli. Cell Microbiol 17(12):1766–1778. https://doi.org/10.1111/cmi.12469

Grabe GJ, Zhang Y, Przydacz M, Rolhion N, Yang Y, Pruneda JN, Komander D, Holden DW, Hare SA (2016) The Salmonella effector SpvD is a cysteine hydrolase with a serovar-specific polymorphism influencing catalytic activity, suppression of immune responses, and bacterial virulence. J Biol Chem 291(50):25853–25863. https://doi.org/10.1074/jbc.M116.752782

Grishin AM, Cherney M, Anderson DH, Phanse S, Babu M, Cygler M (2014) NleH defines a new family of bacterial effector kinases. Structure 22(2):250–259. https://doi.org/10.1016/j.str.2013.11.006

Guiney DG, Lesnick M (2005) Targeting of the actin cytoskeleton during infection by Salmonella strains. Clin Immunol 114(3):248–255. https://doi.org/10.1016/j.clim.2004.07.014

Gunster RA, Matthews SA, Holden DW, Thurston TL (2017) SseK1 and SseK3 type III secretion system effectors inhibit NF-kappaB signaling and Necroptotic cell death in salmonella-infected macrophages. Infect Immun 85(3):e00010–e00017. https://doi.org/10.1128/IAI.00010-17

Haneda T, Ishii Y, Shimizu H, Ohshima K, Iida N, Danbara H, Okada N (2012) Salmonella type III effector SpvC, a phosphothreonine lyase, contributes to reduction in inflammatory response during intestinal phase of infection. Cell Microbiol 14(4):485–499. https://doi.org/10.1111/j.1462-5822.2011.01733.x

Haraga A, Miller SI (2006) A Salmonella type III secretion effector interacts with the mammalian serine/threonine protein kinase PKN1. Cell Microbiol 8(5):837–846. https://doi.org/10.1111/j.1462-5822.2005.00670.x

Hardt WD, Galan JE (1997) A secreted Salmonella protein with homology to an avirulence determinant of plant pathogenic bacteria. Proc Natl Acad Sci U S A 94(18):9887–9892

Heiskanen P, Taira S, Rhen M (1994) Role of rpoS in the regulation of Salmonella plasmid virulence (spv) genes. FEMS Microbiol Lett 123(1–2):125–130

Hensel M (2000) Salmonella pathogenicity island 2. Mol Microbiol 36(5):1015–1023

Hochmann H, Pust S, von Figura G, Aktories K, Barth H (2006) Salmonella enterica SpvB ADP-ribosylates actin at position arginine-177-characterization of the catalytic domain within the SpvB protein and a comparison to binary clostridial actin-ADP-ribosylating toxins. Biochemistry 45(4):1271–1277. https://doi.org/10.1021/bi051810w

Hodgson A, Wier EM, Fu K, Sun X, Yu H, Zheng W, Sham HP, Johnson K, Bailey S, Vallance BA, Wan F (2015) Metalloprotease NleC suppresses host NF-kappaB/inflammatory responses by cleaving p65 and interfering with the p65/RPS3 interaction. PLoS Pathog 11(3):e1004705. https://doi.org/10.1371/journal.ppat.1004705

Jones RM, Wu H, Wentworth C, Luo L, Collier-Hyams L, Neish AS (2008) Salmonella AvrA coordinates suppression of host immune and apoptotic defenses via JNK pathway blockade. Cell Host Microbe 3(4):233–244. https://doi.org/10.1016/j.chom.2008.02.016

Kaniga K, Trollinger D, Galan JE (1995) Identification of two targets of the type III protein secretion system encoded by the inv and spa loci of Salmonella typhimurium that have homology to the Shigella IpaD and IpaA proteins. J Bacteriol 177(24):7078–7085

Kaniga K, Uralil J, Bliska JB, Galan JE (1996) A secreted protein tyrosine phosphatase with modular effector domains in the bacterial pathogen Salmonella typhimurium. Mol Microbiol 21(3):633–641

Keestra AM, Winter MG, Klein-Douwel D, Xavier MN, Winter SE, Kim A, Tsolis RM, Baumler AJ (2011) A Salmonella virulence factor activates the NOD1/NOD2 signaling pathway. MBio 2(6):e00266–e00211. https://doi.org/10.1128/mBio.00266-11

Kelly M, Hart E, Mundy R, Marches O, Wiles S, Badea L, Luck S, Tauschek M, Frankel G, Robins-Browne RM, Hartland EL (2006) Essential role of the type III secretion system effector NleB in colonization of mice by Citrobacter rodentium. Infect Immun 74(4):2328–2337. https://doi.org/10.1128/IAI.74.4.2328-2337.2006

Kenny B, DeVinney R, Stein M, Reinscheid DJ, Frey EA, Finlay BB (1997) Enteropathogenic E. coli (EPEC) transfers its receptor for intimate adherence into mammalian cells. Cell 91(4):511–520

Keszei AF, Tang X, McCormick C, Zeqiraj E, Rohde JR, Tyers M, Sicheri F (2014) Structure of an SspH1-PKN1 complex reveals the basis for host substrate recognition and mechanism of activation for a bacterial E3 ubiquitin ligase. Mol Cell Biol 34(3):362–373. https://doi.org/10.1128/MCB.01360-13

Kim J, Thanabalasuriar A, Chaworth-Musters T, Fromme JC, Frey EA, Lario PI, Metalnikov P, Rizg K, Thomas NA, Lee SF, Hartland EL, Hardwidge PR, Pawson T, Strynadka NC, Finlay BB, Schekman R, Gruenheid S (2007) The bacterial virulence factor NleA inhibits cellular protein secretion by disrupting mammalian COPII function. Cell Host Microbe 2(3):160–171. https://doi.org/10.1016/j.chom.2007.07.010

Knodler LA, Celli J, Hardt WD, Vallance BA, Yip C, Finlay BB (2002) Salmonella effectors within a single pathogenicity island are differentially expressed and translocated by separate type III secretion systems. Mol Microbiol 43(5):1089–1103

Lara-Tejero M, Kato J, Wagner S, Liu X, Galan JE (2011) A sorting platform determines the order of protein secretion in bacterial type III systems. Science 331 (6021):1188–1191. https://doi.org/10.1126/science. 1201476

Le Negrate G (2012) Subversion of innate immune responses by bacterial hindrance of NF-kappaB pathway. Cell Microbiol 14(2):155–167. https://doi.org/10. 1111/j.1462-5822.2011.01719.x

Le Negrate G, Faustin B, Welsh K, Loeffler M, Krajewska M, Hasegawa P, Mukherjee S, Orth K, Krajewski S, Godzik A, Guiney DG, Reed JC (2008) Salmonella secreted factor L deubiquitinase of Salmonella typhimurium inhibits NF-kappaB, suppresses IkappaBalpha ubiquitination and modulates innate immune responses. J Immunol 180(7):5045–5056

Li S, Zhang L, Yao Q, Li L, Dong N, Rong J, Gao W, Ding X, Sun L, Chen X, Chen S, Shao F (2013) Pathogen blocks host death receptor signalling by arginine GlcNAcylation of death domains. Nature 501 (7466):242–246. https://doi.org/10.1038/nature12436

Lin SL, Le TX, Cowen DS (2003) SptP, a Salmonella typhimurium type III-secreted protein, inhibits the mitogen-activated protein kinase pathway by inhibiting Raf activation. Cell Microbiol 5(4):267–275

Litvak Y, Sharon S, Hyams M, Zhang L, Kobi S, Katsowich N, Dishon S, Nussbaum G, Dong N, Shao F, Rosenshine I (2017) Epithelial cells detect functional type III secretion system of enteropathogenic Escherichia coli through a novel NF-kappaB signaling pathway. PLoS Pathog 13(7):e1006472. https://doi.org/10.1371/journal.ppat.1006472

Liu X, Lu R, Wu S, Sun J (2010) Salmonella regulation of intestinal stem cells through the Wnt/beta-catenin pathway. FEBS Lett 584(5):911–916. https://doi.org/10. 1016/j.febslet.2010.01.024

Liu T, Zhang L, Joo D, Sun SC (2017) NF-kappaB signaling in inflammation. Signal Transduction Targeted Ther 2:e17023. https://doi.org/10.1038/sigtrans.2017. 23

Luperchio SA, Newman JV, Dangler CA, Schrenzel MD, Brenner DJ, Steigerwalt AG, Schauer DB (2000) Citrobacter rodentium, the causative agent of

transmissible murine colonic hyperplasia, exhibits clonality: synonymy of C. rodentium and mouse-pathogenic Escherichia coli. J Clin Microbiol 38 (12):4343–4350

Ma KW, Ma W (2016) YopJ family effectors promote bacterial infection through a unique acetyltransferase activity. Microbiol Mol Biol Rev 80(4):1011–1027. https://doi.org/10.1128/MMBR.00032-16

Marches O, Covarelli V, Dahan S, Cougoule C, Bhatta P, Frankel G, Caron E (2008) EspJ of enteropathogenic and enterohaemorrhagic Escherichia coli inhibits opsono-phagocytosis. Cell Microbiol 10 (5):1104–1115. https://doi.org/10.1111/j.1462-5822. 2007.01112.x

Marcus SL, Brumell JH, Pfeifer CG, Finlay BB (2000) Salmonella pathogenicity islands: big virulence in small packages. Microbes Infect 2(2):145–156

Mazurkiewicz P, Thomas J, Thompson JA, Liu M, Arbibe L, Sansonetti P, Holden DW (2008) SpvC is a Salmonella effector with phosphothreonine lyase activity on host mitogen-activated protein kinases. Mol Microbiol 67(6):1371–1383. https://doi.org/10.1111/j. 1365-2958.2008.06134.x

McClelland M, Sanderson KE, Spieth J, Clifton SW, Latreille P, Courtney L, Porwollik S, Ali J, Dante M, Du F, Hou S, Layman D, Leonard S, Nguyen C, Scott K, Holmes A, Grewal N, Mulvaney E, Ryan E, Sun H, Florea L, Miller W, Stoneking T, Nhan M, Waterston R, Wilson RK (2001) Complete genome sequence of Salmonella enterica serovar Typhimurium LT2. Nature 413(6858):852–856. https://doi.org/10. 1038/35101614

McDaniel TK, Jarvis KG, Donnenberg MS, Kaper JB (1995) A genetic locus of enterocyte effacement conserved among diverse enterobacterial pathogens. Proc Natl Acad Sci U S A 92(5):1664–1668

Medzhitov R, Janeway CA Jr (1997) Innate immunity: the virtues of a nonclonal system of recognition. Cell 91 (3):295–298

Miao EA, Miller SI (2000) A conserved amino acid sequence directing intracellular type III secretion by Salmonella typhimurium. Proc Natl Acad Sci U S A 97 (13):7539–7544

Murli S, Watson RO, Galan JE (2001) Role of tyrosine kinases and the tyrosine phosphatase SptP in the interaction of Salmonella with host cells. Cell Microbiol 3 (12):795–810

Newell DG, La Ragione RM (2018) Enterohaemorrhagic and other Shiga toxin-producing Escherichia coli (STEC): where are we now regarding diagnostics and control strategies? Transbound Emerg Dis 65:49–71. https://doi.org/10.1111/tbed.12789

Newton HJ, Pearson JS, Badea L, Kelly M, Lucas M, Holloway G, Wagstaff KM, Dunstone MA, Sloan J, Whisstock JC, Kaper JB, Robins-Browne RM, Jans DA, Frankel G, Phillips AD, Coulson BS, Hartland EL (2010) The type III effectors NleE and NleB from enteropathogenic E. coli and OspZ from Shigella block nuclear translocation of NF-kappaB p65. PLoS Pathog

6(5):e1000898. https://doi.org/10.1371/journal.ppat. 1000898

Norris FA, Wilson MP, Wallis TS, Galyov EE, Majerus PW (1998) SopB, a protein required for virulence of Salmonella Dublin, is an inositol phosphate phosphatase. Proc Natl Acad Sci U S A 95(24):14057–14059

Ochman H, Soncini FC, Solomon F, Groisman EA (1996) Identification of a pathogenicity island required for Salmonella survival in host cells. Proc Natl Acad Sci U S A 93(15):7800–7804

Odendall C, Rolhion N, Forster A, Poh J, Lamont DJ, Liu M, Freemont PS, Catling AD, Holden DW (2012) The Salmonella kinase SteC targets the MAP kinase MEK to regulate the host actin cytoskeleton. Cell Host Microbe 12(5):657–668. https://doi.org/10. 1016/j.chom.2012.09.011

Pallett MA, Berger CN, Pearson JS, Hartland EL, Frankel G (2014) The type III secretion effector NleF of enteropathogenic Escherichia coli activates NF-kappaB early during infection. Infect Immun 82(11):4878–4888. https://doi.org/10.1128/IAI.02131-14

Patel JC, Galan JE (2006) Differential activation and function of Rho GTPases during Salmonella-host cell interactions. J Cell Biol 175(3):453–463. https://doi.org/10.1083/jcb.200605144

Pearson JS, Giogha C, Ong SY, Kennedy CL, Kelly M, Robinson KS, Lung TW, Mansell A, Riedmaier P, Oates CV, Zaid A, Muhlen S, Crepin VF, Marches O, Ang CS, Williamson NA, O'Reilly LA, Bankovacki A, Nachbur U, Infusini G, Webb AI, Silke J, Strasser A, Frankel G, Hartland EL (2013) A type III effector antagonizes death receptor signalling during bacterial gut infection. Nature 501(7466):247–251. https://doi.org/10.1038/nature12524

Pearson JS, Giogha C, Muhlen S, Nachbur U, Pham CL, Zhang Y, Hildebrand JM, Oates CV, Lung TW, Ingle D, Dagley LF, Bankovacki A, Petrie EJ, Schroeder GN, Crepin VF, Frankel G, Masters SL, Vince J, Murphy JM, Sunde M, Webb AI, Silke J, Hartland EL (2017) EspL is a bacterial cysteine protease effector that cleaves RHIM proteins to block necroptosis and inflammation. Nat Microbiol 2:16258. https://doi.org/10.1038/nmicrobiol.2016.258

Pham TH, Gao X, Singh G, Hardwidge PR (2013) Escherichia coli virulence protein NleH1 interaction with the v-Crk sarcoma virus CT10 oncogene-like protein (CRKL) governs NleH1 inhibition of the ribosomal protein S3 (RPS3)/nuclear factor kappaB (NF-kappaB) pathway. J Biol Chem 288(48):34567–34574. https://doi.org/10.1074/jbc.M113.512376

Pilar AV, Reid-Yu SA, Cooper CA, Mulder DT, Coombes BK (2012) GogB is an anti-inflammatory effector that limits tissue damage during Salmonella infection through interaction with human FBXO22 and Skp1. PLoS Pathog 8(6):e1002773. https://doi.org/10.1371/journal.ppat.1002773

Poh J, Odendall C, Spanos A, Boyle C, Liu M, Freemont P, Holden DW (2008) SteC is a Salmonella kinase required for SPI-2-dependent F-actin

remodelling. Cell Microbiol 10(1):20–30. https://doi.org/10.1111/j.1462-5822.2007.01010.x

Pollard DJ, Berger CN, So EC, Yu L, Hadavizadeh K, Jennings P, Tate EW, Choudhary JS, Frankel G (2018) Broad-Spectrum regulation of nonreceptor tyrosine kinases by the bacterial ADP-Ribosyltransferase EspJ. MBio 9(2):e00170–e00118. https://doi.org/10.1128/mBio.00170-18

Pollock GL, Oates CV, Giogha C, Wong Fok Lung T, Ong SY, Pearson JS, Hartland EL (2017) Distinct roles of the Antiapoptotic effectors NleB and NleF from Enteropathogenic Escherichia coli. Infect Immun 85 (4):e01071–e01016. https://doi.org/10.1128/IAI.01071-16

Quezada CM, Hicks SW, Galan JE, Stebbins CE (2009) A family of Salmonella virulence factors functions as a distinct class of autoregulated E3 ubiquitin ligases. Proc Natl Acad Sci U S A 106(12):4864–4869. https://doi.org/10.1073/pnas.0811058106

Rolhion N, Furniss RC, Grabe G, Ryan A, Liu M, Matthews SA, Holden DW (2016) Inhibition of nuclear transport of NF-kB p65 by the Salmonella type III secretion system effector SpvD. PLoS Pathog 12(5):e1005653. https://doi.org/10.1371/journal.ppat.1005653

Ruchaud-Sparagano MH, Muhlen S, Dean P, Kenny B (2011) The enteropathogenic E. coli (EPEC) Tir effector inhibits NF-kappaB activity by targeting TNFalpha receptor-associated factors. PLoS Pathog 7(12):e1002414. https://doi.org/10.1371/journal.ppat.1002414

Ruter C, Hardwidge PR (2014) 'Drugs from bugs': bacterial effector proteins as promising biological (immune-) therapeutics. FEMS Microbiol Lett 351(2):126–132. https://doi.org/10.1111/1574-6968.12333

Rytkonen A, Poh J, Garmendia J, Boyle C, Thompson A, Liu M, Freemont P, Hinton JC, Holden DW (2007) SseL, a Salmonella deubiquitinase required for macrophage killing and virulence. Proc Natl Acad Sci U S A 104(9):3502–3507. https://doi.org/10.1073/pnas.0610095104

Salomon D, Guo Y, Kinch LN, Grishin NV, Gardner KH, Orth K (2013) Effectors of animal and plant pathogens use a common domain to bind host phosphoinositides. Nat Commun 4:2973. https://doi.org/10.1038/ncomms3973

Scott NE, Giogha C, Pollock GL, Kennedy CL, Webb AI, Williamson NA, Pearson JS, Hartland EL (2017) The bacterial arginine glycosyltransferase effector NleB preferentially modifies Fas-associated death domain protein (FADD). J Biol Chem 292(42):17337–17350. https://doi.org/10.1074/jbc.M117.805036

Silberger DJ, Zindl CL, Weaver CT (2017) Citrobacter rodentium: a model enteropathogen for understanding the interplay of innate and adaptive components of type 3 immunity. Mucosal Immunol 10:1108–1117. https://doi.org/10.1038/mi.2017.47

Song T, Li K, Zhou W, Zhou J, Jin Y, Dai H, Xu T, Hu M, Ren H, Yue J, Liang L (2017) A type III effector NleF from EHEC inhibits epithelial inflammatory cell death by targeting Caspase-4. Biomed Res Int 2017:4101745. https://doi.org/10.1155/2017/4101745

Steele-Mortimer O (2008) The Salmonella-containing vacuole: moving with the times. Curr Opin Microbiol 11 (1):38–45. https://doi.org/10.1016/j.mib.2008.01.002

Stender S, Friebel A, Linder S, Rohde M, Mirold S, Hardt WD (2000) Identification of SopE2 from Salmonella typhimurium, a conserved guanine nucleotide exchange factor for Cdc42 of the host cell. Mol Microbiol 36(6):1206–1221

Sun J, Hobert ME, Rao AS, Neish AS, Madara JL (2004) Bacterial activation of beta-catenin signaling in human epithelia. Am J Physiol Gastrointest Liver Physiol 287 (1):G220–G227. https://doi.org/10.1152/ajpgi.00498. 2003

Sun SC, Chang JH, Jin J (2013) Regulation of nuclear factor-kappaB in autoimmunity. Trends Immunol 34 (6):282–289. https://doi.org/10.1016/j.it.2013.01.004

Sun H, Kamanova J, Lara-Tejero M, Galan JE (2016) A family of Salmonella type III secretion effector proteins selectively targets the NF-kappaB signaling pathway to preserve host homeostasis. PLoS Pathog 12(3):e1005484. https://doi.org/10.1371/journal.ppat. 1005484

Suresh R, Mosser DM (2013) Pattern recognition receptors in innate immunity, host defense, and immunopathology. Adv Physiol Educ 37(4):284–291. https://doi.org/10.1152/advan.00058.2013

Wan F, Weaver A, Gao X, Bern M, Hardwidge PR, Lenardo MJ (2011) IKKbeta phosphorylation regulates RPS3 nuclear translocation and NF-kappaB function during infection with Escherichia coli strain O157:H7. Nat Immunol 12(4):335–343. https://doi.org/10.1038/ ni.2007

Wei P, Wong WW, Park JS, Corcoran EE, Peisajovich SG, Onuffer JJ, Weiss A, Lim WA (2012) Bacterial virulence proteins as tools to rewire kinase pathways in yeast and immune cells. Nature 488(7411):384–388. https://doi.org/10.1038/nature11259

Wu S, Ye Z, Liu X, Zhao Y, Xia Y, Steiner A, Petrof EO, Claud EC, Sun J (2010) Salmonella typhimurium infection increases p53 acetylation in intestinal epithelial cells. Am J Physiol Gastrointest Liver Physiol 298 (5):G784–G794. https://doi.org/10.1152/ajpgi.00526. 2009

Xu X, Hensel M (2010) Systematic analysis of the SsrAB virulon of Salmonella enterica. Infect Immun 78 (1):49–58. https://doi.org/10.1128/IAI.00931-09

Yang Z, Soderholm A, Lung TW, Giogha C, Hill MM, Brown NF, Hartland E, Teasdale RD (2015) SseK3 is a Salmonella effector that binds TRIM32 and modulates the Host's NF-kappaB Signalling activity. PLoS One 10(9):e0138529. https://doi.org/10.1371/journal.pone. 0138529

Yao Q, Zhang L, Wan X, Chen J, Hu L, Ding X, Li L, Karar J, Peng H, Chen S, Huang N, Rauscher FJ 3rd, Shao F (2014) Structure and specificity of the bacterial cysteine methyltransferase effector NleE suggests a novel substrate in human DNA repair pathway. PLoS Pathog 10(11):e1004522. https://doi.org/10.1371/jour nal.ppat.1004522

Ye Z, Petrof EO, Boone D, Claud EC, Sun J (2007) Salmonella effector AvrA regulation of colonic epithelial cell inflammation by deubiquitination. Am J Pathol 171(3):882–892. https://doi.org/10.2353/ajpath.2007. 070220

Yen H, Ooka T, Iguchi A, Hayashi T, Sugimoto N, Tobe T (2010) NleC, a type III secretion protease, compromises NF-kappaB activation by targeting p65/RelA. PLoS Pathog 6(12):e1001231. https://doi. org/10.1371/journal.ppat.1001231

Yen H, Sugimoto N, Tobe T (2015) Enteropathogenic Escherichia coli uses NleA to inhibit NLRP3 Inflammasome activation. PLoS Pathog 11(9): e1005121. https://doi.org/10.1371/journal.ppat. 1005121

Yombi JC, Martins L, Vandercam B, Rodriguez-Villalobos H, Robert A (2015) Clinical features and outcome of typhoid fever and invasive non-typhoidal salmonellosis in a tertiary hospital in Belgium: analysis and review of the literature. Acta Clin Belg 70 (4):265–271. https://doi.org/10.1179/2295333715y. 0000000016

Young JC, Clements A, Lang AE, Garnett JA, Munera D, Arbeloa A, Pearson J, Hartland EL, Matthews SJ, Mousnier A, Barry DJ, Way M, Schlosser A, Aktories K, Frankel G (2014) The Escherichia coli effector EspJ blocks Src kinase activity via amidation and ADP ribosylation. Nat Commun 5:5887. https:// doi.org/10.1038/ncomms6887

Zhang L, Ding X, Cui J, Xu H, Chen J, Gong YN, Hu L, Zhou Y, Ge J, Lu Q, Liu L, Chen S, Shao F (2012) Cysteine methylation disrupts ubiquitin-chain sensing in NF-kappaB activation. Nature 481(7380):204–208. https://doi.org/10.1038/nature10690

Zhou D, Mooseker MS, Galan JE (1999a) An invasion-associated Salmonella protein modulates the actin-bundling activity of plastin. Proc Natl Acad Sci U S A 96(18):10176–10181

Zhou D, Mooseker MS, Galan JE (1999b) Role of the S. typhimurium actin-binding protein SipA in bacterial internalization. Science 283(5410):2092–2095

Adv Exp Med Biol - Protein Reviews (2019) 20: 219–240
https://doi.org/10.1007/5584_2018_297
© Springer Nature Switzerland AG 2018
Published online: 24 November 2018

New Techniques to Study Intracellular Receptors in Living Cells: Insights Into RIG-I-Like Receptor Signaling

M. J. Corby, Valerica Raicu, and David N. Frick

Abstract

This review discusses new developments in Förster resonance energy transfer (FRET) microscopy and its application to cellular receptors. The method is based on the kinetic theory of FRET, which can be used to predict FRET not only in dimers, but also higher order oligomers of donor and acceptor fluorophores. Models based on such FRET predictions can be fit to observed FRET efficiency histograms (also called FRET spectrograms) and used to estimate intracellular binding constants, free energy values, and stoichiometries. These "FRET spectrometry" methods have been used to analyze oligomers formed by various receptors in cell signaling pathways, but until recently such studies were limited to receptors residing on the cell surface. To study complexes residing inside the cell, a technique called Quantitative Micro-Spectroscopic Imaging (Q-MSI) was developed. Q-MSI combines determination of quaternary structure from pixel-level apparent FRET spectrograms with the determination of both donor and acceptor concentrations at the organelle level. This is done by resolving and analyzing the spectrum of a third fluorescent marker, which does not participate in FRET. Q-MSI was first used to study the interaction of a class of cytoplasmic receptors that bind viral RNA and signal an antiviral response via complexes formed mainly on mitochondrial membranes. Q-MSI revealed previously unknown RNA mitochondrial receptor orientations, and the interaction between the viral RNA receptor called LGP2 with the RNA helicase encoded by the hepatitis virus. The biological importance of these new observations is discussed.

Keywords

Antiviral response · ATPase · FRET · Hepatitis C virus · Innate immunity · LGP2 · MDA5 · RIG-I · RNA helicase

M. J. Corby
Department of Chemistry and Biochemistry, University of Wisconsin-Milwaukee, Milwaukee, WI, USA

V. Raicu (✉)
Departments of Physics and Biological Sciences, University of Wisconsin-Milwaukee, Milwaukee, WI, USA
e-mail: vraicu@uwm.edu

D. N. Frick (✉)
Department of Chemistry and Biochemistry, University of Wisconsin-Milwaukee, Milwaukee, WI, USA
e-mail: frickd@uwm.edu

Abbreviations

CARD	Capsase activation and recruitment domain
CFP	Cyan fluorescent protein
FP	Fluorescent protein
FRET	Förster resonance energy transfer
FSI	Fully quantitative spectral imaging
GFP	Green fluorescent protein

HCV	Hepatitis C virus
LGP2	Laboratory of genetics and physiology-2
MAVS	Mitochondrial antiviral signaling protein
MDA5	Melanoma differentiated antigen-5
NS3	Non-structural protein 3
NS4A	Non-structural protein 4A
PAMP	Pathogen associated molecular pattern
Poly (I:C)	Polyinosinic: polycytidylic acid
Q-MSI	Quantitative micro-spectroscopic imaging
RIG-I	Retinoic inducible gene-I
RLR	RIG-I like receptor
ROI	Region of interest
YFP	Yellow fluorescent protein

1   Introduction

Förster resonance energy transfer (FRET) (Forster 1946) has been used in microscopy for many years to study molecular interactions. FRET assays monitor the transfer of energy from a "donor" fluorophore to an "acceptor" fluorophore, which need to be within 10 nm (100 Å). Because FRET efficiency depends on the sixth power of the distance between the donor and acceptor (Stryer and Haugland 1967), FRET microscopy has been used to estimate the relative distances between two macromolecules, typically by fusing each to a different fluorescent protein (FP). Introduction of the kinetic theory of FRET (Raicu 2007) allowed one to predict FRET efficiencies in situations where more than one donor or acceptor participates in FRET and has expanded the amount of information that can be obtained using FRET-based microscopy (Patowary et al. 2015; King et al. 2017). For example, donor and acceptor concentrations, and pixel-level FRET efficiencies can be determined in various subcellular locations (Raicu et al. 2009). These data can be used with the Law of Mass Action to estimate intracellular binding constants, free energy values, and stoichiometries, and provide clues to the quaternary structure of oligomeric complexes (King et al. 2016; Raicu

and Schmidt 2017). This review discusses such FRET techniques, and how they have been used to analyze oligomers formed by various receptors in cell signaling pathways. Since much of this work with cell surface receptors was recently reviewed elsewhere (Raicu and Singh 2013; Raicu and Schmidt 2017), the focus here is mainly on the latest FRET methods designed to analyze receptors located in the cytoplasm or intracellular organelles, in particular. The receptors discussed here are called RIG-I like receptors (RLRs).

RLRs bind cytoplasmic nucleic acids known to be pathogen associated molecular patterns (PAMPs). Once activated, RLRs change conformation and signal through downstream enzymes to initiate transcription of interferons and other antiviral proteins (Fig. 1). RLRs locate their cytoplasmic ligands using motor domains resembling those found in helicases that separate the DNA double helix by disrupting Watson-Crick base pairs in an ATP-fueled reaction. The RLR helicases do not separate complementary strands, but instead scan cytoplasmic RNA to identify "non-self" motifs indicative of a viral infection.

The FRET methods discussed rely on the analysis of image stacks acquired at a series of emission wavelengths such that a spectrum is resolved at each pixel (Biener et al. 2013). These methods, termed "optical micro-spectroscopy," can be performed using commercially available two-photon (Biener et al. 2013) and possibly confocal microscopes with spectral resolution (Zimmerman et al. 2003). The main focus here involves the recent development of a technique called Quantitative Micro-Spectroscopic Imaging (Q-MSI) (Stoneman et al. 2017; Mishra et al. 2016; Corby et al. 2017). Q-MSI combines determination of quaternary structure from pixel-level apparent FRET efficiency histograms (also known as FRET spectrograms (Raicu and Singh 2013)) with the determination of both donor and acceptor concentrations at the organelle level (King et al. 2016). Q-MSI enables the ability to visualize the sub-cellular locations of both donors and acceptors by resolving and analyzing the spectrum of a third fluorescent marker, which does not participate in FRET.

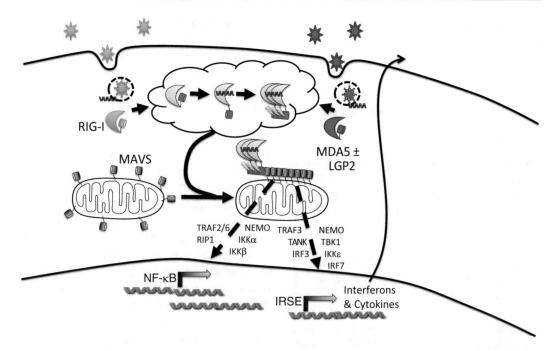

Fig. 1 Oligomer formation in RLR signaling. Various RNA viruses (red, blue) activate antiviral genes after being detected through the RLRs (RIG-I, MDA5, & LGP2) which all converge on MAVS (grey). Ligand binding leads to a conformational change exposing CARDs (purple) that are ubiquitinated (green) and seed oligomers formed by RLRs and a mitochondrial antiviral signaling protein (MAVS)

2   Overview of FRET

FRET between FPs fused to cellular proteins has been measured using microscopes for over 20 years, and numerous comprehensive reviews on the topic are available (Jares-Erijman and Jovin 2003; Piston and Kremers 2007; Shaner et al. 2005; Padilla-Parra and Tramier 2012; Bajar et al. 2016). Briefly, successful FRET experiments depend on the selection of an appropriate donor/acceptor FP pair. For example if *Aequorea victoria* GFP derivatives are used, the donor FP could be a green (GFP, or GFP2) or cyan (Cerulean or Sapphire) variant, while the acceptor could be a yellow variant, such as YFP, Citrine, or Venus. The donor should have a high quantum yield, while the excitation spectrum of the acceptor FP should have the highest spectral overlap possible with the donor emission spectrum, to ensure that it is able to receive the largest number of excitations from the donor. In most cases, it is best if the overlap between the donor emission spectrum and the acceptor absorption spectrum is optimized while minimizing any direct excitation by the acceptor at the donor excitation wavelength. In addition, FPs must be fused to target proteins so that they are not diametrically opposed in a complex, and so that they do not block key motifs needed for protein-protein interactions. FRET can be detected using various methods including filter sets designed to detect donor and acceptor fluorescence at one or two excitation wavelengths (Xia and Liu 2001; Hoppe et al. 2002), acceptor photobleaching (Tramier et al. 2006), fluorescence lifetime imaging (Lakowicz et al. 1992), changes in florescence polarization (Mattheyses et al. 2004), or by spectral imaging (Zimmermann et al. 2002). The strengths and weaknesses of each of these techniques have been rigorously discussed elsewhere (Jares-Erijman and Jovin 2003; Piston and Kremers 2007; Shaner et al. 2005; Padilla-Parra and Tramier 2012; Leavesley et al. 2013; Bajar et al. 2016; Raicu and Singh 2013). The methods below are based exclusively on spectral FRET imaging microscopy (Chen et al. 2007; Raicu et al. 2005, 2009).

2.1 Optical Micro-spectroscopy

The techniques described below were designed for use with two-photon microscopes capable of acquiring emission spectra of donor and acceptor molecules using either one or two excitation wavelengths (Raicu et al. 2009; Raicu and Singh 2013). In theory, however, they could be used with any system capable of acquiring emission spectra at pixel level. Two-photon excitation is preferred because it reduces bleaching of fluorophores located below or above the focal plane, allowing delicate, biological samples to be scanned multiple times with minimal damage. It also allows for the excitation of FPs at a wavelength far red-shifted from their emission spectrum, thus reducing bleed-through of the light used to excite the donor and reducing scattering.

FRET experiments may be performed using either one or two excitation wavelength scans. Experiments using a single excitation wavelength scan require carefully chosen donor and acceptor fluorophores, such that the acceptor does not fluoresce in the absence of the donor. Wild-type GFP and a yellow fluorescent protein are one example of a pair which meet this criteria. In addition, a detection method with spectral resolution is needed in order to separate the donor fluorescence from that of acceptor fluorescence stimulated by energy transfer from the donor (Raicu et al. 2009; Raicu and Singh 2013). It is essential that the fluorescence is captured simultaneously for all emission wavelengths and within a time shorter than the time it takes diffusion to move the molecules and their complexes in or out of the excitation volume. This ensures that the signals captured at each image pixel for all the emission wavelengths originate from the same donors and acceptors. Under those conditions, a single excitation scan gives the option to compute the FRET efficiency at a pixel level (Raicu et al. 2009). By contrast, with methods that need two

excitation scans to determine the FRET efficiency, the length of time (~60 s) needed to collect those scans leaves times for molecular diffusion between scans to scramble the molecular makeup of the sample at the point of excitation and hence the FRET efficiency may only be calculated as an average over many pixels (i.e., an entire area in an image). In our Q-MSI method described herein, a second excitation scan is used to excite the acceptors, such that their concentration is determined, in addition to the donors concentration and the energy of energy transfer, which can both be determined from the same scan as described above. As discussed below, a second scan also allows for the correction of any direct acceptor excitation at the first wavelength, which is sometimes referred to as "acceptor spectral bleed through" (Chen et al. 2007). As a result, donor concentrations and acceptor concentrations can also be calculated, though only donor concentration may be computed with pixel-level resolution at this time.

As already mentioned above, acquisition of a spectrum rather than recording fluorescence intensities at peak emission wavelengths allows separation of the donor and acceptor signals when FRET occurs. Composite spectra acquired from samples expressing both the donor and acceptor are unmixed using spectra obtained from cells expressing either donor alone or acceptor alone. By assuming that the mixed spectrum is a linear combination of the contributions of each individual spectrum, the values of scaling factors (k^l) can be adjusted to fit the measured composite spectrum (S^m):

$$S^m(\lambda_{ex,i}\ \lambda_{em}) = \Sigma_l \left[k^l(\lambda_{ex,i}) s^l(\lambda_{em}) \right] \quad (1)$$

In Eq. 1, i is the excitation wavelength index and l is the summation index for each spectrum present in the composite spectrum (donor (s^D) or acceptor, s^A). If only donors and acceptors exist (i.e. there no other fluorescent species present), the coefficients represent donor in the presence of acceptor (k^{DA}), acceptor in the presence of donor (k^{AD}), and are defined as the contributions of each fluorescent species present in the composite spectrum. They can be extracted using a least squares minimization method (Raicu et al. 2005; Biener et al. 2013; Patowary et al. 2015).

Unmixing fluorescence spectra on the image pixel level generates 2D intensity maps of total fluorescence (or number of photons) of the donor in the presence of acceptor (F^{DA}) and total fluorescence of the acceptor in the presence of donor (F^{AD}) (Raicu et al. 2005, 2009; Raicu and Singh 2013). F^{DA} is connected to a quantity k^{DA} to determine the total number of donor photons emitted per unit time by integrating the fluorescence spectra, $I^{DA}(\lambda_{em}) = k^{DA}i^{D}(\lambda_{em})$ over all emission wavelengths. The mathematical relation thus obtained is:

$$F^{DA} = k^{DA} \int i^{D}(\lambda_{em})d\lambda_{em} = k^{DA}w^{D} \quad (2)$$

In Eq. 2, $i^{D}(\lambda_{em})$ is the emission intensity of the donor spectrum measured in a sample only expressing donor and normalized to its maximum value, and w^{D} is the integral of the elementary spectrum of the donor.

Similarly, the k^{AD} value extracted from unmixing can be used to determine the total number of photons emitted by the acceptor per unit time by integrating the fluorescence spectrum $I^{AD}(\lambda_{em}) = k^{AD}i^{A}(\lambda_{em})$ over all emission wavelengths.

$$F^{AD} = k^{AD} \int i^{A}(\lambda_{em})d\lambda_{em} = k^{AD}w^{A} \quad (3)$$

In Eq. 3, $i^{A}(\lambda_{em})$ is the emission intensity of the acceptor spectrum measured in a sample only expressing acceptor and normalized to the maximum value and w^{A} is the integral of the elementary spectra of the acceptor.

Apparent FRET efficiency (E_{app}), or the average efficiency per donor in a mixture of donors and acceptors, can be then calculated on a pixel level using F^{DA} values:

$$E_{app} = 1 - \frac{F^{DA}}{F^{D}} \quad (4)$$

In Eq. 4, F^{D} represents the total fluorescence of the donor in the absence of acceptor and F^{DA} is

defined as the fluorescence of the donor in the presence of acceptor. F^{D} can be calculated assuming conservation of energy. In other words, F^{D} is equal to the fluorescence of the donor in the presence of the acceptor (F^{DA}) plus the energy lost to FRET through the interaction of the donor and the acceptor ($F^{D}(FRET)$).

$$F^{D} = F^{DA} + F^{D}(FRET) \quad (5)$$

$F^{D}(FRET)$ can be calculated using the quantum yield of the donor (Q^{D}) and the total number of excitations transferred via FRET, N^{FRET}, according to the equation:

$$N^{FRET}Q^{D} = F^{D}(FRET) \quad (6)$$

However, only a fraction of the transferred excitations will be emitted by the acceptor, as defined by the quantum yield of the acceptor, Q^{A}. Thus,

$$N^{FRET}Q^{A} = F^{A}(FRET) \quad (7)$$

Combining Eqs. 7 and 8 yields:

$$F^{D}(FRET) = \frac{Q^{D}}{Q^{A}} F^{A}(FRET) \quad (8)$$

$F^{A}(FRET)$ is related to the fluorescence of the acceptor in the presence of donor if there is no direct excitation of the acceptor:

$$F^{AD} = F^{A}(FRET) \quad (9)$$

Because F^{AD} is obtained from unmixing, $F^{D}(FRET)$ can be calculated from a combination of Eqs. 8 and 9. By combining terms, F^{D} can also be calculated:

$$F^{D} = k^{DA}w^{D} + \frac{Q^{D}}{Q^{A}} F^{AD} \quad (10)$$

E_{app} can therefore be calculated by combining Eqs. 4, 5, and 10:

$$E_{app} = 1 - \frac{k^{DA}w^{D}}{k^{DA}w^{D} + \frac{Q^{D}}{Q^{A}} k^{AD}w^{A}} \quad (11)$$

which can be simplified as:

$$E_{app} = \frac{1}{1 + \frac{Q^A k^{DA} w^D}{Q^D k^{AD} w^A}} \quad (12)$$

2.2 FRET Spectrometry

Pixel level FRET calculation allows for the collection of a large number of E_{app} values per unit area, and thus a more detailed analysis of precise sub-cellular locations where oligomeric complexes form. In addition, and perhaps more importantly, E_{app} values may be organized into histograms based on selected regions of interest. The histograms may have several unique peaks which can indicate the presence of varying combinations of donors and acceptors within a larger oligomeric complex. Therefore, the distributions of FRET efficiencies shown by such histograms contain much more information than simple averages over regions of "interest (which is the older and more widespread method of analysis of FRET data) and allows" one to extract the number and relative distances between protomers within an oligomer, or the *quaternary structure* (Raicu and Schmidt 2017). When the oligomers associate transiently and have a short lifetime relative to the integration time of the light detector, or if multiple combinations of donors and acceptors within the same oligomer are not visible within the cell, it is more convenient to collect the position of the predominant peaks in each cell-level histogram and then build a histogram of such peak positions, or a "meta-histogram." This meta-histogram represents the dominant oligomeric configurations within the larger population (Singh et al. 2013). Whether a histogram or a meta-histogram is used for data analysis, extraction of quaternary structure information proceeds the same way: Models of oligomers of different sizes and geometry are used to compute peak positions corresponding to each combination of donors and acceptors within a certain oligomer, and these are then fitted to the experimental E_{app} distributions by adjusting a set of fitting parameters that control the amplitudes and widths of the peaks and the

strength of the energy transfer, embodied within a quantity termed "pairwise FRET efficiency." The model that best fits the experimental data is taken as the most likely quaternary structure of the protein of interest. Because of the existence of distinct peaks that are visible in both the experimental and theoretical E_{app} histograms, the histograms resemble spectrograms as seen in many spectrometric methods (such as NMR, mass-spec, etc.), and this method is called "FRET spectrometry."

Previously, FRET spectrometry (Raicu et al. 2009) has been used mainly to study receptors that exist on the cell surface or are an integral part of the cellular membranes, the most recent examples being G-protein coupled receptors (Mishra et al. 2016; Stoneman et al. 2017), protein kinases (Mannan et al. 2013), ABC transporters (Singh et al. 2013), sigma-1 receptors (Mishra et al. 2015), and the receptor-like kinase EMS1 (Huang et al. 2017). An important advantage of this method is its ability to distinguish between biologically significant interactions, which lead to formation of protein complexes with possible biological functions, and random encounters between molecules caused by molecular crowding (King et al. 2017), which lead to the existence of the so-called "stochastic FRET." Specifically, while FRET methods based on averages over regions of interest do not distinguish between stochastic and functional interactions (because both effects are combined into a single average E_{app} value), FRET spectrograms reveal clear peaks corresponding to longer-lived oligomers and a broad distribution of FRET efficiency values corresponding to stochastic FRET (Singh and Raicu 2010).

2.3 Fully Quantitative Spectral Imaging (FSI) Using Two Excitation Wavelengths

FSI expanded on the FRET spectrometry technique to include the use of a second excitation wavelength chosen to excite the acceptor but not the donor (Stoneman et al. 2017; King et al. 2016). The second excitation scan allows for the correction of E_{app} calculations for direct acceptor excitation when present at the first excitation wavelength, and the calculation of acceptor and donor concentrations at each position in the image. As mentioned above, the calculation of the fluorescence emission of the donor in the absence of acceptor (F^D) requires a correction for any direct excitation of the acceptor. Thus, a second excitation scan allows for the calculation of E_{app} between a donor and acceptor even when there is significant direct excitation of the acceptor. We will explain this procedure next.

Although, Eq. 10 is not valid under conditions where both the donor and acceptor are excited at the same wavelength ($\lambda_{ex,1}$), direct excitation of the acceptor can be calculated from acceptor fluorescence at a second wavelength using the known ratio of the acceptor fluorescence at each wavelength ($\Gamma^{\lambda ex1,A}$ and $\Gamma^{\lambda ex2,A}$) and the following relationship:

$$F^A(\lambda_{ex,1}) = \frac{\Gamma^{\lambda ex,1,A}}{\Gamma^{\lambda ex2,A}} F^A(\lambda_{ex,2}) \qquad (13)$$

If a "gamma ratio" is defined as the ratio between the emission of the acceptor at the excitation wavelength (λ_{ex1}) and the second excitation wavelength (λ_{ex2}),

$$\rho^A \frac{\Gamma^{\lambda ex,1,A}}{\Gamma^{\lambda ex2,A}}, \qquad (14)$$

then F^{AD} values at each excitation wavelength, $\lambda_{ex,1}$ and $\lambda_{ex,2}$, can be used to calculate the donor and acceptor fluorescence in the absence of FRET for samples co-expressing both:

$$F^D(\lambda_{ex,1}) = F^{DA}(\lambda_{ex,1})$$
$$+ \frac{Q^D}{Q^A} \left[F^{AD}(\lambda_{ex,1}) - \rho^A F^{AD}(\lambda_{ex,2}) \right] \qquad (15)$$

$$F^A(\lambda_{ex,2}) = \frac{\rho^D F^{AD}(\lambda_{ex,2}) - F^{AD}(\lambda_{ex,1})}{\rho^D - \rho^A} \qquad (16)$$

By inserting Eq. (15) into Eq. (4), the average FRET efficiency (E_{app}) over regions of interest then can be calculated. In this way, both the FRET efficiency and the concentrations of the molecules of interest are determined. By plotting the average FRET efficiency against either the sum or the ratio of the donor and acceptor concentrations, it is possible to extract the proportion of associated and unassociated molecules in a sample. Using standard mathematical procedures, it is then possible to determine the dissociation constants and hence the binding energies of the oligomers (King et al. 2016).

Assuming that the fluorescence of a compound will increase linearly with concentration, a standard curve can be constructed which relates fluorescence intensity to concentration. Such a standard curve can be used to calculate the concentrations of the donor and acceptors within each regions of interest (ROI).

Although the ability to correct for direct excitation of the acceptor expands the application of FSI to systems where the donor and acceptor spectrum significantly overlap, it also comes with a tradeoff. If two excitation wavelengths are used in the calculation of E_{app} then two excitation scans must be collected to calculate FRET. If two excitation scans are acquired, then the time it takes to acquire those two scans allows molecular diffusion to occur such that the molecules imaged in the first scan are in different positions in the second scan (Patowary et al. 2015; King et al. 2016).

3 Q-MSI with Mito-Tracker Staining

Q-MSI (Corby et al. 2017) expands FSI and FRET spectrometry (Raicu and Singh 2013; King et al. 2016) by incorporating the use of a fluorescent marker to identify sub-cellular structures. The third fluorescent marker is also unmixed from the composite spectrum and generates a 2D fluorescence intensity map of the desired sub-cellular structure. Sub-cellular ROIs can then be selected within the unmixed fluorescence intensity maps representing the location of the sub-cellular structure being analyzed. The ROIs selected on the fluorescent marker intensity map are applied as a mask to the 2D fluorescent intensity maps generated through the unmixing of the composite spectrum to

Fig. 2 Quantitative Micro-Spectroscopic Imaging (Q-MSI). Q-MSI is a variant of FRET spectrometry, in which pixel-level spectra are used to calculate the FRET efficiency (E_{app}) at each pixel and the concentrations of donor (D) and acceptor (A) in each region of interest (ROI). Spectral unmixing is used to determine the fluorescence of the donor in the presence of acceptor (F^{DA}) and the fluorescence of the acceptor in the presence of donor (F^{AD}), at two different excitation wavelengths ($\lambda_{ex,1}$ and $\lambda_{ex,2}$). These values are used to calculate donor and acceptor concentrations from previously prepared standard curves, and E_{app} from the equation shown, where Q^D and Q^A are the quantum yields of the donor and acceptor and ρ^D and ρ^A the ratios of fluorescence intensities observed upon excitation of the donor and acceptor at $\lambda_{ex,1}$ and $\lambda_{ex,2}$. The difference between Q-MSI and previous quantitative FRET techniques is that it combines the analysis of a third spectrum used as a marker for a sub-cellular compartment. In the above example, MitoTracker is used to identify mitochondria

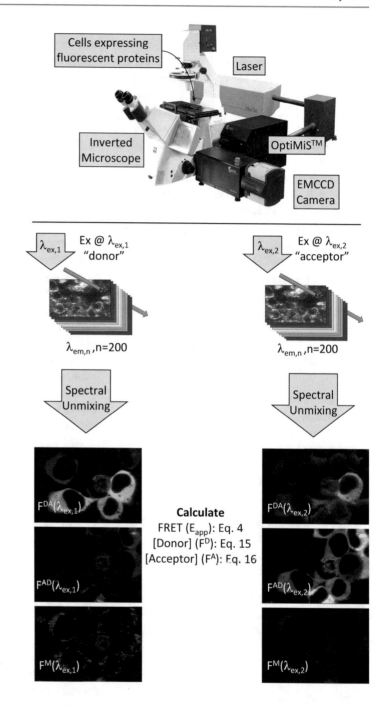

specifically calculate E_{app} within the sub-cellular region of interest (Fig. 2).

To generate the composite spectrum, a "FRET" scan is performed on the cells expressing the combination of CFP and YFP tagged proteins using the peak excitation of the CFP donor ($\lambda_{ex,1} = 840$ nm). To generate an image primarily exciting and targeting the acceptor fluorescence in the presence of donor, a second scan is taken at the acceptor peak excitation ($\lambda_{ex,2} = 960$ nm).

Typically, 15–30 fields of view at 100× magnification are collected for each well plate analyzed. The FRET scan and the acceptor scan are taken sequentially for each field of view and the image acquisition time is about 10 s for each excitation scan and about 40 s to change the wavelength between the FRET and acceptor scans. The total image acquisition time is therefore about 60 s to collect two sets of images each containing 440 x 300 pixels. The average laser power used during the measurements is about 200 mW per line (Biener et al. 2013) for both the FRET and acceptor scans and the power is held constant (Corby et al. 2017; Stoneman et al. 2017). The composite spectra are then unmixed (Raicu et al. 2005) at pixel level using the assumption that the composite spectrum is a linear combination of each of the elementary spectra per Eq. 1. The contributing spectra are donor in the presence of acceptor ($l = DA$), acceptor in the presence of donor ($l = AD$), and Mitotracker ($l = M$).

The coefficients equal to the fluorescent contributions of each of the above mentioned fluorescent species (k^{DA}, k^{AD}, k^M) are extracted using a least-squares minimization (Patowary et al. 2015), and E_{app} (FRET) is calculated as described above from abundance coefficients (k^{DA}, k^{AD}, k^M) for CFP (donor), YFP (acceptor) multiplied by the spectral integrals to calculate F^{DA}, F^{AD}, and F^M, and displayed as intensity maps. These maps are then used to generate regions of interest focused on the fluorescence intensity of the protein or marker of interest.

To calculate the total concentration of the donor-tagged proteins and the acceptor-tagged proteins in the dual-expressing cell samples, the values for F^D ($\lambda_{ex,1}$) and F^A ($\lambda_{ex,2}$) are applied to a standard curve of fluorescence intensity. The standard curve is generated using a dilution series of purified CFP (donor) and YFP (acceptor) which were imaged under the exact same conditions as the cells. Calibration curves are generated by plotting F^D ($\lambda_{ex,1}$) vs. CFP concentration and F^A($\lambda_{ex,2}$) vs. YFP concentration (Corby et al. 2017).

4 Q-MSI Analysis of Interactions Between RIG-I Like Receptors

In the first published Q-MSI study (Corby et al. 2017), HEK293T cells were transfected with various vector combinations to express CFP or YFP fused to one of the biologically active RLRs. The purpose was to determine if FRET spectrometry could be used to study intra-cellular receptors, specifically those that bound to RNA ligands.

4.1 RIG-I Like Receptors (RLRs)

The prototype RLR is RIG-I (retinoic acid-inducible gene 1), which consists of tandem N-terminal caspase activation, and recruitment domains (CARDs), a helicase domain, and a C-terminal ligand recognition domain (Schlee 2013). RIG-I recognizes viral RNA such as poly(U) tracks in RNA (Schnell et al. 2012), or uncapped RNA that still contain an intact triphosphate (Saito et al. 2008; Wang et al. 2010; Linehan et al. 2018). These RNAs bind RLRs causing CARDs to interact with their downstream effector MAVS. Like RIG-I, MDA5 (Melanoma Differentiation-Associated protein 5) also has tandem CARDs that regulate signaling and are modified by ubiquitination and phosphorylation. However, MDA5 and RIG-I recognize different PAMPS, with MDA5 sensing longer, mainly double-stranded RNA (Loo et al. 2008). When activated by an RNA pattern, MDA5 forms long oligomers on its target RNA (Berke and Modis 2012). LGP2 (laboratory of genetics and physiology 2) (Cui et al. 2001) differs from RIG-I and MDA5 in that it lacks CARDs needed to activate MAVS. Without CARDs, LGP2 signals PAMP detection by a path that involves other proteins, like MDA5 (Bruns et al. 2014). LGP2 interacts with MDA5 to prime the formation of oligomers on RNA, exposing CARDs on MDA5 that interact with CARDs on MAVS to initiate the formation of a signaling complex linked to the mitochondrion (Hei and Zhong 2017; Bruns and Horvath 2015).

The RLRs detect a diverse array of viruses. RIG-I detects rhabdoviruses (*e.g.* vesicular stomatitis virus and rabies virus), paramyxoviruses (*e.g.* Sendai virus and measles virus), orthomyxoviruses (*e.g.* influenza A and B), hepaciviruses (hepatitis C virus), and filoviruses (*e.g.* Ebola virus). MDA5 detects picornaviruses (*e.g.* polio and hepatitis A viruses). Both RIG-I and MDA5 are needed to respond to reoviruses and flaviviruses like, dengue, West Nile and perhaps Zika virus (Loo et al. 2008). The role for LGP2 is less clear, but it seems to also assist in the detection of several viruses. For example, Hei and Zhong reported that LGP2 helps the innate immune system respond to HCV infection. They used CRISPR-Cas to knock out the LGP2 gene and found that the cells produced less interferon upon HCV infection. LGP2 also enhances the ability of MDA5 to bind HCV RNA (Hei and Zhong 2017). All the viruses listed above elicit different antiviral responses via RIG-I or MDA5, however, so the nature of one or more of the components in the signaling pathway must differ in each case.

The oligomers formed by the CARDs of MAVS, RIG-I and/or MDA5 recruit a variety of other proteins to initiate interferon production. In one well-characterized pathway, MAVS activates TBK1 (tank binding kinase 1) and IKKε to phosphorylate IRF3 (interferon regulatory factor 3), which in turn dimerizes, migrates to the nucleus and activates transcription from Interferon Stimulated Response Elements (ISREs). Another pathway leads from MAVS through IKKα and IKKβ to NF-κB (Zhang et al. 2013; Belgnaoui et al. 2011) (Fig. 1).

Mutations in genes encoding RLRs are linked to diseases related to abnormal interferon production. For example, Aicardi-Goutières syndrome is linked to mutations in the gene encoding MDA5 (Oda et al. 2014), Singleton-Merten syndrome is linked to mutations in both MDA5 and RIG-I (Jang et al. 2015), and systemic lupus erythematosus is linked to MDA5 mutations (Cunninghame Graham et al. 2011).

The RLRs are part of the DExD/H-box protein family, which is named after the amino acid sequence of a motif in the conserved Walker-type nucleotide-binding site (Walker et al. 1982) where ATP binds to fuel helicase movements on RNA (Frick et al. 2007). DExD/H-box proteins can translocate as monomers, but they typically move more rapidly when associated with themselves or other proteins as oligomers on RNA.

RIG-I (Peisley et al. 2013), MDA5 (Berke and Modis 2012) and LGP2 (Murali et al. 2008; Bruns et al. 2014) all form filaments on duplex or single stranded RNA that have been observed with electron microscopy, x-ray crystallography, and immunofluorescence using either proteins isolated from cell lysates or recombinant proteins purified from *E. coli* or insect cells. When RIG-I and MDA5 bind activating RNA ligands and form oligomers, they change conformation to expose their CARDs, associate with K63-polyubiquitin chains (Zeng et al. 2010) and undergo E3-ligase tripartite motif-containing 25 (TRIM25)-dependent K63 polyubiquitination (Gack et al. 2007). The RLR CARDs then bind a CARD on the MAVS complex (MAVS is also known as IPS-1, CARDIF and VISA). MAVS then forms a long oligomer that can extend the length of a mitochondrion or bridge two mitochondria.

Unlike canonical helicases, RLRs do not separate RNA duplexes. Instead, they use ATP to locate RNA ligands that activate the interferon response. When their ability to cleave ATP is abolished, the receptors fail to induce interferon production (Myong et al. 2009) (reviewed in (Errett and Gale 2015)). The RLR helicase domains contain two tandem RecA-like motor domains. Unlike related proteins that are not RLRs, RIG-I's helicase motor domains are separated by an insertion domain at the N-terminal of hel2, called hel2i, and the RNA binding site can accommodate an RNA duplex rather than only single stranded RNA (Jiang et al. 2011; Kowalinski et al. 2011; Luo et al. 2011).

In addition to these modified motor domains, RLR's possess a C-terminal domain (CTD) that is also needed for RNA interactions. The CTD, is a Zn^{2+} containing regulatory domain that engages ligands in a recognition groove. Variability in the CTDs seems to account for some of the different ligand specificity of the RLRs (Cui et al. 2008).

For example, the RIG-I CTD binds $5'$ ppp dsRNA, ssRNA, and blunt-ended dsRNA more tightly than $5'$ ppp dsRNA (Lu et al. 2010). Whereas, MDA5 and LGP2 CTDs preferentially bind duplex RNA and do not differentiate between RNAs with modified $5'$ ends (Pippig et al. 2009) (reviewed in (Bruns and Horvath 2014)).

The CTDs of RLRs are connected to RLR motor domains by a bridging "pincer", which consists of two tandem alpha helices and connects the CTD to the hel2 domain. The pincer is needed for the interferon response, and deletion of portions of the pincer significantly reduces the ATPase activity in response to RNA binding (Luo et al. 2011; Kowalinski et al. 2011).

The tandem N-terminal CARDs present in RIG-I and MDA5 are responsible for downstream signaling. The CARDs belong to the "Death Domain" superfamily, and consist of six anti-parallel alpha helices. In other proteins, CARDs are needed for apoptotic signaling and inflammatory responses. When the CARDs of RIG-I oligomerize, they form a helical tetrameric unit (reviewed in (Ferrao and Wu 2012) and (Wu and Hur 2015)). The RIG-I tandem CARDs have been crystallized in conjunction with MAVS to reveal that CARDs create a stable tetramer that seeds the MAVS filament formation necessary for downstream signaling (Wu et al. 2013, 2014).

In sum, present data support a model where the RLR helicase domains form a base that is covered by two flaps, the CARDs and the C-terminal domain. In the absence of an RNA ligand, the CARDs block the helicase domain from interacting with "self" RNAs through a direct interaction. Recognition of "non-self" ligands (or PAMPS) causes a conformation change where the CTD swings away from helicase domains exposing CARDs for oligomerization to initiate downstream signaling (Ferrage et al. 2012). This topic has been recently reviewed (Schlee 2013, Goubau et al. 2013). Distinct ligand preferences of each RLR are discussed briefly below.

In vitro, RIG-I binds most tightly to 7–10 base pair duplex RNA with di- or triphosphates at a $5'$-end. Such RNAs are not normally present in the cytosol of healthy cells because host mRNAs are capped before exiting the nucleus after transcription. Other cellular RNAs either have $5'$-monophosphate ends or are masked by ribonuclear complexes. RIG-I ligand preference is tightly regulated to prevent self-RNA recognition (Marques et al. 2006). Recent evidence suggest that RIG-I might also differentiate subtle differences in RNA $5'$ caps. For example, RIG-I can bind to m-7-guanosine capped RNA but it cannot accommodate 2'O-methylation of the first $5'$ nucleotide and therefore self-capped RNAs are excluded from RIG-I recognition (Devarkar et al. 2016).

Unlike RIG-I, MDA5 does not discriminate among RNA $5'$-ends. Instead, MDA5 binds with highest affinity to double-stranded RNAs greater than 2,000 base pairs that exhibit perfect complementarity. Such long duplex RNAs are rare in the cytoplasm, and ATP mediates the increased binding stability of MDA5 for long duplex RNA (Louber et al. 2015). ATP hydrolysis shifts the stable MDA5 filament formation towards a preference for long duplex RNA and increases the likelihood that MDA5 will dissociate from the short filaments (Peisley et al. 2011).

LGP2 differs from RIG-I and MDA5 in that it lacks the CARDs needed to activate MAVS. This suggests that LGP2 plays a unique role in the innate immune system by signaling non-self RNA detection through a path that does not involve CARDs, but instead involves an interaction with MDA5 (Satoh et al. 2010). It has been shown that LGP2 can form active dimers (Murali et al. 2008) and oligomers that interact with MDA5 filaments (reviewed in (Bruns and Horvath 2014)). In addition, LGP2 negatively regulates RIG-I signaling, possibly through substrate sequestration (Rodriguez et al. 2014). LGP2 favors double-stranded RNA ligands and binds with highest affinity to the ends of RNA, however, there is no difference in binding affinity for RNA ligands with or without $5'$-triphosphates. It is possible that LGP2's preference for the ends of dsRNA is in direct competition with RIG-I's RNA end preference and this competition is what causes the observed hindrance of RIG-I signaling when LGP2 is upregulated. LGP2 enhances

MDA5 filament formation and thus MDA5 signaling activity (Bruns et al. 2014; Errett and Gale 2015; Uchikawa et al. 2016).

4.2   Q-MSI Analysis of RLR Interactions

To better understand how the RLRs interact with each other and RNA ligands in cells, the first Q-MSI study (Corby et al. 2017) was initiated to examine cells exposed to polyinosinic: polycytidylic acid (poly(I:C)), which elicits RLR accumulation at mitochondria and subsequent antiviral signaling (Corby et al. 2017). In that study, Q-MSI was used to observe high FRET in CFP-LGP2:YFP-MDA5, CFP-MDA5:YFP-LGP2, and CFP-LGP2:YFP-LGP2, and lower FRET in CFP-RIG-I:YFP-RIG-I and CFP-MDA5:YFP-MDA5 complexes (Table 1), demonstrating that the former interactions are not artifacts due to CFP binding to YFP. FRET also changed significantly in the presence of the RLR ligand poly(I:C).

The key new observations facilitated by Q-MS1 in the Corby *et al.* study (Corby et al. 2017), concerned the nature of the LGP2:MDA5 complex, which had been known to form in cells (Bruns and Horvath 2015). Prior work suggested that LGP2 simply helps prime MDA5 filaments, suggesting that more MDA5 is present in oligomers than LGP2. Q-MSI did not support this model. Instead, Q-MSI data supported a model where more LGP2 was present at the mitochondria, that LGP2:LGP2 complexes dissociated in the presence of an RLR ligand, and that the new complexes formed with MDA5 formed in the presence of poly(I:C) contained more LGP2 than MDA5 (Corby et al. 2017).

The first Q-MSI experiments with RLRs were performed using commercially available CFP and YFP proteins (Corby et al. 2017). These FPs are not ideal for FRET studies, however, because CFP has a relatively low quantum yield, and the CFP two-photon excitation spectrum overlaps the excitation spectrum of YFP, which leads to inadvertent acceptor excitation by laser light. The RLR open reading frames were therefore subcloned into vectors expressing other GFP variants that can be used in experiments requiring only a single excitation scan so E_{app} values could be determined at pixel level resolution. One of the best FRET donors is a protein called "GFP$_2$" (Zimmermann et al. 2002; Stoneman et al. 2017) and one of the best acceptors is a protein called "Venus" (Nagai et al. 2002). Unlike CFP, GFP$_2$ excites at a lower $\lambda_{ex,1}$, and it displays a larger Stokes shift, with a higher peak excitation than CFP. Unlike YFP, Venus has less excitation at lower wavelengths. Variants of GFP$_2$ and Venus used also incorporate an A206K mutation to minimize non-specific dimerization seen with many GFP derivatives (von Stetten et al. 2012). These new fusion proteins have allowed the extraction of pixel-level E_{app} directly from the "FRET scan" without the need to perform corrections for acceptor direct excitation.

Table 1 FRET efficiencies observed at mitochondria regions of interest (RLRs) in HEK293T cells expressing various RLR fusion proteins in the presence and absence of the RLR ligand poly(I:C)

Donor (CFP)	Acceptor (YFP)	Poly(I:C)	ROIs analyzed (#)	E_{app} (Ave ± SD)
RIG-I	RIG-I	−	132	2 ± 4%
		+	132	4 ± 17%
MDA5	MDA5	−	200	3 ± 6%
		+	200	6 ± 10%
LGP2	LGP2	−	594	11 ± 12%
		+	594	7 ± 6%
LGP2	MDA5	−	210	2 ± 3%
		+	187	3 ± 4%
MDA5	LGP2	−	436	**8 ± 1%**
		+	430	**21 ± 27%**

Data from Corby et al. (2017)

Pixel level, fully quantitative FRET analysis was used to examine the interaction between GFP$_2$-LGP2 and Venus-MDA5 at mitochondrial membranes. Plasmids expressing recombinant GFP$_2$-LGP2 and Venus-MDA5 were transfected into HEK293T cells. The cells were allowed to incubate and express the fluorescently tagged proteins for 20 h. Each set was then transfected with either poly(I:C) (0.36 μg/mL) or vehicle, incubated for 4 h, and stained with MitoTracker. After elementary spectra were collected using cells solely expressing GFP$_2$-LGP2, Venus-MDA5, or only stained with MitoTracker Red, composite spectra were unmixed, to calculate E_{app} at every pixel in the image. A map depicting the 2D spatial distribution of pixel-level E_{app} values was also generated (Fig. 3).

The k^m (MitoTracker) map was then used to identify the mitochondrial regions of the cell (Fig. 4a) and this mitochondrial region mask was applied to the E_{app} map (Fig. 4b) to select only the mitochondrial regions. For each mitochondrial region of interest, a histogram of E_{app} peaks was created. Predominant peaks (modes) were then selected in the collection of E_{app} histograms and compiled into a meta-histogram (Fig. 4c). The E_{app} range was between 0–30%, which was similar to what was seen before with CFP-LGP2 and YFP-MDA5 (Corby et al. 2017). When poly(I:C) was added to the cells, a sub-set

of cells exhibited E_{app} values larger than 30% and up to a maximum of 40% (Fig. 4a), again reproducing results with CFP-LGP2 and YFP-MDA5 (Corby et al. 2017).

To analyze LGP2:MDA5 complexes in the cytoplasm, another set of ROIs using the k^{DA} values (i.e. LGP2) to create a mask in images of cells expressing both GFP$_2$-LGP2 and Venus-MDA5 (Fig. 4c). The predominant peaks were selected from each of the E_{app} histograms and compiled into a meta-histogram (Fig. 4d). E_{app} values for the cells not treated with poly(I:C) were between 0–50% with two clear populations (Fig. 4e), in agreement with what was seen previously with CFP-LGP2 and YFP-MDA5 (Corby et al. 2017). However, the population of higher E_{app} values (30–50%), was significantly smaller when the cells were transfected with poly(I:C) (Fig. 4e).

4.3 Significance of the Q-MSI Observations with RLRs

The above data reveal new insights into how MDA5 and LGP2 interact upon ligand binding, confirming predictions made in Corby et al. (Corby et al. 2017). Past models of RLR function predicted that MDA5 and LGP2 interact in the cytoplasm and jointly recognize RNA ligands,

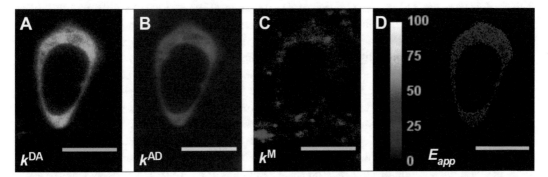

Fig. 3 Pixel-level FRET determined from spectrally resolved GFP2-LGP2, Venus-MDA5, and Mitotracker. Cells were co-transfected with plasmids expressing GFP$_2$-LGP2 and Venus-MDA5. The mitochondria were stained with MitoTracker-Red and FRET was calculated at a pixel level. (**a**) The spectrally-resolved 2D fluorescence intensity map for GFP$_2$-LGP2 (k^{DA}). (**b**) The 2D fluorescence intensity map for Venus-MDA5 (k^{AD}). (**c**) The 2D fluorescence intensity map for MitoTracker (k^M). (**d**) The FRET intensity distribution for cells co-expressing GFP$_2$-LGP2 and Venus-MDA5 (bars = 10 μm)

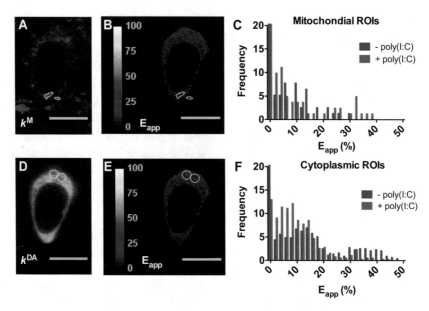

Fig. 4 **Pixel level FRET analysis of the GFP$_2$-LGP2: Venus-MDA5 interaction.** (a–c) Selection and analysis of E_{app} in mitochondria regions of cells co-expressing GFP$_2$-LGP2 and Venus-MDA5 using MitoTracker. The selected mitochondrial ROIs (a) then used as a mask and applied to the E_{app} intensity map (b), and E_{app} values in each mitochondrial region were analyzed to identify the most common E_{app} value within that ROI, which were plotted on a meta-histogram (c) compiling all the peak E_{app} values across all mitochondrial ROIs selected. (d–f) The same set of images used for analysis in (a–c) were re-analyzed by selecting cytoplasmic regions using a consistent circle comprising 146 pixels on k^{DA} intensity maps such that most of the cell's cytoplasm was selected but none of the circles were over-lapping (d). The mask created was applied to the E_{app} intensity map (e) to select random cytoplasmic regions of a consistent size and most common E_{app} values in each region was used to generate a meta-histograms comprising all peak E_{app} values selected (f). Meta histograms in (c) and (f) compare the results of cells analyzed in the presence (red) or absence (blue) of poly(I:C) RNA (bars = 10 μm)

forming oligomeric complexes before interacting with MAVs at the mitochondria (Rodriguez et al. 2014; Bruns and Horvath 2015; Bruns et al. 2014). However, the Q-MSI data reveal more complex details regarding how these interactions change when the LGP2:MDA5 complex migrates to the mitochondria to prime oligomerization of the MAVS complex.

In all experiments with RLRs, the highest FRET efficiencies observed were between LGP2 and MDA5. When YFP-LGP2 was co-expressed with CFP-MDA5 in the absence of poly I:C, there was a small population of E_{app} values as high as 10%, but with the addition of poly I:C, the E_{app} values shifted above 20%, with some reaching as high as 80–90% (Table 1). When the tags were switched, the overall FRET efficiency decreased but there was still an overall shift to higher E_{app} values with the addition of poly I:C. In addition to the shift in E_{app} values, the molar fraction of acceptor shifted when the tags were switched between LGP2 and MDA5 (Corby et al. 2017). When the acceptor was tagged to LGP2 and the donor attached to MDA5, a shift towards higher molar fractions of acceptor was visible. In contrast, when the donor was LGP2 and the acceptor was MDA5, the molar fraction of acceptor shifted lower. All this data combined suggests that LGP2 does not just prime RNA filaments for MDA5 interaction, but rather is the dominant protein in the MDA5: LGP2 oligomer (Corby et al. 2017). Addition of an RNA ligand likely leads to accumulation of LGP2 at mitochondria through interaction with either endogenous or exogenous MDA5. The interaction of LGP2:LGP2 at the mitochondria in response to poly(I:C) contained

Fig. 5 Models for LGP2: MDA5 oligomers. (a) Proposed model for LGP2: MDA5 oligomer where LGP2 functions as an endcap or primer for the MDA5 filament. **(b)** Proposed model for the LGP2:MDA5 oligomer where LGP2 is the predominant protomer in the overall MDA5:LGP2 oligomer

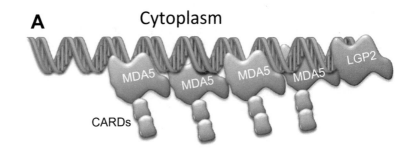

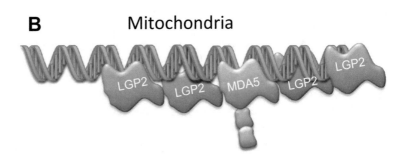

overall larger E_{app} values than the self-interaction of either MDA5 or RIG-I with E_{app} reaching values as high as 30% both with and without poly(I:C). Interestingly, the FRET efficiency for the LGP2 self-interaction was highest in the absence of poly(I:C) and then decreased when poly(I:C) was added. This possibly indicates that the presence of a PAMP encourages LGP2 interaction with MDA5 to aid in the PAMP recognition, thus decreasing the self LGP2:LGP2 interaction (Fig. 5).

This MAVS interaction might explain the difference in FRET efficiencies observed in the cytoplasm and mitochondria. If the oligomeric complexes formed in the cytoplasm between MDA5 and LGP2 is solely between those two proteins and the RNA ligand, it's possible that the E_{app} in the cytoplasm would be overall higher than the when the complex forms at the mitochondria, where MDA5 also interacts with the MAVS. The interaction of the MAVs CARDs may shift the overall E_{app} value for the complex at the mitochondria to be slightly lower and therefore, the overall population displays a slightly lower predominant E_{app} peak.

5 The Hepatitis C Virus Helicase Interacts with the Human Pattern Recognition Receptor LGP2

Many viruses that establish chronic infections often do so by disrupting detection by RLRs. One common technique is for viral proteins to cleave adaptor proteins. For example, the hepatitis C virus (HCV) encodes a protease that cleaves MAVS (Li et al. 2005; Sumpter et al. 2005) and TRIF (Li et al. 2005), thus cutting off pattern receptor signaling and allowing the virus to establish a persistent infection (Li et al. 2005). The HCV protease catalyzing these events is part of a multifunctional protein that has a helicase domain homologous to the RLR helicase domains. A Q-MSI project was therefore initiated to examine if this HCV protease/helicase (called NS3) interacts with one of the RLRs.

HCV is a blood borne pathogen that infects human hepatocytes. First identified in 1988 (Choo et al. 1989), HCV slowly destroys the liver, and if left untreated causes cirrhosis, hepatocellular carcinoma, and liver failure. HCV is in the *Flaviviridae* family, the members of which have a positive-sense RNA genome with one open reading frame encoding an approximately 3,000 amino acid-long polyprotein. The HCV polyprotein is processed by host and viral proteases into structural (core, E1, and E2) and nonstructural proteins (p7, NS2, NS3, NS4A, NS4B, NS5A, and NS5B) (Gu and Rice 2013). The main protein in this set involved in RLR evasion is NS3.

NS3 contains covalently linked protease and helicase domains. Such a linkage has only been observed to date in *Flaviviridae* NS3 proteins, and the biological advantage for linking the two activities is unclear. In vitro, NS3 protease activity is higher when the helicase domain is present, and the activity of the helicase is higher when the protease domain is present (Frick et al. 2004; Beran et al. 2007; Beran and Pyle 2008; Aydin et al. 2013).

The NS3 protease is most active upon binding of the NS4A peptide and it cleaves several sites in the HCV poly-protein: NS3/NS4A (*in cis*), NS4A/NS5B, NS4B/NS5A, and NS5A/NS5B (*in trans*) (Bartenschlager et al. 1994). NS4A binding to NS3 influences the position of the catalytic triad (His[57], Asp[81], and Ser[139]) in the active site to activate the serine protease function of NS3 (Tomei et al. 1996; Kim et al. 1996). The polyprotein processing occurs in a membranous replication web closely associated with the ER membrane and containing vesicles with HCV non-structural proteins, HCV RNA, and lipid droplets. Outside this web, NS3:NS4A disrupts RLR signaling by cleaving MAVS (Loo et al. 2006) and it also cleaves TRIF (Li et al. 2005). Toll-like receptors, signal through TRIF, which initiates down-stream signaling and the production of interferon. This second cleavage effectively abrogates Toll-like receptor signaling in the cell (Akira et al. 2006; Liu and Gale 2010; Errett and Gale 2015).

Like the RLRs, the NS3 helicase is a DExD/H-box protein in helicase super family 2 (Gorbalenya and Koonin 1993; Byrd and Raney 2012). ATP binds NS3 between two RecA-like motor domains (Walker et al. 1982)

near a phosphate binding loop and a catalytic base (Glu[291]) that activates the water molecule needed for ATP hydrolysis. The activated water attacks the γ-phosphate on the ATP (Frick et al. 2007).

5.1 RLRs and HCV Detection

HCV is recognized by the innate immune system by both TLRs (Li et al. 2005) and RLRs (Loo et al. 2006). Defects in RLRs facilitate robust HCV replication in cell culture (Sumpter et al. 2005). For example, in the cell line most permissive to HCV replication there is a mutation that prevents proper RIG-I CARD ubiquitination (reviewed in (Liu and Gale 2010)). MDA5 has also been implicated in HCV recognition (Cao et al. 2015), possibly by binding duplex RNA replication intermediates formed by the HCV RNA-dependent RNA polymerase (reviewed in (Errett and Gale 2015)). LGP2 is likely a regulator of HCV infection because cell lines with low LGP2 levels express less interferon response to HCV infection. LGP2 likely exerts its effect upstream of MDA5 and the presence of HCV RNA enhances the interaction between MDA5 and LGP2, which also enhances MDA5 interaction with RNA (Hei and Zhong 2017).

HCV's regulation of the RLR pathway is clearly mediated by the NS3:NS4A complex, which cleaves MAVS, liberating it from the mitochondrial surface and abrogating its downstream signaling which leads to activation of IFN and other cytokines. Cleavage of MAVS lowers RLR signaling, and the prevention of MAVS cleavage through mutation of the MAVS cleavage site restores the RIG-I like receptor function (Foy et al. 2003).

5.2 Observation of NS3:RLR Interactions with Q-MSI

Given the key role of NS3:NS4A in RLR evasion, a Q-MSI experiment was designed to test the possibility that NS3 helicase aids in the cleavage of MAVS by helping to localize the NS3 protease domain to MAVS through interaction with one of the RLRs. All the RLRs and HCV NS3 possess similar helicase domains known to oligomerize, and which have a preference for HCV RNA (Banerjee and Dasgupta 2001; Schnell et al. 2012). This interaction might aid in localizing NS3 near MAVS for cleavage and provide one possible explanation for linking the protease and helicase functions in HCV.

To examine the interaction of NS3 with the RLRs, cells were designed to express NS3 proteins tagged with CFP and various RLRs tagged with YFP. The elementary spectrum of the CFP and the YFP were used to unmix composite CFP:YFP spectra. Q-MSI was first used to analyze cells transfected with CFP-NS3:NS4A and either YFP-RIG-I, YFP-MDA5, or YFP-LGP2. In this case, the regions of interest were whole cells that displayed both a CFP and YFP signal. This experiment only revealed an interaction between CFP-NS3:NS4A and YFP-LGP2 (Corby et al. 2017).

To explore which NS3 domains were needed for the NS3:LGP2 interaction, CFP was attached to truncated versions of NS3:NS4A. A CFP-NS3: NS4A construct expressed the full length NS3, including both the helicase and protease domains, and the full length NS4A protein including both the protease co-factor and membrane anchors. In a second construct, the NS4A membrane anchors were absent (CFP-scNS3-4A). In a third, all of NS4A was absent (CFP-NS3). In the fourth, the NS3 protease region was absent (CFP-NS3h), and in the fifth the helicase was absent (CFP-NS3p). Clear FRET was observed with YFP-LGP2 and each truncated protein except CFP-NS3p, suggesting the interaction is mediated by the NS3 helicase domain (Corby et al. 2017).

Because other studies showed that LGP2 responds to poly(I:C) in cells (Childs et al. 2013), Q-MSI was also used to examine the effect of poly(I:C) on the NS3:LGP2 interaction. When whole cells were analyzed, E_{app} values increased with the addition of poly(I:C), indicating more NS3 bound LGP2 in the cytoplasm. In addition, a shift in CFP and YFP protein concentration was observed with the addition of poly(I:C). When poly(I:C) was added, a noticeable population of cells expressed less of each fusion protein, possibly indicating that poly(I:C) treatment caused an interferon-induced repression of protein translation. When only mitochondrial regions were analyzed however, E_{app} values at the mitochondria did not increase significantly upon poly(I:C) addition (Corby et al. 2017).

5.3 Significance of the LGP2:NS3 Interaction

The NS3:LGP2 interaction might be biologically relevant if it helps NS3 locate cleavage targets, such as MAVS. This is a key new discovery because prior work had not revealed a role for the HCV helicase in assisting RLR evasion. For example, ectopic expression of an NS3 mutant lacking the ability to unwind RNA blocks IRF-3 activation down stream of MAVs in response to Sendai virus infection (Foy et al. 2003), and NS3 lacking the helicase domain cleaves MAVs *in vitro* (Horner et al. 2012). These prior findings, however, do not rule out the possibility that LGP2 positions NS3 in the cell near MAVS. The ability of LGP2 to shuttle NS3 to sites with high MAVS concentrations might also be enhanced by the LGP2 interaction with MDA5. LGP2 is known to bind MDA5 to aid in its recognition and binding of viral substrates. MDA5 then, in turn, exposes its CARDs to interact with the CARDs on MAVS. NS3 might therefore bind to the LGP2:MDA5 complex, with MDA5 forming a bridge between NS3 and MAVS (Fig. 6).

6 Conclusions and Future Directions

Q-MSI allows for the determination of FRET and fluorescent protein concentrations within specific sub-cellular regions in living cells under different conditions. The method was made possible by the incorporation of a third fluorescent marker to identify mitochondria. Q-MSI showed oligomerization of RIG-I and MDA5, and that these interactions increased with the addition of poly (I:C). The increase in FRET efficiency was

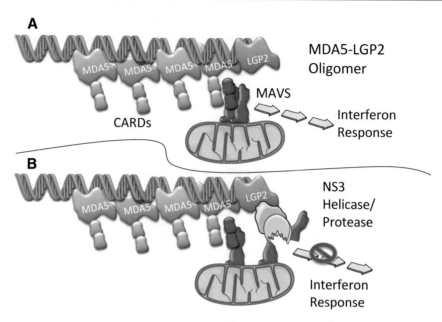

Fig. 6 Model for HCV NS3 interacting with an RLR oligomer to facilitate the cleavage of MAVS. (a) Helicase domains of RLRs scan the cytoplasm for viral PAMPs like duplex RNA. Upon PAMP binding, RLRs change conformation to expose CARDs, which bind to CARDs on MAVS (green) to form a signaling complex. This triggers a kinase cascade leading to phosphorylation of IRF-3 and transcription of pro-inflammatory cytokine and interferon genes. (b) The viral NS3 helicase (grey), which interacts with LGP2 (red), is covalently tethered to a protease (yellow), which cleaves MAVS and other key proteins needed to initiate the interferon response

greater overall for MDA5 than for RIG-I. The largest FRET efficiency increase in response to RNA, however, was seen between LGP2 and MDA5 (Table 1). This interaction was previously studied and it is known that LGP2 and MDA5 interact on RNA ligands (reviewed in (Bruns and Horvath 2015)). When fluorescently tagged LGP2 and MDA5 were co-expressed there was a small population of FRET efficiencies greater than 10%, but upon addition of the RNA ligand, the FRET efficiencies increased dramatically, with some reaching as high as 80–90% (Corby et al. 2017). These high FRET values suggest that LGP2 might not simply "prime" RNA filaments for MDA5 assembly as has been previously proposed. Instead, LGP2 might be the dominant protomer in the LGP2:MDA5 oligomer (Fig. 5). Q-MSI also was used to identify a new, possibly biologically relevant, interaction between LGP2 and the HCV NS3 helicase. LGP2 was the only RLR which showed any significant interaction with HCV NS3 and domain deletion mutants of NS3 showed that the interaction of NS3 and LGP2 is due, at least in part, to the helicase domain.

These studies show that Q-MSI could be useful for studying protein: protein and protein: ligand interactions in living cells at a subcellular level. It is important to note that proteins could be localized using a wide variety of markers besides MitoTracker. For example, there is presently evidence that the RLRs assemble not only at the mitochondria but also at peroxisomes (Dixit et al. 2010) and mitochondrial-associated endoplasmic reticulum membranes (MAMs) (Horner et al. 2011). Q-MSI could be used to quantify interactions in these regions using probes like red fluorescent fusions of PSS-1 (for MAMs), PMP70 (for peroxisomes) and TOM20 (for mitochondria).

Q-MSI is therefore now a powerful technique that could be applied to study almost any protein: protein interaction in live cells and determine fluorescent protein concentrations in sub-cellular regions of interest. The next challenge will be to enable FRET spectrograms to be generated in regions other than those at the cellular surface or the plasma membrane. In the past, FRET spectrometry (Raicu et al. 2009) has been used mainly to study receptors that exist on the cell surface or are integral parts of cellular membranes, the most recent examples being G-protein coupled receptors (Mishra et al. 2016; Stoneman et al. 2017), protein kinases (Mannan et al. 2013), ABC transporters (Singh et al. 2013), and sigma-1 receptors (Mishra et al. 2015). The challenge now is to refine Q-MSI to permit similar studies to be performed with protein complexes that reside in other parts of the cell.

Acknowledgements Funding for this research was provided by the National Institutes of Health (Grant No. RO1AI088001 awarded to DNF), UWM Research Growth Innitiative (Grant No. 101X333 awarded to DNF and VR) and the National Science Foundation, Major Research Instrumentation Program (Grants No. PHY-1126386 and PHY-1626450 awarded to V.R.).

References

Akira S, Uematsu S, Takeuchi O (2006) Pathogen recognition and innate immunity. Cell 124:783–801

Aydin C, Mukherjee S, Hanson AM, Frick DN, Schiffer CA (2013) The interdomain interface in bifunctional enzyme protein 3/4A (NS3/4A) regulates protease and helicase activities. Protein Sci 22:1786–1798

Bajar BT, Wang ES, Zhang S, Lin MZ, Chu J (2016) A guide to fluorescent protein FRET Pairs. Sensors (Basel) 16:1488. https://doi.org/10.3390/s16091488

Banerjee R, Dasgupta A (2001) Specific interaction of hepatitis C virus protease/helicase NS3 with the 3′-terminal sequences of viral positive- and negative-strand RNA. J Virol 75:1708–1721

Bartenschlager R, Ahlborn-Laake L, Mous J, Jacobsen H (1994) Kinetic and structural analyses of hepatitis C virus polyprotein processing. J Virol 68:5045–5055

Belgnaoui SM, Paz S, Hiscott J (2011) Orchestrating the interferon antiviral response through the mitochondrial antiviral signaling (MAVS) adapter. Curr Opin Immunol 23:564–572

Beran RK, Pyle AM (2008) Hepatitis C viral NS3-4A protease activity is enhanced by the NS3 helicase. J Biol Chem 283:29929–29937

Beran RK, Serebrov V, Pyle AM (2007) The serine protease domain of hepatitis C viral NS3 activates RNA helicase activity by promoting the binding of RNA substrate. J Biol Chem 282:34913–34920

Berke IC, Modis Y (2012) MDA5 cooperatively forms dimers and ATP-sensitive filaments upon binding double-stranded RNA. EMBO J 31:1714–1726

Biener G, Stoneman MR, Acbas G, Holz JD, Orlova M, Komarova L, Kuchin S, Raicu V (2013) Development and experimental testing of an optical microspectroscopic technique incorporating true line-scan excitation. Int J Mol Sci 15:261–276

Bruns AM, Horvath CM (2014) Antiviral RNA recognition and assembly by RLR family innate immune sensors. Cytokine Growth Factor Rev 25:507–512

Bruns AM, Horvath CM (2015) LGP2 synergy with MDA5 in RLR-mediated RNA recognition and antiviral signaling. Cytokine 74:198–206

Bruns AM, Leser GP, Lamb RA, Horvath CM (2014) The innate immune sensor LGP2 activates antiviral signaling by regulating MDA5-RNA interaction and filament assembly. Mol Cell 55:771–781

Byrd AK, Raney KD (2012) Superfamily 2 helicases. Front Biosci (Landmark Ed) 17:2070–2088

Cao X, Ding Q, Lu J, Tao W, Huang B, Zhao Y, Niu J, Liu YJ, Zhong J (2015) MDA5 plays a critical role in interferon response during hepatitis C virus infection. J Hepatol 62:771–778

Chen Y, Mauldin JP, Day RN, Periasamy A (2007) Characterization of spectral FRET imaging microscopy for monitoring nuclear protein interactions. J Microsc 228:139–152

Childs KS, Randall RE, Goodbourn S (2013) LGP2 plays a critical role in sensitizing mda-5 to activation by double-stranded RNA. PLoS One 8:e64202

Choo QL, Kuo G, Weiner AJ, Overby LR, Bradley DW, Houghton M (1989) Isolation of a cDNA clone derived from a blood-borne non-A, non-B viral hepatitis genome. Science 244:359–362

Corby MJ, Stoneman MR, Biener G, Paprocki JD, Kolli R, Raicu V, Frick DN (2017) Quantitative microspectroscopic imaging reveals viral and cellular RNA helicase interactions in live cells. J Biol Chem 292:11165–11177

Cui Y, Li M, Walton KD, Sun K, Hanover JA, Furth PA, Hennighausen L (2001) The Stat3/5 locus encodes novel endoplasmic reticulum and helicase-like proteins that are preferentially expressed in normal and neoplastic mammary tissue. Genomics 78:129–134

Cui S, Eisenächer K, Kirchhofer A, Brzózka K, Lammens A, Lammens K, Fujita T, Conzelmann KK, Krug A, Hopfner KP (2008) The C-terminal regulatory domain is the RNA 5′-triphosphate sensor of RIG-I. Mol Cell 29:169–179

Cunninghame Graham DS, Morris DL, Bhangale TR, Criswell LA, Syvänen AC, Rönnblom L, Behrens TW, Graham RR, Vyse TJ (2011) Association of NCF2, IKZF1, IRF8, IFIH1, and TYK2 with systemic lupus erythematosus. PLoS Genet 7:e1002341

Devarkar SC, Wang C, Miller MT, Ramanathan A, Jiang F, Khan AG, Patel SS, Marcotrigiano J (2016) Structural basis for m7G recognition and 2′-O-methyl discrimination in capped RNAs by the innate immune receptor RIG-I. Proc Natl Acad Sci U S A 113:596–601

Dixit E, Boulant S, Zhang Y, Lee AS, Odendall C, Shum B, Hacohen N, Chen ZJ, Whelan SP, Fransen M, Nibert ML, Superti-Furga G, Kagan JC (2010) Peroxisomes are signaling platforms for antiviral innate immunity. Cell 141:668–681

Errett JS, Gale M (2015) Emerging complexity and new roles for the RIG-I-like receptors in innate antiviral immunity. Virol Sin 30:163–173

Ferrage F, Dutta K, Nistal-Villán E, Patel JR, Sánchez-Aparicio MT, De Ioannes P, Buku A, Aseguinolaza GG, García-Sastre A, Aggarwal AK (2012) Structure and dynamics of the second CARD of human RIG-I provide mechanistic insights into regulation of RIG-I activation. Structure 20:2048–2061

Ferrao R, Wu H (2012) Helical assembly in the death domain (DD) superfamily. Curr Opin Struct Biol 22:241–247

Forster T (1946) Energiewanderung und fluoreszenz. Naturwissenschaften 33:166–175

Foy E, Li K, Wang C, Sumpter R, Ikeda M, Lemon SM, Gale M (2003) Regulation of interferon regulatory factor-3 by the hepatitis C virus serine protease. Science 300:1145–1148

Frick DN, Rypma RS, Lam AM, Gu B (2004) The nonstructural protein 3 protease/helicase requires an intact protease domain to unwind duplex RNA efficiently. J Biol Chem 279:1269–1280

Frick DN, Banik S, Rypma RS (2007) Role of divalent metal cations in ATP hydrolysis catalyzed by the hepatitis C virus NS3 helicase: magnesium provides a bridge for ATP to fuel unwinding. J Mol Biol 365:1017–1032

Gack MU, Shin YC, Joo CH, Urano T, Liang C, Sun L, Takeuchi O, Akira S, Chen Z, Inoue S, Jung JU (2007) TRIM25 RING-finger E3 ubiquitin ligase is essential for RIG-I-mediated antiviral activity. Nature 446:916–920

Gorbalenya AE, Koonin EV (1993) Helicases: amino acid sequence comparisons and structure-function relationships. Curr Opin Struct Biol 3:419–429

Goubau D, Deddouche S, Reis E Sousa C (2013) Cytosolic sensing of viruses. Immunity 38:855–869

Gu M, Rice CM (2013) Structures of hepatitis C virus nonstructural proteins required for replicase assembly and function. Curr Opin Virol 3:129–136

Hei L, Zhong J (2017) Laboratory of genetics and physiology 2 (LGP2) plays an essential role in hepatitis C virus infection-induced interferon responses. Hepatology 65:1478–1491

Hoppe A, Christensen K, Swanson JA (2002) Fluorescence resonance energy transfer-based stoichiometry in living cells. Biophys J 83:3652–3664

Horner SM, Liu HM, Park HS, Briley J, Gale M (2011) Mitochondrial-associated endoplasmic reticulum membranes (MAM) form innate immune synapses and are targeted by hepatitis C virus. Proc Natl Acad Sci U S A 108:14590–14595

Horner SM, Park HS, Gale M (2012) Control of innate immune signaling and membrane targeting by the Hepatitis C virus NS3/4A protease are governed by the NS3 helix $\alpha 0$. J Virol 86:3112–3120

Huang J, Li Z, Biener G, Xiong E, Malik S, Eaton N, Zhao CZ, Raicu V, Kong H, Zhao D (2017) Carbonic anhydrases function in anther cell differentiation downstream of the receptor-like kinase EMS1. Plant Cell 29:1335–1356

Jang MA, Kim EK, Now H, Nguyen NT, Kim WJ, Yoo JY, Lee J, Jeong YM, Kim CH, Kim OH, Sohn S, Nam SH, Hong Y, Lee YS, Chang SA, Jang SY, Kim JW, Lee MS, Lim SY, Sung KS, Park KT, Kim BJ, Lee JH, Kim DK, Kee C, Ki CS (2015) Mutations in DDX58, which encodes RIG-I, cause atypical Singleton-Merten syndrome. Am J Hum Genet 96:266–274

Jares-Erijman EA, Jovin TM (2003) FRET imaging. Nat Biotechnol 21:1387–1395

Jiang F, Ramanathan A, Miller MT, Tang GQ, Gale M, Patel SS, Marcotrigiano J (2011) Structural basis of RNA recognition and activation by innate immune receptor RIG-I. Nature 479:423–427

Kim JL, Morgenstern KA, Lin C, Fox T, Dwyer MD, Landro JA, Chambers SP, Markland W, Lepre CA, O'Malley ET, Harbeson SL, Rice CM, Murcko MA, Caron PR, Thomson JA (1996) Crystal structure of the hepatitis C virus NS3 protease domain complexed with a synthetic NS4A cofactor peptide. Cell 87:343–355

King C, Stoneman M, Raicu V, Hristova K (2016) Fully quantified spectral imaging reveals in vivo membrane protein interactions. Integr Biol (Camb) 8:216–229

King C, Raicu V, Hristova K (2017) Understanding the FRET signatures of interacting membrane proteins. J Biol Chem 292:5291–5310

Kowalinski E, Lunardi T, McCarthy AA, Louber J, Brunel J, Grigorov B, Gerlier D, Cusack S (2011) Structural basis for the activation of innate immune pattern-recognition receptor RIG-I by viral RNA. Cell 147:423–435

Lakowicz JR, Szmacinski H, Nowaczyk K, Berndt KW, Johnson M (1992) Fluorescence lifetime imaging. Anal Biochem 202:316–330

Leavesley SJ, Britain AL, Cichon LK, Nikolaev VO, Rich TC (2013) Assessing FRET using spectral techniques. Cytometry A 83:898–912

Li K, Foy E, Ferreon JC, Nakamura M, Ferreon AC, Ikeda M, Ray SC, Gale M, Lemon SM (2005) Immune evasion by hepatitis C virus NS3/4A protease-mediated cleavage of the Toll-like receptor 3 adaptor protein TRIF. Proc Natl Acad Sci U S A 102:2992–2997

Linehan MM, Dickey TH, Molinari ES, Fitzgerald ME, Potapova O, Iwasaki A, Pyle AM (2018) A minimal RNA ligand for potent RIG-I activation in living mice. Sci Adv 4:e1701854

Liu HM, Gale M (2010) Hepatitis C virus evasion from RIG-I-dependent hepatic innate immunity. Gastroenterol Res Pract 2010:548390

Loo YM, Owen DM, Li K, Erickson AK, Johnson CL, Fish PM, Carney DS, Wang T, Ishida H, Yoneyama M, Fujita T, Saito T, Lee WM, Hagedorn CH, Lau DT, Weinman SA, Lemon SM, Gale M (2006) Viral and therapeutic control of IFN-beta promoter stimulator 1 during hepatitis C virus infection. Proc Natl Acad Sci U S A 103:6001–6006

Loo YM, Fornek J, Crochet N, Bajwa G, Perwitasari O, Martinez-Sobrido L, Akira S, Gill MA, García-Sastre A, Katze MG, Gale M (2008) Distinct RIG-I and MDA5 signaling by RNA viruses in innate immunity. J Virol 82:335–345

Louber J, Brunel J, Uchikawa E, Cusack S, Gerlier D (2015) Kinetic discrimination of self/non-self RNA by the ATPase activity of RIG-I and MDA5. BMC Biol 13:54

Lu C, Xu H, Ranjith-Kumar CT, Brooks MT, Hou TY, Hu F, Herr AB, Strong RK, Kao CC, Li P (2010) The structural basis of 5′ triphosphate double-stranded RNA recognition by RIG-I C-terminal domain. Structure 18:1032–1043

Luo D, Ding SC, Vela A, Kohlway A, Lindenbach BD, Pyle AM (2011) Structural insights into RNA recognition by RIG-I. Cell 147:409–422

Mannan MA, Shadrick WR, Biener G, Shin BS, Anshu A, Raicu V, Frick DN, Dey M (2013) An ire1-phk1 chimera reveals a dispensable role of autokinase activity in endoplasmic reticulum stress response. J Mol Biol 425:2083–2099

Marques JT, Devosse T, Wang D, Zamanian-Daryoush M, Serbinowski P, Hartmann R, Fujita T, Behlke MA, Williams BR (2006) A structural basis for discriminating between self and nonself double-stranded RNAs in mammalian cells. Nat Biotechnol 24:559–565

Mattheyses AL, Hoppe AD, Axelrod D (2004) Polarized fluorescence resonance energy transfer microscopy. Biophys J 87:2787–2797

Mishra AK, Mavlyutov T, Singh DR, Biener G, Yang J, Oliver JA, Ruoho A, Raicu V (2015) The sigma-1 receptors are present in monomeric and oligomeric forms in living cells in the presence and absence of ligands. Biochem J 466:263–271

Mishra AK, Gragg M, Stoneman MR, Biener G, Oliver JA, Miszta P, Filipek S, Raicu V, Park PS (2016) Quaternary structures of opsin in live cells revealed by FRET spectrometry. Biochem J 473:3819–3836

Murali A, Li X, Ranjith-Kumar CT, Bhardwaj K, Holzenburg A, Li P, Kao CC (2008) Structure and function of LGP2, a DEX(D/H) helicase that regulates the innate immunity response. J Biol Chem 283:15825–15833

Myong S, Cui S, Cornish PV, Kirchhofer A, Gack MU, Jung JU, Hopfner KP, Ha T (2009) Cytosolic viral sensor RIG-I is a 5′-triphosphate-dependent translocase on double-stranded RNA. Science 323:1070–1074

Nagai T, Ibata K, Park ES, Kubota M, Mikoshiba K, Miyawaki A (2002) A variant of yellow fluorescent protein with fast and efficient maturation for cell-biological applications. Nat Biotechnol 20:87–90

Oda H, Nakagawa K, Abe J, Awaya T, Funabiki M, Hijikata A, Nishikomori R, Funatsuka M, Ohshima Y, Sugawara Y, Yasumi T, Kato H, Shirai T, Ohara O, Fujita T, Heike T (2014) Aicardi-Goutières syndrome is caused by IFIH1 mutations. Am J Hum Genet 95:121–125

Padilla-Parra S, Tramier M (2012) FRET microscopy in the living cell: different approaches, strengths and weaknesses. BioEssays 34:369–376

Patowary S, Pisterzi LF, Biener G, Holz JD, Oliver JA, Wells JW, Raicu V (2015) Experimental verification of the kinetic theory of FRET using optical microspectroscopy and obligate oligomers. Biophys J 108:1613–1622

Peisley A, Lin C, Wu B, Orme-Johnson M, Liu M, Walz T, Hur S (2011) Cooperative assembly and dynamic disassembly of MDA5 filaments for viral dsRNA recognition. Proc Natl Acad Sci U S A 108:21010–21015

Peisley A, Wu B, Yao H, Walz T, Hur S (2013) RIG-I forms signaling-competent filaments in an ATP-dependent, ubiquitin-independent manner. Mol Cell 51:573–583

Pippig DA, Hellmuth JC, Cui S, Kirchhofer A, Lammens K, Lammens A, Schmidt A, Rothenfusser S, Hopfner KP (2009) The regulatory domain of the RIG-I family ATPase LGP2 senses double-stranded RNA. Nucleic Acids Res 37:2014–2025

Piston DW, Kremers GJ (2007) Fluorescent protein FRET: the good, the bad and the ugly. Trends Biochem Sci 32:407–414

Raicu V (2007) Efficiency of resonance energy transfer in homo-oligomeric complexes of proteins. J Biol Phys 33:109–127

Raicu V, Schmidt WF (2017) Advanced microscopy techniques. In: Herrick-Davis K, Milligan G, Di Giovanni G (eds) G-protein-coupled receptor dimers. Humana Press, Cham

Raicu V, Singh DR (2013) FRET spectrometry: a new tool for the determination of protein quaternary structure in living cells. Biophys J 105:1937–1945

Raicu V, Jansma DB, Miller RJ, Friesen JD (2005) Protein interaction quantified in vivo by spectrally resolved fluorescence resonance energy transfer. Biochem J 385:265–277

Raicu V, Stoneman MR, Fung R, Melnichuk M, Jansma DB, Pisterzi LF, Rath S, Fox M, Wells JW, Saldin DK (2009) Determination of supramolecular structure and spatial distribution of protein complexes in living cells. Nat Photonics 3:107–113

Rodriguez KR, Bruns AM, Horvath CM (2014) MDA5 and LGP2: accomplices and antagonists of antiviral signal transduction. J Virol 88:8194–8200

Saito T, Owen DM, Jiang F, Marcotrigiano J, Gale M (2008) Innate immunity induced by composition-dependent RIG-I recognition of hepatitis C virus RNA. Nature 454:523–527

Satoh T, Kato H, Kumagai Y, Yoneyama M, Sato S, Matsushita K, Tsujimura T, Fujita T, Akira S, Takeuchi O (2010) LGP2 is a positive regulator of RIG-I- and MDA5-mediated antiviral responses. Proc Natl Acad Sci U S A 107:1512–1517

Schlee M (2013) Master sensors of pathogenic RNA - RIG-I like receptors. Immunobiology 218:1322–1335

Schnell G, Loo YM, Marcotrigiano J, Gale M (2012) Uridine composition of the poly-U/UC tract of HCV RNA defines non-self recognition by RIG-I. PLoS Pathog 8:e1002839

Shaner NC, Steinbach PA, Tsien RY (2005) A guide to choosing fluorescent proteins. Nat Methods 2:905–909

Singh DR, Raicu V (2010) Comparison between whole distribution- and average-based approaches to the determination of fluorescence resonance energy transfer efficiency in ensembles of proteins in living cells. Biophys J 98:2127–2135

Singh DR, Mohammad MM, Patowary S, Stoneman MR, Oliver JA, Movileanu L, Raicu V (2013) Determination of the quaternary structure of a bacterial ATP-binding cassette (ABC) transporter in living cells. Integr Biol (Camb) 5:312–323

Stoneman MR, Paprocki JD, Biener G, Yokoi K, Shevade A, Kuchin S, Raicu V (2017) Quaternary structure of the yeast pheromone receptor Ste2 in living cells. Biochim Biophys Acta 1859:1456–1464

Stryer L, Haugland RP (1967) Energy transfer: a spectroscopic ruler. Proc Natl Acad Sci U S A 58:719–726

Sumpter R, Loo YM, Foy E, Li K, Yoneyama M, Fujita T, Lemon SM, Gale M (2005) Regulating intracellular antiviral defense and permissiveness to hepatitis C virus RNA replication through a cellular RNA helicase, RIG-I. J Virol 79:2689–2699

Tomei L, Failla C, Vitale RL, Bianchi E, De Francesco R (1996) A central hydrophobic domain of the hepatitis C virus NS4A protein is necessary and sufficient for the activation of the NS3 protease. J Gen Virol 77:1065–1070

Tramier M, Zahid M, Mevel JC, Masse MJ, Coppey-Moisan M (2006) Sensitivity of CFP/YFP and GFP/mCherry pairs to donor photobleaching on FRET determination by fluorescence lifetime imaging microscopy in living cells. Microsc Res Tech 69:933–939

Uchikawa E, Lethier M, Malet H, Brunel J, Gerlier D, Cusack S (2016) Structural analysis of dsRNA binding to anti-viral pattern recognition receptors LGP2 and MDA5. Mol Cell 62:586–602

von Stetten D, Noirclerc-Savoye M, Goedhart J, Gadella TW, Royant A (2012) Structure of a fluorescent protein from Aequorea victoria bearing the obligate-monomer mutation A206K. Acta Crystallogr Sect F Struct Biol Cryst Commun 68:878–882

Walker JE, Saraste M, Runswick MJ, Gay NJ (1982) Distantly related sequences in the alpha- and beta-subunits of ATP synthase, myosin, kinases and other ATP-requiring enzymes and a common nucleotide binding fold. EMBO J 1:945–951

Wang Y, Ludwig J, Schuberth C, Goldeck M, Schlee M, Li H, Juranek S, Sheng G, Micura R, Tuschl T, Hartmann G, Patel DJ (2010) Structural and functional insights into 5′-ppp RNA pattern recognition by the innate immune receptor RIG-I. Nat Struct Mol Biol 17:781–787

Wu B, Hur S (2015) How RIG-I like receptors activate MAVS. Curr Opin Virol 12:91–98

Wu B, Peisley A, Richards C, Yao H, Zeng X, Lin C, Chu F, Walz T, Hur S (2013) Structural basis for dsRNA recognition, filament formation, and antiviral signal activation by MDA5. Cell 152:276–289

Wu B, Peisley A, Tetrault D, Li Z, Egelman EH, Magor KE, Walz T, Penczek PA, Hur S (2014) Molecular imprinting as a signal-activation mechanism of the viral RNA sensor RIG-I. Mol Cell 55:511–523

Xia Z, Liu Y (2001) Reliable and global measurement of fluorescence resonance energy transfer using fluorescence microscopes. Biophys J 81:2395–2402

Zeng W, Sun L, Jiang X, Chen X, Hou F, Adhikari A, Xu M, Chen ZJ (2010) Reconstitution of the RIG-I pathway reveals a signaling role of unanchored polyubiquitin chains in innate immunity. Cell 141:315–330

Zhang HX, Liu ZX, Sun YP, Zhu J, Lu SY, Liu XS, Huang QH, Xie YY, Zhu HB, Dang SY, Chen HF, Zheng GY, Li YX, Kuang Y, Fei J, Chen SJ, Chen Z, Wang ZG (2013) Rig-I regulates NF-κB activity through binding to Nf-κb1 3′-UTR mRNA. Proc Natl Acad Sci U S A 110:6459–6464

Zimmermann T, Rietdorf J, Girod A, Georget V, Pepperkok R (2002) Spectral imaging and linear un-mixing enables improved FRET efficiency with a novel GFP2-YFP FRET pair. FEBS Lett 531:245–249

Zimmermann T, Rietdorf J, Pepperkok R (2003) Spectral imaging and its applications in live cell microscopy. FEBS Lett 546:87–92

Adv Exp Med Biol - Protein Reviews (2019) 20: 241
https://doi.org/10.1007/5584_2019_390
© Springer Nature Switzerland AG 2019
Published online: 19 July 2019

Correction to: Polyphosphoinositide-Binding Domains: Insights from Peripheral Membrane and Lipid-Transfer Proteins

Joshua G. Pemberton and Tamas Balla

Correction to:
Chapter "Polyphosphoinositide-Binding Domains: Insights from Peripheral Membrane and Lipid-Transfer Proteins" in:
J. G. Pemberton and T. Balla, Adv Exp Med Biol - Protein Reviews,
https://doi.org/10.1007/5584_2018_288

This chapter was inadvertently published with an incorrect copyright holder. It has now been updated as

"This is a U.S. government work and not under copyright protection in the U.S.; foreign copyright protection may apply 2019".

The updated version of this chapter can be found at https://doi.org/10.1007/5584_2018_288

Adv Exp Med Biol - Protein Reviews (2019) 20: 243–249
https://doi.org/10.1007/978-3-030-14339-8

Index

A

Aagaard, K., 9
Aakko, J., 10, 11
Actinobacteria, 5, 13
Acute infectious diarrhea (AID)
    AMP-dependent chloride channel, 110
    clinical features of, 109–110
    large-scale immunization programs, 110
    LGG, 112
    prevention of, 113–114
    probiotic metabolites, 111
    probiotic strains, 117
    RV infection, 110
    secretory diarrhea, 111
    treatment of, 115–116
    Ussing chamber technique, 111
Adhesive-invasive *E. coli* (AIEC), 73
Aitoro, R., 42, 43, 61
Akedo, I., 75, 76
Akobeng, A.K., 75
Alard, J., 105
Albenberg, L., 102
Alber, D.G., 9
Alcohol dehydrogenase (ADH), 90
Alisi, A., 98
Alterio, T., 85–98
Amarasekara, R., 9
Amarri, S., 72
Amniotic fluid, 11
Amogan, H.P., 11
Amoroso, A., 61
Andrew, H., 75
Annese, V., 75
Antigen-presenting cells (APC), 52
Antony, K.M., 9
Apical sodium-dependent bile acid transporter (ASBT), 87
Arakawa, Y., 75, 76
Arboleya, S., 13
Atopic disease, 13

B

Bacterial exopolysaccharides, 41
Bacterial vaginosis (BV), 15
Bacteria-produced SCFAs, 63

Bacteroides, 4, 12
Bacteroidetes, 5, 102
Bahrami, B., 75
Baldassano, R.N., 75, 104
Balk, F., 61
Banaszkiewicz, A., 130
Baranowski, J.R., 25–33
Barbato, M., 72
Barbera, C., 72
Barbosa, T., 42
Barera, G., 72
Barrett, H.L., 10
Bassols, J., 10
Basson, A.R., 102
Bauserman, M., 129
Bellantoni, A., 72
Benitez-Paez, A., 75
Bercik, P., 54
Berezin, I., 54
Berni Canani, R., 57–65
Bibiloni, R., 75
Bifidobacteria, 51
Bifidobacteria-fermented milk (BFM), 75
Bifidobacterium, 12
Bifidobacterium breve, 74, 116
Bifidobacterium breve BBG-01, 31
Bifidobacterium longum, 92
Bile salt export pump (BSEP), 87
Blackett, K.L., 75
β-lactoglobulin (BLG), 62
Bonnaud, E., 73
Borchers, A.T., 61
Borthakur, A., 61
Bousvaros, A., 73
Boutillier, D., 105
Braat, H., 61
Brandimarte, G., 75
Breton, J., 105
Briggs, C.M., 10
Bronsky, J., 103
Bruno, C., 57–65
Buccigrossi, V., 109–117
Buhimschi, C.S., 11
Bu, L.N., 130

Buono, A., 57–65
Butterworth, A.D., 75
Butyrate-producing bacteria, 51

C
Calvo, C., 75
Camarca, A., 75
Cammarota, G., 104
Campieri, M., 75
Cao, B., 9, 10
Capilla, A., 75
Cardile, S., 85–98
Carreras-Badosa, G., 10
Castellaneta, S., 72
Castellano, E., 72
Castillejo, G., 75
Catassi, C., 72
Cavallo, N., 75
Celiac disease (CeD)
    CDGEMM study, 76–77
    microbiome composition, manipulation of, 73–74
    microbiome, role of, 72
    pathogenesis of, 70–72
    prevention and/or treatment, 74–75
Centers for Disease Control and Prevention (CDC), 27
Chang, H.-Y., 30
Chang, J.-H., 30
Chardon, P., 73
Chau, K., 126
Chen, J.-H., 30
Cheon, J.H., 102
Che, X., 102
Choi, Y.O., 61
Choline, 88
Choline deficient diet (CDD), 88
Chronic intestinal diseases, 75
Claud, E.C., 25–33
Clostridium difficile (C. diff)
    clinical symptoms, 139
    management of, 140–141
    nucleic acid amplification testing, 140
    prevention of, 141–143
Coccorullo, P., 130
Cognitive behavioral therapy (CBT), 128
Collado, M.C., 3–17
Collins, S.M., 54
Cominelli, F., 102
Community state types (CSTs), 4
Conlon, B., 9
Constipation, 130–133
Corynebacterium, 14
Cow's milk allergy (CMA), 59
Cox, S.B., 103
Crohn's disease (CD), 70, 101
Cucchiara, S., 75
Cummings, J.H., 75

D
Dal Bello, F., 75
D'amico, T., 75
Danese, S., 75
Dargenio, V.N., 49–54
De Angelis, M., 75
Debruyn, J., 104
Delaney, M.L., 9
Del Chierico, F., 90
Della Gatta, G., 57–65
De Palma, G., 75
Der, T., 54
Derwa, Y., 104
De Simone, C., 75
Devaney, K.L., 73
Diaz, H., 104
Di Cagno, R., 75
Di Costanzo, M., 57–65
DiGiulio, D.B., 11
Di Giulio, E., 75
Dimova, T., 10
Di Nardo, G., 75
Dinleyici, M., 139–144
Di Rosa, S., 75
Di Scala, C., 57–65
Dissanayake, V.H.W., 9
Djerov, L., 10
Donat, E., 75
Dong, X.D., 9
Dore, J., 73
Dowd, S.E., 103
Doyle, R.M., 9, 10
DuBois, A.M., 9
Du, N., 11
Dysbiosis, 26

E
Eftekhari, K., 129
Ehara, A., 75, 76
Elisei, W., 75
Endotoxins, 51
Engering, A., 61
Enterobacteriaceae, 28, 102
Enterococcus, 5
Enterococcus faecium, 11
Escherichia, 13
Extensively hydrolyzed casein formula (EHCF), 63

F
Famouri, F., 98
Farland, Mc., 142
Fasano, A., 69–77
Fecal microbiota transplantation (FMT), 96–98, 141
Fedele, M.C., 109–117
Fedorak, R., 104
Fedorak, R.N., 75

Feeding method, 26
Ferrari, F., 75
Firmicutes, 5, 102
Food allergy (FA)
  allergen-specific and allergen non-specific strategies, 59
  atopic disease manifestations, 58
  CMA, 59
  dietary fat, 58
  EHCF, 63–64
  gut microbiota, 59
  immune system dysfunction, 58
  milk and egg allergy, 60
  national and geographical variations, 58
  PBMCs, 63
  preclinical evidences, 60–61
  SCFAs, 62–63
  T cell-mediated immunomodulation, 60
  3D co-culture model, 61
  time trend analysis, 58
  vitamin D insufficiency, 58
Ford, A.C., 104
Fork head box P3 (FOXP3) messenger RNA, 52
Forti, G., 75
Francavilla, R., 49–54, 72, 75, 129
Frangeul, L., 73
Fructooligosaccharides (FOS), 32
Fujimori, S., 75, 76
Fujiwara, S., 61
Functional abdominal pain (FAP), 128–130
Functional gastrointestinal disorders (FGID), 50
  constipation, 130–133
  definition of, 121
  FAP/IBS, 128–130
  infantile colic
    inadequate maternal-infant bonding, 125
    inconsolable crying, 124, 125
    irritability and fussing, 124
    maternal anxiety, 125
    nervous/digestive system, 125
    treatment of, 126–128
  Rome IV criteria, 121–124
Furrie, E., 75
Fusobacterium nucleatum, 8

G
Galactooligosaccharides (GOS), 32
García-Mantrana, I., 3–17
Garrote, J.A., 75
Gasbarrini, A., 75, 104
Gawronska, A., 129
GE, J., 61
Gerardi, V., 104
Gershwin, M.E., 61
Gestational diabetes mellitus (GDM), 15–16
Gevers, D., 73
Ghadimi, D., 61
Giannetti, E., 75, 104, 129
Gianni, M.L., 37–43
Giglio, A., 75

Gill, R.K., 61
Gionchetti, P., 75
Giordano, P., 49–54
Giorgetti, G.M., 75
Girgis, S., 104
Gleinser, M., 42
Gloux, K., 73
Gobbetti, M., 75
Gomez-Arango, L.F., 10
Gómez-Gallego, C., 3–17
G-protein-coupled receptors (GPCRs), 63
Gracie, D.J., 104
Grangette, C., 105
Guandalini, S., 75, 101–105, 129
Guan, L., 104
Guarino, A., 109–117
Guariso, G., 72
Gudis, K., 75, 76
Guerra, P.V., 130
Gupta, P., 75
Gut-associated lymphoid tissue (GALT), 52, 70
Gut-liver axis, 87
Gut microbiota, 7, 37–38, 60, 102–103

H
Haileselassie, Y., 43
Hamlin, P.J., 104
Harris, K., 10
Hassan, C., 75
Hasunuma, O., 75, 76
Hecht, J.L., 9
Helwig, U., 61
Herfarth, H.H., 75
High-fat diet (HFD), 97
Hoepner, L., 13
Hojsak, I., 121–133
Holowacz, S., 105
Holtz, L.R., 11
Huizina, J.D., 54
Huizinga, J.D., 54
Human microbiome, 3, 26
Human milk oligosaccharides (HMOs), 32, 141
Hurme, M., 61
Huttenhower, C., 73
Huynh, H.Q., 104

I
Iliev, I.D., 42
Imaoka, A., 75, 76
Imeneo, M., 75
Indrio, F., 49–54
Infant colic, 50
  crying, 49
  definition, 50
  epidemiology, 50
  FGID, 50
    inadequate maternal-infant bonding, 125
    inconsolable crying, 124, 125
    irritability and fussing, 124

Infant colic (*cont.*)
    maternal anxiety, 125
    nervous/digestive system, 125
    treatment of, 126–128
  infant colic, 50
  microbiota, etiology and role of, 50–52
  probiotics, 52–54
Infant gut microbiota, 38
Infant microbiome, 26–27
Inflammatory bowel diseases (IBD)
  CDGEMM study, 76–77
  crohn's disease, 101
  gut microbiota, 102–103
  microbiome composition, manipulation of, 73–74
  microbiome, role of, 72–73
  mucosal homeostasis/chronic inflammation, 101
  pathogenesis of, 70–72
  prevention and/or treatment, 75–76
  ulcerative colitis, 103–104
Interferon gamma (IFNγ), 71
Interleukin-1β (IL-1β), 87
Interleukin-15 (IL15), 70
Interleukin-18 (IL18), 70
Interleukin-21 (IL21), 70
Intestinal homeostasis, 38
Intestinal lactobacilli, 51
In vitro fertilization (IVF), 14
Irritable bowel syndrome (IBS), 128–130
Ishii, Y., 75, 76
Ishikawa, H., 75, 76
Isolauri, E., 61
Iwabuchi, N., 61
Iwasaki, A., 75, 76

J
Jadresin, O., 129
Jarrin, C., 73
Jayasekara, R.W., 9
Jones, H.E., 9

K
Kalina, W.V., 61
Kamiza, S., 10
Kanda, Y., 61
Kannan, P.S., 8, 10, 12
Kaplan, J.L., 103
Kato, K., 75, 76
Kawamura, T., 61
Keen, C.L., 61
Kellermayer, R., 103
Kennedy, A., 75
Kim, D.H., 102
Kim, J.H., 102
Kim, J.Y., 61
Kim, S., 102
Kim, S.W., 102
Kim, T.I., 102
Kim, W.H., 102
Kirschner, B.S., 75
Kishida, T., 75, 76
Kitazawa, H., 61

Kobayashi, T., 75, 76
Korhonen, H., 61
Korzenik, J., 73
Kurihara, R., 75, 76
Kuylle, S., 105

L
Lactic acid bacteria (LAB) strains, 60
Lactobacillus, 4
Lactobacillus acidophilus (LA1), 102–103, 116
Lactobacillus rhamnosus GG (LGG), 74, 112
Lactobacillus spp., 4
Lam, M., 102
Landeau, M., 9
Larussa, T., 75
Lauder, A.P., 10
Lee, J.H., 102
Leleiko, N., 73
Li, M., 104
Lim, E.S., 11
Limongelli, M.G., 72
Linsalata, M., 74
Lionetti, E., 72
Lipopolysaccharide (LPS), 87
Liu, Y., 42
Li, X.R., 9
Lødrup Carlsen, K.C., 11, 12
Lopetuso, L.R., 104
Lo Vecchio, A., 109–117
Luan, J.J., 9
Luzza, F., 75

M
Maassen, C.B., 61
Macfarlane, G.T., 75
Macfarlane, S., 75
Madsen, K., 104
Madsen, K.L., 75
Ma, H.W., 102
Ma, J., 8–10, 12
Malaguarnera, M., 90
Malin M., 61
Mallardo, S., 75
Mangiola, F., 104
Manichanh, C., 73
Maragkoudaki, M., 129
Marcos, A., 75
Marteau, P., 73
Maternal diet, 26
Maternal-foetal-neo-natal microbiota, 4
Matteoli, G., 42
McElrath, T.F., 9
McIntyre, H.D., 10
McKenzie, C., 63
Microbiome dysbiosis, 4
Miele, E., 75, 104
Mi, G.L., 126
Mikkelsen, H.B., 54
Mileti, E., 42
Miniero, R., 53
Mir, S.A., 103

Mitchell, K.B., 9
Mitsui, K., 75, 76
Mizuno, S., 75, 76
Modeo, M.E., 75
Mohamadzadeh, M., 61
Molloy, E., 42
Morgan, X.C., 73
Morini, S., 75
Mosca, A., 85–98
Mosca, F., 37–43
Mueller, N.T., 13
Mysorekar, I.U., 9

N
Nagy-Szakal, D., 103
Nalin, R., 73
Navis, M., 43
Necrotizing enterocolitis (NEC)
    administration of probiotics, 29–30
    bacterial colonization patterns, 27–28
    complex echo-system, 37
    definition, 25
    dysbiosis, 26
    epidemiology and pathogenic factors, 38–40
    gut microbiota, 37–38
    homeostasis, 41
    human microbiota, 26
    immunomodulation activity, 42–43
    infant microbiome, 26–27, 32–33
    intestinal barrier, enhancement of, 42
    intestinal epithelial cells, 41
    intestinal homeostasis, 38
    microbiome and, 28–29
    preventive strategies, 40–41
    risks of administering probiotics, premature and
        VLBW infants, 30–31
    vaginal microbiota, 26
Neisseria, 28
Neonatal intensive care units (NICUs), 25
Ng, S.C., 102
Niers, L.E., 61
Ni, J., 102
Nobili, V., 85–98
Nocerino, R., 42, 43, 57–65
Nogacka, A.M., 13
Nonalcoholic fatty liver disease (NAFLD)
    bariatric surgery techniques, 86
    FMT, 96–98
    gut-liver axis, 87
    intestinal microbiome, 86–87
    intestinal microbiota, 87–89
    lifestyle and pharmacological interventions, 86
    microbiota, 89–91
    microbiota, trials on, 93–94
    obesity, trial in, 95–96
    probiotics and symbiotics, 91–92
    type 1 NAFLD, 86
    type 2 NAFLD, 86
Nonalcoholic steatohepatitis (NASH), 86
Non-communicable diseases (NCDs), 4
Non-obstructive azoospermia, 5

Notarnicola, M., 74
Nova, E., 75

O
Oggero, R., 53
Ohnishi, E., 11
Olivares, M., 75
Oliva, S., 75
Olson, S., 61
Onderdonk, A.B., 9
O'Neil, D.A., 75
Oral food immunotherapy (OIT), 64
Oral microbiota, 7–8
Orlando, A., 74
Orsini, F., 64
Ortigosa, L., 75
Otani, T., 75, 76
Otsuka, M., 75, 76

P
Palau, F., 75
Palumeri, E., 53
Papa, A., 75, 104
Paparo, L., 42, 43, 57–65
Parnell, L.A., 10
Partty, A., 126
Pascarella, F., 75, 104
Patrizi, G., 75
Peanut OIT (PPOIT), 64
Pelle, E., 53
Pellegrino, S., 72
Pelletier, E., 73
Penna, G., 42
Peripheral blood mononuclear cells (PBMCs), 63
Petito, V., 104
Peucelle, V., 105
Pigneur, B., 43
Pistoia, M.A., 75
Pizzoferrato, M., 104
Polanco, I., 75
Polloni, C., 72
Polyunsaturated fatty acids (PUFAs), 86
Ponsonby, A.L., 64
Poscia, A., 104
Pot, B., 105
Prevotella, 4, 5, 12
Prince, A.L., 8, 10, 12
Propionibacterium, 5, 14
Proteobacteria, 5, 128
Pseudomonas, 5
Pulvirenti, A., 72
Putignani, L., 85–98

Q
Quaglietta, L., 75, 104

R
Randomized controlled trials (RCTs), 16, 30
Rath, H.C., 75
Rautava, S., 10, 11
Reddy, S., 103

Rehbinder, E.M., 11, 12
Rescigno, M., 37–43
Reyes, J.A., 73
Reynolds, N., 75
Ribes-Koninckx, C., 75
Rifaximin, 140
Rigottier-Gois, L., 73
Rijkers, G.T., 61
Risk, C.D., 72
Rizzello, C.G., 75
Roca, J., 73
Roche, A.M., 10
Rodino, S., 75
Rodriguez, C., 11
Romano, C., 129
Romero, R., 11
Rossi, P., 75
Rotavirus (RV) infection, 110
Russo, F., 74

S
Sacca, N., 75
Sadeghzadeh, M., 130
Sakamoto, C., 75, 76
Salazar, N., 13
Salminen, S., 3–17
Sands, B.E., 73
Sansotta, N., 101–105
Sanz, Y., 75
Sartor, R.B., 75
Savino, F., 53, 126
Sawamura, S., 61
Scaldaferri, F., 104
Schiavi, E., 42
Schrezenmeir, J., 61
Schultz, M., 75
Sebkova, L., 75
Sekita, Y., 75, 76
Selma-Royo, M., 3–17
Senanayake, H., 9
Seo, D.H., 102
Seo, T., 75, 76
Serena, G., 69–77
Serino, M., 10
Shah, S.A., 73
Shaken baby syndrome, 50
Shaping microbiota
   dietary and nutritional counselling, 14
   host's environmental characteristics, 4
   human microbiome, 3
   IVF, 14
   maternal-foetal-neo-natal microbiota, 4
   maternal microbiota, during pregnancy
      gut microbiota, 7
      oral microbiota, 7–8
      vaginal and uterine microbiota, 6–7
   microbiome dysbiosis, 4
   NCDs, 4
   NEC, effect on, 41–42

postnatal bacterial exposition, 13–14
probiotics
   during gestation, 15–16
   during postnatal period, 16
reproductive microbiotas, 4–6
step-wise neonatal microbiota colonization, 4
in utero colonization hypothesis
   amniotic fluid, 11
   amniotic fluid microbiome, 11, 12
   atopic disease, 13
   bacterial DNA, 12
   blood microbiome, 8
   enterococcus faecium, 11
   fusobacterium nucleatum, 8
   haematogenous route, 8
   maternal diet, 13
   placental microbiome, 9–10
   proteobacteria phylum, 13
   vagina/gut microbiota, 8
   vaginal infections, 14–15
Sherrill-Mix, S., 10
Shi, W., 11
Short-chain fatty acids (SCFAs), 52, 62, 88
Sidhu, A., 42
Simeoli, R., 61
Single Ig IL-1-related receptor (SIGIRR), 30
Smits, H.H., 61
Snapper, S.B., 73
Sokol, H., 43, 73
Staff, A.C., 11, 12
Staiano, A., 75, 104
Staphylococcus, 5, 13, 28
Steed, H., 75
Stojanovic, J., 104
Stout, M.J., 9
Streptococcus, 4, 28
Stronati, L., 75
Suárez, M., 13
Suda, W., 11
Sudo, N., 61
Sugitani, M., 75, 76
Sung, V., 126
Sun, Y., 103
Sütas, Y., 61
Syrjänen, S., 10
Szajewska, H., 126

T
Tabbers, M.M., 130
Takahashi, N., 61
Takahashi, S., 61
Tanaka, K., 61
Tanaka, R., 75, 76
Tang, H., 62
Tang, M.L., 64
Tan, J., 63
Tannock, G.W., 75
Tarrazó, M., 3–17
Tatsuguchi, A., 75, 76

Temoin, S., 11
Terrin, G., 43
Terzieva, A., 10
Thomas, A.G., 75
Tickle, T.L., 73
Timmer, A., 75
Timmerman, H.M., 61
Toll-interacting protein (TOLLIP), 30
Toll-like receptor 4 (TLR4), 30
Toll-like receptors (TLRs), 39, 51, 87
Tomov, V.T., 102
Tonutti, E., 72
Torii, A., 61
Torii, S., 61
Tsilingiri, K., 42
Tumor necrosis factor-alpha (TNFα), 71
Tuominen, H., 10
Turner, J., 104
Tursi, A., 75
Tutino, V., 74
Tyagi, S., 61

U
Ughi, C., 72
Ulcerative colitis (UC), 75, 101, 103–104
Umesaki, Y., 75, 76
Urinary tract infections (UTIs), 15
Urushiyama, D., 11
Ussing chamber technique, 111

V
Vaginal and uterine microbiota, 6–7
Vaginal microbiome, 27
Vajro, P., 98
van den Brande, J., 61
Vandenplas, Y., 139–144
Vanderhoof, J.A., 75
van der Kleij, D., 61
van Holten-Neelen, C., 61
van Tol, E., 61
Varea, V., 75
Veillonella, 4

Verronen, P., 61
Verrucomicrobia, 74
Very low birth weight (VLBW) infants, 30
Very low density lipoproteins (VLDLs), 88
Vibrio cholerae, 110
Vuillermin, P.J., 63
Vulvovaginal candidiasis (VVC), 15
Vu, N., 43

W
Walsh, S.V., 75
Wang, J., 11, 62
Wang, X., 11
Wang, X.-Y., 54
Wang, Y., 42
Ward, D.V., 73
Watterlot, L., 43
Weaning, S.W.G.O., 72
Wegner, A., 130
Weizman, Z., 129
Wessel, M.A., 50
Western diet, 94
Whyatt, R., 13
Winter, H.S., 103
Wojtyniak, K., 130
Wu, G.D., 102

X
Xavier, R.J., 73
Xiao, X., 9, 10
Xiao, X.-H., 10
Xu, X., 11

Z
Zhang, C., 62
Zhang, Q., 9, 10
Zhang, Y., 11
Zheng, J., 9–11
Zhou, D., 97
Zuin, G., 72
Zuo, T., 102

Printed in the United States
by Baker & Taylor Publisher Services